This book is an introduction to the arrow of time in thermodynamics and cosmology, and develops a new quantum measurement theory in which the foregoing concepts play an essential role.

The first chapter is an overview and 'route map' and is followed by an exposition of irreversibility, the expansion of the universe and other arrows of time. The author examines the thesis that the thermodynamic arrow follows the cosmological one, and in doing so extends traditional statistical mechanics. The second part of the book presents a new theory of quantum measurement and possible experimental tests. This theory incorporates the extended statistical mechanics in an essential way. The last chapter discusses open experimental and theoretical issues. Written in a lively and accessible style, the text is liberally sprinkled with exercises. Each chapter ends with a resources section that includes notes, further reading, and technical appendices.

All graduate students and researchers in theoretical physics and philosophy of physics will find this a stimulating and thought-provoking volume.

TIME'S ARROWS AND QUANTUM MEASUREMENT

Time's arrows and quantum measurement

L. S. SCHULMAN

CAMBRIDGE
UNIVERSITY PRESS

CAMBRIDGE UNIVERSITY PRESS
Cambridge, New York, Melbourne, Madrid, Cape Town, Singapore,
São Paulo, Delhi, Dubai, Tokyo, Mexico City

Cambridge University Press
The Edinburgh Building, Cambridge CB2 8RU, UK

Published in the United States of America by
Cambridge University Press, New York

www.cambridge.org
Information on this title: www.cambridge.org/9780521567756

© Cambridge University Press 1997

First published 1997

A catalogue record for this publication is available from the British Library

Library of Congress Cataloguing in Publication Data

Schulman, L. S. (Lawrence S.),
Time's arrows and quantum measurement / L. S. Schulman.
p.   cm. — (Cambridge monographs on mathematical physics)
Includes bibliographical references and index.
ISBN 0 521 56122 1 (he). – ISBN 0 521 56775 0 (pbk.)
1. Thermodynamics. 2. Cosmology. 3. Time measurements.
4. Quantum theory. 5. Quantum statistics. 6. Mathematical physics.
I. Title.
QC311.S4   1997
530.1–dc20   95–47482   CIP

ISBN 978-0-521-56122-8 Hardback
ISBN 978-0-521-56775-6 Paperback

This book is dedicated
to my mother
Anna Schulman
and to my wife
Claire Schulman

Contents

Preface

Where is the frontier of physics? Some would say 10^{-33} cm, some 10^{-15} cm and some 10^{+28} cm. My vote is for 10^{-6} cm. Two of the greatest puzzles of our age have their origins at this interface between the macroscopic and microscopic worlds. The older mystery is the thermodynamic arrow of time, the way that (mostly) time-symmetric microscopic laws acquire a manifest asymmetry at larger scales. And then there's the superposition principle of quantum mechanics, a profound revolution of the twentieth century. When this principle is extrapolated to macroscopic scales, its predictions seem wildly at odds with ordinary experience.

This book deals with both these 'mysteries,' the foundations of statistical mechanics and the foundations of quantum mechanics. It is my thesis that they are related. Moreover, I have teased the reader with the word 'foundations,' a term that many of our hardheaded colleagues view with disdain. I think that new experimental techniques will soon subject these 'foundations' to the usual scrutiny, provided the right questions and tests can be formulated. Historically, it is controlled observation that transforms philosophy into science, and I am optimistic that the time has come for speculations on these two important issues to undergo that transformation.

My central theme is a modification of the fundamental principle of statistical mechanics, the experimentally unverified postulate that gives equal weight to all microscopic states consistent with the perceived macroscopic state. I come to this proposal from critical examination of the following idea, first suggested by Thomas Gold in 1958: the local thermodynamic arrow of time is a consequence of the expansion of the universe. This Preface is not the place to elaborate on the physics, and in Chapter 1 (Section 1.1) I give an overview that shows how the conclusions of my critical examination modify the notion of causality. At the classical level my proposal already has potential observational consequences; in particular we will see how certain statistical features of 'large scale structure' (of the galaxies) could indicate an impending big crunch (so the frontiers are interrelated).

At the quantum level, this extension of statistical mechanics permits an understanding of the measurement process that is at once fully faithful to quantum dynamical laws (meaning nothing ever happens but $\psi \rightarrow \exp(-iHt/\hbar)\psi$), is fully deterministic, and takes place within a single 'world' (in the 'many-worlds' sense, i.e., there is only one world). This theory invokes statistical mechanics in a second way as well: I rely on the complexity of macroscopic systems to provide what I call 'special' states. Again, let Chapter 1 and the book tell the story.

The book also has pedagogical objectives. There is a long chapter on 'irreversibility,' which could be used to supplement a statistical mechanics course, including at the undergraduate level. In fact much of this material is drawn from my own teaching. The original goal was to present the subject so as to make my extensions of the standard theory more understandable and more palatable. But the professorial instinct took over. More advanced readers would skip much of this, and Section 1.2 provides an outline of the contents of the book to help those who want to get into the material quickly.

Anyone with an elementary physics background should be able to handle most of the book. Keeping this promise lies behind further pedagogical material on general relativity and even a bit on organization in nonlinear open systems. There are also exercises within the text. What I don't do is explain *other* quantum measurement theories, except in passing. However, there are tidbits even for advanced students. For example, the Wigner-Araki-Yanase theorem is an important limitation on our usual picture of quantum measurement, and yet few non-specialists are aware of it.

The size scale that I characterized as the frontier of physics—10^{-6} cm— is actually a common frontier with biology and chemistry. In all these fields the advancing technology of observation and manipulation promises new understanding of the emergence of collective effects. Whether or not my proposals are correct in detail, I believe that their consideration will lead to a deeper comprehension of Nature.

Each chapter contains a multipurpose 'Notes and sources' section. Its first intent is to provide citations, more or less as in a journal article. There are also references for peripheral and background material—a recommendation for a text on general relativity, a book about algorithmic information theory, an article about winds on Venus. Finally, this is the place for (a higher density of) wisecracks, for brief words on philosophy and for historical and personal remarks about the people whose ideas are presented.

Both the development of scientific ideas and the writing of a book are enterprises for which one is fortunate if there is help and participation by others. In the several journal articles that preceded this book I have thanked many individuals and institutions and it is a pleasure once again to recall those contributions. The list I now give is unweighted, reflecting an awareness of my own limited diplomatic skills. Nevertheless, at the level of specific scientific contributions, including discussions not previously acknowledged, I am happy to mention particularly my colleague Bernard Gaveau. The roles of others on the list range from devil's advocates to scientific counselors to providers of references to TEX advisors to sources of moral support. Here then are individuals to whom I am grateful: Yakir Aharonov, Josi Avron, Dani ben-Avraham, Michael Berry, Iwo Bialynicki-Birula, Roger Bidaux, Christian Borgs, Geoffrey Canright, Sudip Chakravarty, Jennifer Chayes, Netta Cohen, Aaldert Compagner, Jozef Devreese, Charlie Doering, Gabor Forgacs, Jerry Franklin, Christopher Gerry, Larry Glasser, Thomas Gold, Shelly Goldstein, Francesco Guerra, Michael Harrison, Eivind Hauge, Per Hemmer, Ted Jacobson, Steve Kivelson, Gyula Kotek, Charles Kuper, Rolf Landauer, Gordon Lasher, Joel Lebowitz, Jan Adrian Leegwater, Andrew Lenard, Don Lichtenberg, Jon Machta, Ady Mann, Tal Mor, Chuck Newman, Roger Newton, Kåre Olaussen, Marc Parmet, George Patsakos, Phil Pechukas, Asher Peres, Sandu Popescu, Vladimir Privman, Anedio Ranfagni, Michael Revzen, Tom Reynolds, Amiram Ron, Nathan Rosen, Maura Sassetti, Claire Schulman, Leonard Schulman, Phil Seiden, Mike Shlesinger, Lee Smolin, Rafael Sorkin, Birger Stølan, Henk van Beijeren, J.-P. Vigier, Don Weingarten, Uli Weiss, John Wheeler, Arthur Wightman, Ching-Hung Woo, Bill Wootters, Alice Young, Joshua Zak, H.-Dieter Zeh and Wojciech Zurek. I have also enjoyed the hospitality of a number of institutions and would here like to thank those where I did work connected to the quantum measurement theory material reported in this book. They are Columbia University, Istituto di Ricerca sulle Onde Elettromagnetiche del CNR (Florence), Institutt for teoretisk fysikk, Universitetet i Trondheim, University of Paris VI (Université Pierre et Marie Curie), CEA Saclay, University of Stuttgart (Max Planck Institute), University of Utrecht (F.C. Donders Professorship), University of Antwerp, and IBM (Yorktown Heights). As will be clear in the reading, the subjects in this book represent problems on which I have worked much of my scientific life. As such I am pleased to thank those institutions in which I have held permanent positions, and indeed spent many years at each (with overlap). They are Indiana University (Bloomington), the Technion-Israel Institute of Technology (Haifa), and Clarkson University (Potsdam, New York). Financial support has also been forthcoming from several sources and I would like to express my appreciation to the United States National Science Foundation (most recently PHY 93 16681), NATO, jointly with the Italian national science foundation (CNR), the French national science foundation (CNRS) and atomic energy commission (CEA).

1

Introduction

In the next few pages I provide previews of the book: Section 1.1 is a statement of the main ideas. Section 1.2 is a chapter by chapter guide.

There are two principal themes: time's arrows and quantum measurement. In both areas I will make significant statements about what are usually called their foundations. These statements are related, and involve modification of the underlying hypotheses of statistical mechanics. The modified statistical mechanics contains notions that are at variance with certain primitive intuitions, but it is consistent with all known experiments.

I will try to present these ideas as intellectually attractive, but this virtue will be offered as a reason for study, not as a reason for belief. Historically, intellectual satisfaction has not been a reliable guide to scientific truth. For this reason I have striven to provide experimental and observational tests where I could, even where such experiment is not feasible today. The need for this hardheaded, or perhaps, intellectually humble, approach is particularly felt in the two areas that I will address. The foundations of thermodynamics and the foundations of quantum mechanics have been among the most contentious areas of physics; indeed, some would deny their place within that discipline. In my opinion, this situation is a result of the paucity of relevant experiment.

The goal of this book is therefore to present ideas, to stimulate comparisons with Nature and, ultimately, should the comparisons prove favorable, significantly extend our understanding of the physical world. However, given the predilections of the author and the actual state of experiment, what I mostly do in this book is offer intellectual satisfaction. My hope is that those who find these ideas attractive will join me not only in the development of the theory, but in seeking better tests and in performing those tests.

Besides the two sections mentioned above, this chapter contains an appendix that I call 'Propaganda,' and which consists of observations about my ideas and their relation to other theories. This propaganda is not part of the development, but may give perspective. There is also a section of 'Notes and sources' (as there is for every chapter in the book) that lists related books and articles.

1.1 The main ideas

Both major themes in this book require an extension of the usual foundations of statistical mechanics. Our thesis on quantum mechanics has additional elements as well. We first discuss 'arrows of time' and then quantum measurement theory.

By the 'thermodynamic arrow of time' I mean the second law of thermodynamics. There are numerous formulations of this law, statements about the conversion of heat to work or about the increase in entropy in isolated systems. We give another formulation, substantially equivalent to the others: to predict the future of a macroscopic system, average over all microscopic states consistent with its present macroscopic state. By contrast, to estimate its previous condition, make a hypothesis about the earlier macroscopic state and use that hypothesis for prediction. If it agrees with present data, good. If not, it is rejected as a hypothetical prior state. This distinction in turn is equivalent to the kind of causality implicit in the use initial conditions for forward time evolution.

The 'cosmological arrow of time' is the observation that the universe is expanding. In 1958 Thomas Gold proposed that the thermodynamic arrow is a consequence of the cosmological arrow. To me this thesis has the stark power of a great—nearly obvious—truth, and my puzzle is why it took almost 30 years from Hubble's discovery of the expansion before a connection was made. But others consider Gold's idea neither obvious nor true. Moreover, even in my admiration for this idea, I found Gold's argument flawed. It is a subtle flaw, and if the reader goes through my presentation of Gold's argument (in Chapter 4) it is likely that on the first reading the flaw will not be apparent. However, once it is pointed out it is clear that something is amiss. (In fact, this was the response of Gold, when he was kind enough to read a draft of that chapter.) Basically, the trap is that we are easily lulled into the use of initial conditions in our reasoning, but it is precisely the justi-

fication for the use of initial conditions that the argument should provide.[1]

My contention then is that if you want to prove a relation between
the thermodynamic and cosmological arrows, you must formulate your
arguments without prejudicing the outcome through the use of initial
conditions. One way to do this is to give *two-time boundary conditions*.
That is, you make hypotheses about the state of the universe at two re-
motely separated epochs; you make them such that for at least part of
the intervening time there is cosmological expansion, and you then deduce that during the expansion era the thermodynamic arrow is aligned
with the expansion. This kind of argument is given in Chapter 4. It
has two parts. The first says that even though you have a future condition, you can mostly speak as if you didn't. The second gives prosaic
thermodynamic reasoning in which the expansion of the universe causes
large collections of matter to get caught in long lived metastable states
(like stars). The slow equilibration of these systems maintains other far
flung systems out of equilibrium through messengers, such as 6000 kelvin
photons.[2] In such far flung systems, temporary isolation of a subsystem
allows its relaxation, which we call the second law of thermodynamics.

The foregoing line of reasoning uses roughly time-symmetric boundary
conditions. With highly asymmetric remote data you can probably force
any arrow you want, irrespective of expansion. In that case, if you are
trying to explain the thermodynamic arrow you would have to justify the
asymmetry. On the other hand, if it's the effect of expansion you want
to study, it makes sense to take symmetric data, as in a universe facing a
big crunch. As usual though, you must distinguish between mathematics
and physics. The argument outlined above shows that the expansion *can*
drive the thermodynamic arrow; it does not say anything about whether
the universe is or is not facing a big crunch.

Where a successful two-time boundary condition [cosmo ⇒ thermo]
argument does again make contact with physics, is in providing an independent observational framework for discovering our actual cosmology.
That is, as a result of our investigation we can suggest ways in which
a future collapse could influence events today; in particular, if there are
processes (such as the large scale dynamics of galaxies) whose time scale

---

[1] There are all sorts of caveats and clarifications for the foregoing statement, especially
regarding the micro/macro distinction. Trying to say it all in three lines would make
them unreadable. In fact, it is to explain these things that I have written a book.

[2] Of course the stars themselves are also out of equilibrium and the same is true up
until scales that could be affected by the two-time boundary data. At that point,
equilibration no longer has its usual properties.

is comparable to that of the big bang/big crunch time interval, there will be physical consequences and these may be observable.

In making the last statements I have implicitly used one of the main results of the investigations reported below. Suppose your future condition selects a particular small class of microscopic states. Then there will be a relaxation time associated with that restriction. With time-symmetric dynamics, that relaxation time will be the same for time evolution in either direction away from the time at which the restriction is compelled to hold. Now if you use such a restriction as a future boundary condition (say as part of a two-time boundary condition), then if the time of the conditioning is far from the present, *that conditioning will (at present) be invisible and undetectable.* I call microscopic state selection of this sort a *cryptic constraint*, and it is the inclusion of such conditioning that constitutes our extension of statistical mechanics. Naturally, the extension is especially concerned with ways in which the conditioning is not entirely hidden; an example is the possibility of observing big crunch precursors through slow modes, as mentioned in the previous paragraph.

A consequence of the relaxation time argument to which I just alluded is that at the present epoch the usual thermodynamic processes would look normal, despite a remote (on their time scale) future conditioning. Thus, *we would have no clue of any impending future conditioning,* and the historical development of thermodynamics and all its laws would be the same with or without this future conditioning. This is what I meant in my claim that the extension we make to statistical mechanics is consistent with all contemporary experimental information.

This brings us to quantum mechanics. Quantum mechanics is different! Certain kinds of future conditioning never relax. This is related to wave packet spreading and the difference in quantum behavior will be important in providing a rationale for our description of the quantum measurement process. First, however, I will present that description, and then return to connect these themes.

The first problem of quantum measurement theory is the existence of 'grotesque' states. By this I mean wave functions of macroscopic systems with support on more than one macroscopic state: a superposition of a live Schrödinger cat and its dead counterpart. One arrives at such wave functions by applying, at the *macroscopic* level, mathematical operations, in particular, $\psi \to \exp(-iHt/\hbar)\psi$, that are perfectly legal at the *microscopic* level. Every theory of quantum measurement must account for why we do not experience such wave functions. The answers range from interpretive sleights of hand ('now you must talk about "ensembles"') to multiplication of reality ('when you look at the cat there's one branch of the wave function—and universe—in which you saw a living cat and one in which you saw a dead cat'). My answer relies on relatively rare microscopic states. Just as there are microscopic states in which a collection of water molecules gathers together, cools the water around it, and flies into the air (the time reversal of a raindrop hitting a puddle), so there are microscopic states of the cat and its cruel quantum apparatus in which only one or the other macroscopic outcome evolves, by pure quantum evolution, from the initial state. Let me say this again, since it's an important point. Consider a relatively complicated quantum system, perhaps 10^{10} particles, which under pure quantum evolution for a time T (that is, $\psi \rightarrow \exp(-iHT/\hbar)\psi$) would generally find itself in a grotesque state. Thus ψ could describe a small region of a bubble chamber through which an external particle is about to pass and which would have 50% probability of forming a bubble due to the passage. In this language, $\exp(-iHT/\hbar)\psi$ would have significant support on configurations with a bubble and also on configurations without a bubble, with combined norm in each configuration about 0.5. My claim is that *there are particular initial wave functions ψ for which the evolved object, $\exp(-iHT/\hbar)\psi$, has support on only one of the outcomes.* This claim is supported by specific calculations on realistic models and by less rigorous appeal to the complexity of macroscopic systems. As the reader will note, this way to avoid grotesque states uses the richness of macroscopic systems, just as the problem itself arose from applying to macroscopic objects the rules known to work for microscopic ones.

Microscopic states that evolve to a single macroscopic outcome (when most microstates would have become grotesque) are called 'special.' If the initial conditions of an experiment are special, you will not get a grotesque state; in this way the main conundrum of quantum measurement theory can be avoided. I now postulate that this *is* the way that grotesque states are avoided in Nature: *all situations with potentially grotesque outcomes have 'special' initial conditions.* But now we are confronted with new questions. Why should an initial microstate be special?[3] And how can such a comprehensive constraint be maintained indefinitely? If this is indeed the solution of the quantum measurement problem, then every time you do an experiment that could lead to a grotesque state, in fact, in every natural circumstance that could lead to grotesqueness, experiment or not, you would need to have these special initial conditions. True, they are microstates indistinguishable macroscopically from their non-special macroscopic counterparts, but why or how could such a constraint be enforced?

---

[3] I will drop the quotation marks, but continue to use the term in the sense defined.

Here is where our earlier extension of statistical mechanics can provide perspective. We have already introduced the notion of cryptic constraints on microstates—all microstates, initial, final, whatever—based on future conditioning, and asserted that for most purposes those constraints would be unobservable. One way to justify the sweeping hypothesis of our quantum measurement theory would be to find a future condition that particularly mitigates against grotesque states.

We do indeed propose such a condition. It is based on the observation that when a wave function is grotesque the constituent particles—which are many in number—become widely spread in space. A future condition that would be violated by such a state is one in which particle wave functions are well localized. We then argue that in certain cosmologies such localization could be expected. The intent of this line of reasoning is not to advocate a particular cosmology. Rather, we are interested in showing that the ubiquitous coherence that we demand can arise from conditions not all that difficult to imagine. Personally, I give Nature credit for more subtlety than I can imagine; but for those who could see no reason for the selection of special states, this cosmological scenario is at least one possibility.

But even when you are ready to accept the existence of special states, and even if you can overcome the intuitive feeling that one can control initial conditions arbitrarily (which I say you cannot), still there is one important and central quantitative hurdle to overcome. It is this. If every event is fixed by its precise microstate, why is it that if I perform the same experiment many times the probabilities of the various outcomes follow the square of the microscopic system's wave function, as in the Copenhagen interpretation? (Here 'same' means the same macroscopic characterization.) In other words, how does my theory recover the standard probabilities? One could consider this a demand for a quantum extension of the earlier demonstration that the constraint is cryptic, hidden and unobservable.

This leads us to another postulate: quantum probabilities are like classical probabilities. Although the special states are *relatively* rare, in absolute terms we expect there to be many of them. Our postulate is that the relative probability of each outcome of an experiment is proportional to the relative number of special states for that outcome. This is the same rule as for classical statistical mechanics; there, relative number is measured by phase space volume, here by Hilbert space dimension, as dictated by the correspondence principle. (It is also clear that for any macroscopic outcome, the space of special states for that outcome is a Hilbert subspace.) With this postulate one then has the following challenge: show that the special states do indeed come with the proper abundances, and that they do this no matter what apparatus is used to perform the measurement.

Establishing the foregoing has been the technically most difficult part of this work. It is the only part where I was seriously ready to quit. The analysis of special state abundance is performed in terms of kicks that a system might receive, the 'kicks' being deviations from what would happen to the system in most environments. Such kicks would be the result of encountering the rare microscopic 'special' states. What we find is that if such kicks are distributed with the Cauchy distribution, then one recovers the usual quantum probabilities. The random variable X is Cauchy distributed with parameter a if the probability of finding X between x and $x + dx$ is $C_a(x) = (a\,dx/\pi)/[a^2 + x^2]$. There are still unresolved questions in this demonstration, but the meshing of the mathematical pieces has, for me at least, imparted a certain credibility.

A brief characterization of our ideas is that we have not changed the dynamics, only the class of states on which those dynamics act. If this is the case and if all the usual probabilities are recovered (as asserted in the last paragraph), how can this theory be experimentally distinguished from others? How could one prove or disprove it?

We propose several possibilities, with varying degrees of directness. First one can try to deplete the microscopic degrees of freedom of a system until it would be impossible for a particular class of special states to exist. If this can be done while leaving Copenhagen predictions unchanged, an experimental difference could be noted. Less direct would be a search for ancillary effects of the required Cauchy noise. The long-tailed Cauchy distribution may be expected to affect other phenomena. Another possible experiment would involve entanglement. It is this property that lies behind the EPR experiment. By making a reasonable assumption on the nature of the special states involved in spin-EPR experiments, we find physical features not present in the Copenhagen analysis.

1.2 Outline of the book

This book contains both new and old, both classical and quantum. I hope that exposition of the 'old' will be pedagogically useful and that the entire range of physical phenomena will be of interest. Nevertheless, for those whose goals are more focused, it may be helpful to provide a reading guide.

Chapter 2 is mainly pedagogical. Partly it is collected from material I have used in teaching, and others may find it useful for this purpose. **Section 2.6**, however, is a particular restatement of the second law of thermodynamics that is important in understanding later developments (in particular Chapters 4, 5, 6 and 8.). The goal of Chapter 2 is to show how irreversibility enters in time symmetric theories. Along the way, we point out that such a demonstration is *not* the same as accounting for the physical, thermodynamic arrow of time. As can be seen from the Contents, Chapter 2 progresses from discrete statistical models to more complex ones, and finally to irreversibility in quantum systems. At each stage we show where microscopic information is lost and how this can be consistent with abstract theorems showing how it may be recovered (recurrence and reversibility paradoxes).

Chapter 3 discusses a few of the many 'arrows of time' that have been proposed. This is the first of two chapters in which we focus on the important physical 'arrow of time' questions, namely to what extent the arrows are correlated with one another; which are independent and which are consequences of another. Like Chapter 2, much of Chapter 3 is pedagogical and is preparation for Chapter 4 and later material. In particular, we cover enough on each of several subjects so that readers of different backgrounds will be able to go through our arguments in the later work. One fairly comprehensive section (Section 3.3 and an appendix) is on the radiative arrow of time, and is not vital for the sequel. More specific advice on what can be skipped is given near the beginning of that section. Along the way in Chapter 3, I step on everybody's toes—you can read this chapter to find out what's wrong with quantum cosmology, that there is no such thing as a quantum mechanical arrow of time, and learn other ways to lose friends.

Chapter 4 is the heart of the classical arrow of time development. There I show that the thermodynamic arrow of time follows the cosmological one. First I present Gold's argument, criticize it, and present what I consider to be an improved version. Here is where I advocate the use of two-time boundary conditions and begin to explore the consequences for the arrow of time arguments. I also discuss observational tests by which one could find evidence for an impending (on a time scale of billions of years) big crunch.

Chapter 5 is concerned with technical aspects of solving two-time boundary value problems. In the form of doubly constrained stochastic dynamics it presents a systematic formalism for dealing with our extension of statistical mechanics. It also takes up the matter of *quantum* two-time boundary value problems. This is far more difficult than the classical version and is central to many of the technical problems dealt with later in the book, especially in Chapter 7.

In **Chapter 6** I present my quantum measurement theory. This is the *other* heart of the book. I explain the first problem of quantum measurement theory—grotesque states—and present an explicit example of how they can be avoided with special initial conditions. Here too there is a stretch of rather technical development (Section 6.2), which the first time reader should not get bogged down in. At the end of Section 6.1 specific advice is given on how to avoid this bog!

The next three chapters, **Chapter 7**, **Chapter 8** and **Chapter 9** go into important theoretical questions relating to the plausibility and consistency of the theory presented in Chapter 6. Section 6.4 outlines the themes of these chapters and should be adequate for finding one's way. There is also the usual advice about not getting overly involved in details before seeing where a calculation is going. Section 9.2 is a good example of how it could be possible to get lost, although the calculation itself is vital to the theory.

Finally in **Chapter 10** we discuss experimental tests of the theory. Section 10.1 describes the most direct test, an attempt to affect the results of an experiment by eliminating a class of special states. At the moment this is out of the range of experimental feasibility, but not by all that much. Other kinds of test would involve predictions for EPR experiments (Section 10.2) as well as features that follow from study of the detailed mechanisms of specialization studied in Chapter 9.

Chapter 11 is an overview, my own feeling about what needs to be done, what chances there are for success, and what it would mean if the theory is proved true.

Organization

Equations, exercises, and figures all carry chapter and section number. References are gathered in a 'Notes and sources' section near the end of each chapter. Comments on exercises appear after the references. Any appendices for the chapter follow this.

1.3 Notes and sources

The scientific and philosophical literature on the foundations of thermo-dynamics and quantum mechanics is enormous and widely varying in language and goals. In the next paragraphs I will mention a few sources, but I claim neither balance nor comprehensiveness. Especially on matters remote from my scientific interests, some of the sources are those I found on remainder tables in book stores or that were pushed by book clubs, which in rare instances lured me into membership.

In 1962, Bondi and Gold ran a conference at Cornell on the arrow of time which was attended by some of the outstanding scientists of the day. One participant, who had thought he was speaking privately refused to be identified in the proceedings ([GOLD]) and goes by the name 'Mr. X,' lending the volume yet another exotic quality. A lot of solid physics is adduced in pursuit of the goal, but like other discussions on this subject no one quite knows what the goal is.

The topic continued to attract attention in subsequent years; the discovery of CP violation with its attendant implication of T violation stimulated research. A good collection of articles, with this and other directions, is [GAL-OR]. See also [LANDSBERG]. A recent conference on time asymmetry is [HALLIWELL]. Ideas of Boltzmann are presented there and in [LEBOWITZ]. A book of wise and occasionally humorous articles is [FLOOD]. Fiction has long been a way to express ideas on time, with the recent book of [LIGHTMAN] an interesting example. Science fiction deals with such themes, for example, time travel paradoxes in [HEINLEIN]. In the 1970s there was a series of conferences ([FRASER]) exploring issues on time. Many perspectives are given, physical, biological, philosophical. For example, the biological basis of our perception of time is as puzzling as the physical questions we take up here. My own interest in arrow-of-time questions is long-standing and some of my publications in this field are [SCHULMAN 71, 73, 74, 75, 77, 91a].

For those who study the quantization of gravity, there is an extra twist, since only at a semiclassical level does our usual time parameter become meaningful. This and related issues will not be discussed in this book, but a good source is [ZEH]. At a popular level are works by [HAWKING] and [PENROSE]. These works also bear on the foundations of quantum mechanics, and an article where 'decoherence' issues are raised is [GELL-MANN]. More technical works related to this research direction will be cited later.

With respect to the philosophical literature, note that some books already listed contain contributions from philosophers, for example articles by A. Grünbaum appear in both [GOLD] and [GAL-OR]. Other philosophical works of interest are [PRICE] and [SALMON].

Foundational books and articles on quantum mechanics abound. In some cases there is overlap with material relating to time. An excellent source of historically significant contributions is the collection [WHEELER]. A recent text with emphasis on foundations is [PERES]. There is also the early text of [BOHM]. Recent conferences are [ACCARDI], [ANANDAN], [BLACK], [EZAWA], [GORINI], [GREENBERGER], [HALLIWELL], and [MILLER]. A collection of articles in honor of David Bohm, many of which address quantum measurement theory, is [HILEY]. A survey article on Bell's theorem is [CLAUSER] and a collection containing many recent papers with emphasis on that direction is [BALLENTINE]. For Bell's inequality I found the treatment of [HARRISON] to be clear and readable. Many contributors to this field have published in book form, including collections of articles at varying technical levels. A sampling is: [BELL], [HEISENBERG], [JAMMER], [JAUCH], [LEGGETT], [NELSON], and [WIGNER]. And of course there is the seminal work of [VON

NEUMANN], which established the mathematical framework within which much of contemporary discourse takes place. A volume on the 'many-worlds' interpretation is [DEWITT]. A recent review of experimental work is [LAMOREAUX]. Another experimentally oriented paper is [SHERMAN]. A sampling of articles representing additional points of view is [BALIAN], [BENATTI], [GOLDSTEIN], [GRIFFITHS], [KHALFIN], [PEARLE], and [RIMINI]. Some references for my own work are [SCHULMAN 84, 86] and [SCHULMAN 91b].

A philosophical issue that physicists generally believe is impacted by quantum mechanics is 'free will.' As I later comment, we overestimate the significance philosophers attach to this. In any case, a number of books on this and related issues are [O'CONNOR], [DENNETT], and [FEIGL]. The latter includes articles on many subjects, including the foundations of probability. For additional references on philosophical aspects of probability see the notes to Chap. 6.

References

ACCARDI 94	L. Accardi (1994) *The interpretation of quantum theory: where do we stand?*, Enciclopedia Ital., Rome. (Conf. at Columbia Univ., 1992.)
ANANDAN 90	J. S. Anandan (1990) *Quantum Coherence*, World Sci., Singapore. (Conf. Fund. Aspects of Quant. Theory, 1989)
BALIAN 89	R. Balian (1989) On the principles of quantum mechanics and the reduction of the wave packet, *Am. J. Phys.* **57**, 1019.
BALLENTINE 88	L. E. Ballentine (1988) *Foundations of Quantum Mechanics since the Bell Inequalities*, Amer. Assoc. Phys. Teachers, College Park, Md.
BELL 87	J. S. Bell (1987) *Speakable and unspeakable in quantum mechanics*, Cambridge Univ. Press, New York.
BENATTI 87	F. Benatti, G. C. Ghirardi, A. Rimini & T. Weber (1987) Quantum Mechanics with Spontaneous Localization and the Quantum Theory of Measurement, *Nuov. Cim.* **100 B**, 27.
BERGQUIST 86	J. C. Bergquist, R. G. Hulet, W. M. Itano & D. J. Wineland (1986) Observation of Quantum Jumps in a Single Atom, *Phys. Rev. Lett.* **57**, 1699.
BLACK 92	T. D. Black, M. M. Nieto, H. S. Pilloff, M. O. Scully & R. M. Sinclair (1992) *Foundations of Quantum Mechanics*, World Sci., Singapore. (Santa Fe, 1991.)
BOHM 51	D. Bohm (1951) *Quantum Theory*, Prentice-Hall, Englewood Cliffs, New Jersey.
CHUANG 95	I. L. Chuang, R. Laflamme, P. W. Shor and W. H. Zurek (1995) Quantum Computers, Factoring, and Decoherence, *Science* **270** (December 8), 1633.
CLAUSER 78	J. F. Clauser & A. Shimony (1978) Bell's theorem: experimental tests and implications, *Rep. Prog. Phys.* **41**, 1881.
CLOUDSLEY 78	J. L. Cloudsley-Thompson (1978) Biological Clocks and Their Synchronizers, in [FRASER], III.
DENNETT 84	D. C. Dennett (1984) *Elbow Room: The Varieties of Free Will Worth Wanting*, MIT Press, Cambridge, Mass.
DEWITT 73	B. S. DeWitt & N. Graham (1973) *The Many-Worlds Interpretation of Quantum Mechanics*, Princeton Univ. Press, Princeton.

EZAWA
93
H. Ezawa & Y. Murayama (1993) *Quantum Control and Measurement Theory*, North Holland, Amsterdam. (Int. Symp., Found. Quant. Mech. (ISQM), Hitachi, 1992.)

FEIGL
53
H. Feigl & M. Brodbeck (1953) *Readings in the Philosophy of Science*, Appleton-Century-Crofts, New York.

FLOOD
86
R. Flood & M. Lockwood (1986) *The Nature of Time*, Blackwell, Oxford.

FRASER
78
J. T. Fraser, N. Lawrence & D. Park, ed., *The Study of Time III*, Springer, New York. Volume II pub. 1975 and an earlier related conf. pub. in Studium Generale. See [CLOUDSLEY], [LANDSBERG], [MATSUMOTO], [MICHON], [PALMER], and [PARK].

GAL-OR
74
B. Gal-Or (1974) *Modern Developments in Thermodynamics*, Wiley, New York.

GELL-MANN
90
M. Gell-Mann & J. B. Hartle (1990) Quantum mechanics in the light of quantum cosmology, in *Complexity, Entropy, and the Physics of Information*, W. H. Zurek, ed., Addison-Wesley, Redwood City, Calif.

GOLD
67
T. Gold (1967) *The Nature of Time*, Cornell Univ. Press, Ithaca.

GOLDSTEIN
87
S. Goldstein (1987) Stochastic Mechanics and Quantum Theory, *J. Stat. Phys.* **47**, 645.

GORINI
86
V. Gorini & A. Frigerio (1986) *Fundamental Aspects of Quantum Theory*, NATO ASI, Ser. B, Phys., vol. 144, Plenum, New York.

GREENBERGER
86
D. M. Greenberger (1986) *New Techniques and Ideas in Quantum Measurement Theory*, New York Acad. Sci., New York (vol. 480, *Ann. New York Acad. Sci.*)

GRIFFITHS
84
R. B. Griffiths (1984) Consistent Histories and the Interpretation of Quantum Mechanics, *J. Stat. Phys.* **36**, 219.

HALLIWELL
94
J. J. Halliwell, J. Pérez-Mercader & W. H. Zurek (1994) *Physical Origins of Time Asymmetry*, Cambridge Univ. Press, New York. (Conf., Spain, 1991.)

HARRISON
82
D. Harrison (1982) Bell's inequality and quantum correlations, *Am. J. Phys.* **50**, 811.

HARTLE
83
J. B. Hartle & S. W. Hawking (1983) Wave function of the universe, *Phys. Rev.* D **28**, 2960.

HAWKING
88
S. W. Hawking (1988) *A Brief History of Time, from the big bang to black holes*, Bantam, Toronto.

HEINLEIN
58
R. A. Heinlein (1958) All you zombies, in *The Best From Fantasy and Science Fiction: 9th Series*, ed., R. P. Mills, Ace, New York.

HEISENBERG
58
W. Heisenberg (1958) *Physics and Philosophy, The Revolution in Modern Science*, Harper Torchbooks, New York.

HILEY
87
B. J. Hiley & F. D. Peat (1987) *Quantum implications: essays in honour of David Bohm*, Routledge, London.

JAMMER
66
M. Jammer (1966) *The Conceptual Development of Quantum Mechanics*, McGraw-Hill, New York.

JAUCH
68
J. M. Jauch (1968) *Foundations of Quantum Mechanics*, Addison-Wesley, Reading, Penn.

KHALFIN
92
L. A. Khalfin & B. S. Tsirelson (1992) Quantum/Classical Correspondence in the Light of Bell's Inequalities, *Found. Phys.* **22**, 879.

LAMOREAUX
92
S. K. Lamoreaux (1992) A Review of the Experimental Tests of Quantum Mechanics, *Int. J. Mod. Phys.* A **7**, 6691.

LANDSBERG
78
P. T. Landsberg (1978) Thermodynamics, Cosmology, and the Physical Constants, in [FRASER], III.

LANDSBERG 82   P. T. Landsberg (1982) *The Enigma of Time*, Hilger, Bristol.

LEBOWITZ 93   J. L. Lebowitz (1993) Boltzmann's Entropy and Time's Arrow, Phys. Today, Sep., p. 32.

LEGGETT 87   A. J. Leggett (1987) *The Problems of Physics*, Oxford Univ. Press, Oxford.

LIGHTMAN 93   A. Lightman (1993) *Einstein's Dreams*, Warner, New York.

LLOYD 95   S. Lloyd (1995) Quantum Mechanical Computers, *Sci. Amer.* **273** (4,October), 140.

MATSUMOTO 75   M. Matsumoto (1975) Time: Being or Consciousness Alone? —A Realist View, in [FRASER], II.

MERMIN 85   N. D. Mermin (1985) Is the moon there when nobody looks? Reality and the quantum theory, *Phys. Today*, April 1985, 38.

MICHON 75   J. A. Michon (1975) Time Experience and Memory Processes, in [FRASER], II.

MILLER 90   A. I. Miller (1990) *Sixty-Two Years of Uncertainty: Historical, Philosophical and Physical Inquiries into the Foundations of Quantum Mechanics*, NATO ASI Ser. B: Phys., vol. 226, Plenum, New York.

NELSON 67   E. Nelson (1967) *Dynamical Theories of Brownian Motion*, Princeton Univ. Press, Princeton.

O'CONNOR 71   D. J. O'Connor (1971) *Free Will*, Doubleday, Garden City, New York.

PALMER 78   J. D. Palmer (1978) The Living Clocks of Marine Organisms, in [FRASER], III.

PARK 71   D. Park (1971) The Myth of the Passage of Time, *Stud. Gen.* **24**, 19. (At 1st Conf. of Int. Soc. for the Study of Time, Oberwolfach, 1969.) See also D. Park (1975) Laws of Physics and Ideas of Time, in [FRASER], II, and D. Park (1978) The Past and the Future, in [FRASER], III.

PEARLE 94   P. Pearle & E. Squires (1994) Bound State Excitation, Nucleon Decay Experiments, and Models of Wave Function Collapse, *Phys. Rev. Lett.* **73**, 1.

PENROSE 89   R. Penrose (1989) *The Emperor's New Mind*, Penguin, London.

PERES 93   A. Peres (1993) *Quantum Theory: Concepts and Methods*, Kluwer, Dordrecht.

PRICE 94   H. Price (1996) *Time's Arrow and Archimedes' Point: New Directions for the Physics of Time*, Oxford Univ. Press, New York.

RIMINI 92   A. Rimini & M. Roncadelli (1992) A Framework for a Relativistic Theory of State Reduction, *Found. Phys. Lett.* **5**, 499.

SALMON 84   W. Salmon (1984) *Scientific Explanation and the Causal Structure of the World*, Princeton Univ. Press, Princeton.

SCHULMAN 71   L. S. Schulman (1971) Tachyon Paradoxes, *Am. J. Phys.* **39**, 481.

SCHULMAN 73   L. S. Schulman (1973) Correlating Arrows of Time, *Phys. Rev.* D **7**, 2868.

SCHULMAN 74   L. S. Schulman (1974) Some Differential-Difference Equations Containing Both Advance and Retardation, *J. Math. Phys.* **15**, 295.

SCHULMAN 75   L. S. Schulman, R. G. Newton & R. Shtokhamer (1975) Model of Implication in Statistical Mechanics, *Phil. Sci.* **42**, 503.

SCHULMAN 77   L. S. Schulman (1977) Illustration of Reversed Causality with Remarks on Experiment, *J. Stat. Phys.* **16**, 217.

SCHULMAN 84   L. S. Schulman (1984) Definite Measurements and Deterministic Quantum Evolution, *Phys. Lett.* A **102**, 396.

SCHULMAN 86   L. S. Schulman (1986) Deterministic Quantum Evolution through Modification of the Hypotheses of Statistical Mechanics, *J. Stat. Phys.* **42**, 689.

SCHULMAN 91a   L. S. Schulman (1991) Models for Intermediate Time Dynamics with Two-Time Boundary Conditions, *Physica* A **177**, 373.

SCHULMAN 91b   L. S. Schulman (1991) Definite Quantum Measurements, *Ann. Phys.* **212**, 315.

SHERMAN 92   B. Sherman & G. Kurizki (1992) Preparation and detection of macroscopic quantum superpositions by two-photon field-atom interactions, *Phys. Rev.* A **45**, R7674.

VON NEUMANN 55   J. von Neumann (1955) *Mathematical Foundations of Quantum Mechanics*, Princeton Univ. Press, Princeton. (Trans. from German ed. (1932: Springer, Berlin) by R. T. Beyer.)

WHEELER 83   J. A. Wheeler & W. H. Zurek (1983) *Quantum Theory and Measurement*, Princeton Univ. Press, Princeton.

WIGNER 67   E. P. Wigner (1967) *Symmetries and Reflections, Scientific Essays of Eugene P. Wigner*, Indiana Univ. Press, Bloomington.

ZEH 89   H. D. Zeh (1989) *The Physical Basis of the Direction of Time*, Springer, New York.

Appendix 1.1.A: Propaganda

1) Although intellectual satisfaction is not a reliable guide to truth, intellectual *dissatisfaction* does carry a valid, if less specific, message. For quantum mechanics applied to macroscopic systems, this message has been broadcast for more than 60 years.

2) Why quantum measurement has merited recent attention.

   a) Experiments: Bell inequality checks, delayed choice, etc. This is not new, but is a step beyond the formative era of quantum mechanics. See [CLAUSER] and [GREENBERGER].

   b) There is now a quantum description of objects that function as measurement apparatus. For the SQUID, see Leggett, Chakravarty and Tesche in [GREENBERGER]. Similarly, people now manipulate individual atoms in ways that challenge the wave function collapse ideas that were possible when microscopic processes seemed remote from observation. See, for example, [BERGQUIST].

   c) Quantum cosmology. When you talk about the 'wave function of the universe' there is no longer any 'larger' system to 'collapse' your wave function. See [HARTLE] and [GELL-MANN]; also [SCHULMAN 84].

   d) Quantum computing. Fully coherent evolution of a system allows solution of lengthy problems, exponential in the appropriate parameter; for upbeat reports see [LLOYD], [CHUANG]. Physical construction of a device to take advantage of this forces consideration of coherence or the breakdown thereof in moderately large objects. There are laboratory efforts to build large undisturbed systems with good control over all microscopic coordinates. I expect such systems to provide possibilities for experimental tests of the ideas presented in this book.

3) In the text I say what my theory is. Here are comments on what *kind* of theory it is and what it is *not*.

   a) The theory does not change one jot or tittle of *quantum mechanics*. There is no wave function collapse, no special dynamics at the moment 'measurement.' What I change is the fundamental hypothesis of *statistical mechanics*. I do not assign equal probabilities to all microstates consistent with the perceived macrostate. This hypothesis has never been checked experimentally at the level at which I claim it to be violated. Furthermore, if one looks to the size range where the black magic of measurement is supposed to occur, namely 10^{-7} to 10^{-4} cm, there is no reason to expect any deviation from quantum mechanics, *except* for the measurement problem.

b) Another characterization of the theory is that it is *deterministic*. All that ever happens is $\psi \rightarrow \exp(-iHt/\hbar)\psi$. This may be complicated but (assuming no operator pathology) it gives you only one answer for any initial conditions. I imagine Einstein would be pleased[4] that quantum mechanics is now *complete*: everything needed to evolve the system is right up front. Nor are there hidden variables. It is not possible (for us) to predict with certainty because the requisite microscopic information is not accessible. This introduces probabilities in the same way that they appear in classical statistical mechanics. If you don't know where in phase space your system is, you average over the possible microstates consistent with your macroscopic observations. It's the same here.

c) The wave function is where the particle is. If a particle is absorbed, its *entire* wave function evolves to the region of absorption. We do not need the Copenhagen ensemble interpretation.

4) The ideas in this book can be tough on the intuition, especially with respect to causality. But as [MERMIN] emphasizes, *quantum mechanics* is already tough on the intuition. My point is not that misery loves company, but that intuition needs re-education.

5) Statistical mechanics deals with the micro/macro interface. Phrasing the quantum measurement problem as a conflict between the (supposedly) microscopic superposition principle and our macroscopic experience shows that conflict to fall within its scope. Our solution to the quantum problem uses statistical mechanics in two ways: it appeals to the complexity of the macroscopic environment for the existence of 'special' states (which avoid the aforementioned conflict) and it goes into the foundations of statistical mechanics in its understanding of the arrow of time, thereby making natural the notion of cryptic constraints.

6) The use of cosmological two-time boundary conditions is a temporal version of the Cosmological Principle. Not only is our location not special, our thermodynamic time direction isn't either. This can lead to cryptic constraints and non-cryptic constraints. This book proposes that the absence of grotesque states is a notable example of the latter (as explained in Chapter 6, etc.).

7) *On philosophy.* As one who studied the Talmud, I respect the power and subtlety of words, but am also aware of their limitations. In the early stages of scientific development, philosophers enter uncharted fields using words, but later this tool should give way to more powerful ones, for example mathematics, where good notation compresses and sharpens ideas. One analogy likens philosophers to reconnaissance troops with small arms, clearing the way for the tanks (and subject to more danger than

---

[4] I may be overly optimistic. Despite his kind nature, my Technion colleague (and Einstein collaborator), Nathan Rosen, did not go so far as to describe himself as pleased. At a recent celebration (his 86th birthday and the 60th anniversary of EPR) he opined that my ideas 'require some getting used to.'

those that follow—in this case the danger of sounding foolish). These comments were stimulated by attendance at a recent conference at which there was extensive discussion of 'elements of reality' and 'counterfactual' arguments.

8) *On humility.* 'Can you teach a dog calculus?'[5] The point is not that you can't, but that you couldn't explain to the dog what it is missing. The real point is to apply this to the human race. Less than 2% of evolutionary history separate our brains from those of dogs. Why shouldn't there be concepts too subtle for us to comprehend? One can speculate that evolution-like computer programs designed to develop logic from survival experience could discover kinds of logic that elude our capacities. What's amusing is that it's not clear how you'd know when you succeeded. In any case, in the present volume we'll use the mix of observation, experiment and theory that has served science for the past several centuries. This middle road avoids theorizing based on intellectual satisfaction alone (although one could argue that general relativity arose in this way), but also continues to operate in the face of larger uncertainties, such as those suggested by the dog question or by mystics whom one may think of as trying to reach that next level of wisdom.

---

[5] I had thought this question due to Wigner, but have been told that similar questions had been posed before.

2

Irreversibility

Things happen. That may seem obvious, but it has also been maintained that all the 'happening' does not signify change and that the way of the world is periodic and repetitious:

One generation passeth away, and another generation cometh...
The sun... riseth, and the sun goeth down,
And there is nothing new under the sun.
—Ecclesiastes, Chapter 1

I won't discuss the profound aspects of this passage, but I will do post-industrial age nitpicking. Only in the past century has humanity understood a distinction that exists among these cyclic behaviors. For the rising and setting of the sun, there is indeed little that is happening. To a good approximation this is non-dissipative. But as to the coming and going of generations, with the benefit of wisdom gained in building steam engines, we recognize that birth and death can occur only so long as there is a source of negative entropy.

I could continue in this vein and discuss how the failure to distinguish between free and frictional motion confused humanity's greatest minds as they grappled with elementary mechanics. But I wish to begin a technical discussion of irreversibility and only want to draw a lesson of humility from the historical perspective. Until the past few centuries, humanity failed to appreciate the most manifest of time's arrows, the second law of thermodynamics. Unless one realizes that Nature's dynamics are mostly time symmetric one does not know that there is a problem. And even so, it took more than deep thinking to make the next step. It took the technical demands of mining and manufacturing to give the first formulations of efficiency, dissipation and irreversibility.

In this chapter I will review the ways that the modern scientist can express and 'derive' irreversibility. We experience irreversibility in many ways. Rain falls down. Ice melts. Death. A punctured tire goes flat.

A sample of radium decays. To begin to make precise statements, the system should be isolated. Thus one version of the second law of thermodynamics is that the entropy of an isolated system cannot decrease. The requirement of isolation sharpens things, but there is also the need to define entropy. Unlike energy or angular momentum, entropy is not a function of the microscopic state of a system. Historically, entropy was defined in terms of heat transfer, obviating the need for sophisticated definitions. But statistical mechanics has made us sophisticated and we can go beyond the phenomenology. Nevertheless, even when one defines entropy, there remain substantive questions. If you look at the dynamics of some system, is it indeed true that the entropy—however you've defined it—increases? The most famous argument along these lines is the Boltzmann H-theorem, which is fraught with its no less famous paradoxes: reversibility and recurrence. These then are the problems that lie at the foundations of statistical mechanics. The establishing of results like that of Boltzmann and the understanding of the assumptions and limitations of the demonstration.

The goal of this chapter is to acquaint the reader with the foregoing issues. It is not a treatise on the H-theorem, in fact I will hardly touch on that theorem in its original form.[1] Rather I hope to lay out the important concepts and to explain what is known or conjectured about them. We approach this task with examples, models for which the mathematics is easily understood, and work up from these.

Exercise 2.0.1. The story is told that J. A. Wheeler posed the following question at a Princeton Ph.D. final oral exam. He took two coins, placed them on a smooth table and slid one toward the other so that there was a small collision. The candidate was asked whether this represented elastic or inelastic scattering, and was required to justify the answer *based on the experiment*. That is, if he maintained that it was inelastic, he could not argue that there is *always* some friction, *always* some heat generated in a collision. He was required to offer an observation, sensible as well to the committee members, that the collision was inelastic.

2.1 The Kac ring model

Our first example was created by one of the great mathematicians and teachers of our era, Mark Kac. He considered a ring with N sites on it. See Fig. 2.1.1. In every site there is a ball, which can be either black or white. On each time step, every ball moves counterclockwise to the next site. The dynamics are enriched by considering a subset of the sites to be active. Let there be A of these 'active' sites and let them have the

---

[1] In Section 2.7 is a brief statement of the theorem.

following property: when a ball *leaves* an active site, it switches color, turning white if it was black, black if it was white. In the figure $N = 25$ and $A = 9$.

The behavior of this system can be deduced conversationally. Suppose we start with more white balls than black. Let us also suppose that the active sites are randomly distributed. On the first time step the number of balls that go from white to black will be greater than the number going in the opposite direction, simply because there were more white balls to start with. It follows that the system will not settle down until the number of each color is equal, and we see that half black, half white is the equilibrium state of the system. Making this argument quantitative only requires notation. Let the number of white balls at time t be $W(t)$ and the number of black balls $B(t)$. The normalized difference is defined to be $\delta(t) \equiv [W(t) - B(t)]/N$. Let the number of white balls in active sites at time t be $W_A(t)$, and let $B_A(t)$ be correspondingly defined. The

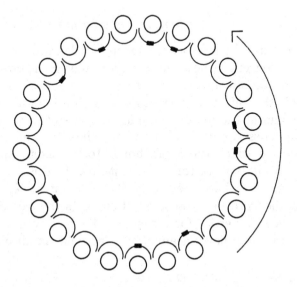

Fig. 2.1.1. Kac ring model. Balls move counterclockwise from site to site. Some sites are active and are marked at their bases. When a ball leaves an active site it changes color. Ball color is not shown.

dynamical scheme implies[2]

$$W(t+1) = W(t) - W_A(t) + B_A(t)$$
$$B(t+1) = B(t) - B_A(t) + W_A(t)$$
(2.1.1)

This is exact. Define $\mu \equiv A/N$, the active site fraction. For randomly distributed active sites it is reasonable to assume that

$$\frac{\text{number of white balls in active sites}}{\text{number of white balls everywhere}} = \frac{\text{number of active sites}}{\text{number of all sites}}$$

Thus $W_A(t)/W(t) = \mu$ and similarly for B. (This is the analog of Boltzmann's molecular chaos assumption, about which more later.) With this assumption Eq. (2.1.1) becomes

$$W(t+1) = W(t) - \mu(W(t) - B(t))$$
$$B(t+1) = B(t) - \mu(B(t) - W(t))$$
(2.1.2)

Subtracting the second equation from the first, the normalized difference on time step $t+1$ is found to be

$$\delta(t+1) = \frac{W(t+1) - B(t+1)}{N} = (1 - 2\mu)\delta(t) \qquad (2.1.3)$$

It follows that $\delta(t) = (1 - 2\mu)^t \delta(0)$. In other words, the system goes gray exponentially rapidly, with oscillation if $\mu > 1/2$.

Entropy in the ring model

We next discuss the increase in entropy in this process. For this model, there is no concept of heat, so our definition of entropy depends on the guidance of statistical mechanics, rather than appeal to experiment. The guidance is this: given a macroscopic description of a system, entropy is the logarithm of the number of microscopic states consistent with that macroscopic description.[3] This guidance has two ambiguities. The first is the base to which the logarithm is to be taken, and represents an overall multiplicative factor in the entropy. In this book we will either work to base 2 or base e or carry the constant as 'k.' The other ambiguity is the word 'macroscopic.' It suggests the existence of collective or large-scale properties for a system. That there are such properties, that entropy could have meaning even without an absolute definition of

---

[2] In words, the first line of Eq. (2.1.1) states: the number of white balls at time $t+1$ is the number that were white at t, reduced by those that were white at t and were in active sites, and increased by those that were black at t and were in active sites.

[3] A caveat for the entire book: pay attention to the i/a distinction. Contrasts between micro and macro will frequently arise. If I could suppress all English language sensibilities, I would write mIcro and mAcro.

'macroscopic,' are central questions. One can view the subject of ergodic theory as a quest for one kind of answer, namely that a dynamical system in equilibrium has only one macroscopic property, energy.[4] However, the interesting situations are those where the system is not fully relaxed, so that even if one is happy with ergodic theory, systems are known to have other macroscopic properties on the way to the relaxed state. A chunk of ice in an isolated glass of water has many macroscopic features: mass, momentum, impurity content, shape. We could describe its melting as a passage through successive macroscopic states, and with a good defini- tion of entropy we expect the entropy of ice plus water to be increasing all along.

However, let's not solve everything just yet. For the Kac ring model there is a natural choice of macroscopic description, the parameter δ. It is a measure of color, $\delta = 1$ being all white, $\delta = -1$ all black and $\delta = 0$ gray. Recall that δ does not take a continuum of values, but is restricted to

$$\delta = \frac{(2W - N)}{N} \qquad \text{for } W = 0, \dots, N \tag{2.1.4}$$

Let the entropy be called S. By the foregoing discussion we define it to be

$$S(\delta) \equiv k \log[\ \#\ \text{microstates consistent with}\ \delta] \tag{2.1.5}$$

The total number of microscopic states available to the balls of the sys- tem is 2^N. Note that the active sites are considered fixed, like quenched impurities in material systems. The number of states of the balls consis- tent with having W of them white is the number of ways of choosing W objects out of N. Thus

$$\exp\left[\frac{1}{k}S(\delta)\right] = \binom{N}{W} \qquad \text{where } \delta = (2W/N) - 1 \tag{2.1.6}$$

Using Stirling's approximation, it is straightforward to conclude that

$$S(\delta) \sim -kN\left[\left(\frac{W}{N}\right)\log\left(\frac{W}{N}\right) + \left(1 - \frac{W}{N}\right)\log\left(1 - \frac{W}{N}\right)\right]$$
$$= -kN\left[\left(\frac{1+\delta}{2}\right)\log\left(\frac{1+\delta}{2}\right) + \left(\frac{1-\delta}{2}\right)\log\left(\frac{1-\delta}{2}\right)\right] \tag{2.1.7}$$

Since we have already deduced the time dependence of δ ($\sim (1 - 2\mu)^t$), we have achieved a complete theory of the relaxation of this ring. In

---

[4] The system is supposed to be pinned to the table; otherwise momentum and angular momentum also persist. The reader may be more familiar with the formulation of the ergodic hypothesis in terms of equality of time and phase space averages. See Appendix B to Section 2.2.

Fig. 2.1.2 are graphs of δ as a function of time (for a system starting all white), along with the corresponding behavior of the entropy.

Remark: We can also express the equilibration as loss of information. Select a particular ball and consider what we know about its color. At any particular time it has probability $(1 + \delta(t))/2$ to be white and one minus that to be black. The information associated with a probability distribution $\{p_\ell\}$ (with ℓ ranging over states of the system) is $I = -k \sum_\ell p_\ell \log(p_\ell)$, for some constant k. For the N balls of our system the total (missing) information is therefore

$$I = -kN \left[\left(\frac{1+\delta}{2} \right) \log \left(\frac{1+\delta}{2} \right) + \left(\frac{1-\delta}{2} \right) \log \left(\frac{1-\delta}{2} \right) \right]$$

If the system starts all white, I begins at 0 and, as $\delta(t)$ decreases, it approaches $kN \log 2$, indicating that we know nothing about the system. Comparing this expression to Eq. (2.1.7), we see that S and I are the same.

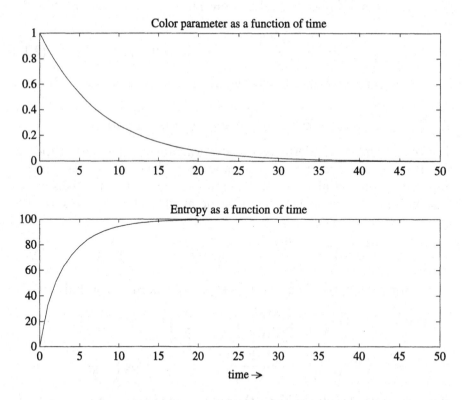

Fig. 2.1.2. Equilibration in the Kac ring model. The upper figure shows the color parameter $\delta(t)$ as a function of time. In this example, for which $\mu = 0.06$, $\delta(t)$ begins at 1 and relaxes exponentially to zero. In the lower figure the entropy, $S(t)$, is plotted, as a function of time. Its value is related to that of $\delta(t)$ by Eq. (2.1.7). It is plotted for $N = 100$ balls and the base of logarithms is taken to be 2.

Remark: As will be shown below, from an initial ball state, this model does not explore $O(2^N)$ configurations, but only $O(N)$. Therefore, as the system evolves there are persistent correlations. An observer capable of perceiving these correlations would not use the product decomposition of the state space implicit in our expression for I, and would find less loss of information than follows from the foregoing estimate. For such an observer, the quantity 'δ' is an inappropriate macroscopic variable. For physical systems, similar discrepancies could arise between observers with different instruments. One observer may believe the system to be in equilibrium while the other finds degrees of freedom that have not fully relaxed. Two examples are the 'negative temperatures' of Ramsey, which depends on nuclear spin degrees of freedom being substantially uncoupled from the lattice, and the spin-echo experiments of Hahn in which energy that has apparently been lost to heat is recovered by manipulation of magnetic fields.

The classical paradoxes and their resolution

The Kac ring model is ideal for developing the two classical paradoxes, paradoxes that cast serious doubt on all of statistical mechanics.

First we consider the reversibility paradox. Suppose you start the system in a state of less than maximum entropy; for example, take 75% of the balls to be white. Let it go for a moderate time T. It becomes grayer and has higher entropy. Now we run the inverse process. For the inverse dynamics the rules are slightly different: balls move *clockwise* and change color when they *enter* active sites, rather than when they leave. However, if you go back to our arguments on the time dependence of δ you will see that nowhere did we use those details. So δ should decrease, and entropy increase, with either rule. Therefore after another T steps with the inverse process the system should be yet grayer, with yet higher entropy. But the inverse process, for T steps, brings us back to exactly the original state, 75% white. Therefore the 'proof' of entropy increase cannot have been correct. For classical mechanics the argument is slightly different. The dynamical laws of Newton are symmetric under reversal of the time parameter. Consider the behavior of a system that at time 0 is started at a state of less than maximum entropy. By the Boltzmann H-theorem, by time T the entropy will have increased. Now run it backwards. For classical mechanics this means reverse all velocities (leaving positions untouched) and use the forward evolution laws. The H-theorem should hold for this new 'initial' state and the entropy again increase. But by the symmetry of the microscopic Newtonian laws, after a time T the system will arrive at its original state of *smaller* entropy.[5]

---

[5] Note that for the ring model we did not require that the inverse dynamics be exactly the same as the forward dynamics. This point is of interest in considering whether *CP*, and presumably *T*, violation could be sources of global time asymme-

The recurrence paradox relies on a more subtle dynamical feature. For Newtonian mechanics it arises from a theorem of Poincaré. A system that is spatially confined returns arbitrarily closely to the point in phase space from which it began. This means that the air from my punctured tire will eventually find its way back in again, very handy if you don't have a spare. The catch is that the amount of time you need to wait before the system 'recurs,' i.e., approaches its initial phase space position, is very long, generally growing exponentially with the number of particles involved. So do carry a spare. Boltzmann could therefore take comfort that this apparent contradiction to his theorem—the return to lower entropy demanded by a recurrence—was not a practical possibility. But in terms of forcing the realization that his proof could not strictly be a proof, the recurrence paradox showed that his assumptions were flawed.

For the Kac ring model recurrence is obvious. Start with all white and consider the state $2N$ time steps later. Every ball has been in every active site twice. Therefore it has switched color an even number of times and is now white again. No matter how you begin, after $2N$ steps, you return to the original state.

It is now our task to examine our 'proof' that $\delta(t) \sim (1 - 2\mu)^t$. This will lead to an approach to statistical mechanics in which our proofs can be made into honest proofs by using ensembles, ensembles of variously defined objects, active site locations, or perhaps ball initial conditions. Unfortunately, there is more notation to deal with. It is convenient to define zero/one-valued variables for the sites and for the balls. Let

$$\epsilon_p \equiv \begin{cases} -1 & \text{if } p \in \mathcal{A} \\ +1 & \text{if } p \notin \mathcal{A} \end{cases} \qquad (2.1.8)$$

where the symbol $\mathcal{A}$ refers to the set of active sites. The dynamical variables are the ball states and these we write as follows

$$\eta_p(t) \equiv \begin{cases} +1 & \text{if the ball in site } p \text{ at time } t \text{ is white} \\ -1 & \text{if the ball in site } p \text{ at time } t \text{ is black} \end{cases} \qquad (2.1.9)$$

Taking the site-numbering to be counterclockwise, the time evolution of the system is given by

$$\eta_p(t) = \epsilon_{p-1}\eta_{p-1}(t - 1) \qquad (2.1.10)$$

from which it follows that $\eta_p(t) = \epsilon_{p-1} \ldots \epsilon_{p-t}\eta_{p-t}(0)$. Addition and subtraction of the index p is modulo N. Recalling the definition of $\delta(t)$

---

try. Clearly, having different dynamics in the two time directions does not to solve the reversibility paradox.

(see Eq. (2.1.4)) we write

$$\delta(t) = \frac{1}{N} \sum_p \eta_p(t) = \frac{1}{N} \sum_p \left[\prod_{\ell=0}^{t-1} \epsilon_{p+\ell} \right] \eta_p(0) \qquad (2.1.11)$$

With this notation we repeat the earlier calculation, so as to pinpoint the assumptions. The key is the adjective that we repeatedly applied to the active site distribution, 'random.' If the sites are random then the site variable ϵ_p will not be correlated with other site variables ϵ_q nor with dynamical variables η_r. The object we wish to evaluate is $\sum_p \left[\prod_{\ell=0}^{t-1} \epsilon_{p+\ell} \right] \eta_p(0)$. In our earlier, casual, discussion the averaging was supposed to be accomplished by the sum over p, but now we explicitly take the site variables to be random variables and include an averaging over *them*. This means that we want the following expectation value, where the brackets refer to the averaging associated with the randomness of the set of active sites,

$$\left\langle \left[\prod_{\ell=0}^{t-1} \epsilon_{p+\ell} \right] \right\rangle \eta_p(0) \qquad (2.1.12)$$

For uncorrelated variables the average of a product is the product of the averages, so that for example $\langle \epsilon_1 \epsilon_2 \epsilon_3 \rangle = \langle \epsilon_1 \rangle \langle \epsilon_2 \rangle \langle \epsilon_3 \rangle$. The ηs average to δ (this will involve the sum over p). The ϵs average to

$$\langle \epsilon \rangle = (+1) \Pr(\epsilon = +1) + (-1) \Pr(\epsilon = -1)$$
$$= (+1) \times \frac{N-A}{N} + (-1) \times \frac{A}{N}$$
$$= 1 - 2\mu$$

We use this in Eqs. (2.1.11) and (2.1.12). Assuming for the moment that the ϵs are uncorrelated, we have that the expectation of the product of the ϵs is the product of their expectation values. Summing over p, it follows that $\delta(t) = (1 - 2\mu)^t \delta(0)$.

What we need for a real proof is thus independence of the variables $\{\epsilon\}$. Statistical mechanics then makes a virtue of necessity. Instead of claiming our equilibration assertion to be true for individual systems, it only makes this claim for ensembles of systems, entire collections of rings each with a different arrangement of active sites. For any particular ring there may be anomalous behavior. But when averaged over the ensemble, the smoothed behavior prevails.

Let us make this precise. Imagine a large collection of rings, with many different subsets, $\mathcal{A}$, selected to be the active sites. These could be generated in the following way: go from site to site, flipping a biased coin. Landing on 'heads' puts the site into $\mathcal{A}$ and the probability of

getting heads is μ. Tails puts the site in the complement of A and has
probability $(1 - \mu)$. All $\{\epsilon_p\}$ are independent of one another.[6] The
expectation used to evaluate Eq. (2.1.11) is not only the sum over sites
p as described in our earlier demonstration, but an averaging over the
ensemble of coin flip-generated rings as well. With this explicit realization
of the independence of the ϵs, the discussion following Eq. (2.1.11) can be
used (with averaging now meaning ensemble averaging) and the formula
for $\delta(t)$ is established.

The proof is now a proof. How have we resolved the paradoxes? First
we address the reversibility paradox. Under the inverse process the dynamical law is not that given in Eq. (2.1.10), but instead is

$$\eta_p(t) = \epsilon_p \eta_{p+1}(t-1) \qquad (2.1.13)$$

Balls change color on entering a site and the motion is clockwise. To see
the implications of this imagine that we evolve for t time steps, arriving
at $\eta_p(t)$, as in Eq. (2.1.10). Now implement the reversed dynamics. For
the next time step we use Eq. (2.1.13), but with $(t+1)$ replacing t. The
result is

$$\eta_p(t+1) = \epsilon_p \eta_{p+1}(t) = \epsilon_p \epsilon_p \eta_p(t-1) \qquad (2.1.14)$$

This equation represents one step with the forward process and one step
with the backward process. The important feature is that ϵ_p appears
twice. The square of ϵ_p is *always* one. Our proof of equilibration used
two essential facts:

$$\begin{aligned}
\langle \epsilon_p \rangle &= 1 - 2\mu \qquad \text{and} \\
\langle \epsilon_p \epsilon_q \rangle &= \langle \epsilon_p \rangle \langle \epsilon_q \rangle
\end{aligned} \qquad (2.1.15)$$

The second fact is the statement that the site variables are uncorrelated.
When you get a particular ϵ_p twice in a row, as in Eq. (2.1.14), they
are correlated, about as correlated as they can get. The expectation of
the product is not that given in Eq. (2.1.15), but is one. In ordinary
language, the ball is retracing its steps. It sees the same environment,
which undoes whatever it did on the first pass. Mathematically, the
argument at the heart of the reversibility paradox fails because for the
motion on the reverse leg of the journey uses *correlated* site variables and
the equilibration proof requires the site variables to be uncorrelated.

The resolution of the recurrence paradox is similar. In Eq. (2.1.11) the
product expression for $\eta_p(t)$ will involve different ϵ_qs so long as $t < N$,

---

[6] Kac also considered another way to generate the ensemble of active sites. He fixed
the total number, but not their positions. This left a correlation, albeit a small
one. For this ensemble, more elaborate reasoning is needed. The simpler procedure
is analogous to what one has in going from the canonical ensemble to the grand
canonical ensemble.

but once $t \geq N$, some site variables ϵ_q will appear more than once in the product.

In both cases, it's a breakdown in the randomness assumption in the environment that invalidates the 'proof' of equilibration. In this, Kac's example is beautifully adapted to its intent. The suspect part of Boltzmann's H-theorem proof was the assumption of molecular chaos, or, in the scientific lingua franca of the day, the *Stosszahlansatz*. So long as you could assume randomness, things went to equilibrium. For the time reversal paradox, that assumption failed, because the reversed motion was highly correlated with the original motion. The breakdown in randomness leading to the recurrence paradox is more subtle and has to do with the difficulty of realizing randomness in a finite deterministic context.

Exercise 2.1.1. The curves in Fig. 2.1.2 are idealized, large N, approximations. In any actual system, with 'random' site locations, there will be fluctuations of δ on its way to equilibrium and around its equilibrium value. Compare Fig. 2.1.3. Evaluate, as a function of N and A, the expected equilibrium entropy per site. It should be less than $\log 2$ by an amount that goes to zero as $N \to \infty$.

Exercise 2.1.2. Find a variant of the Kac ring dynamics that is symmetric in time.

Exercise 2.1.3. For $t = N$ the entropy recurs although no ϵ_p repeats. Explain.

'Deriving' the thermodynamic arrow

Boltzmann's H-theorem is a demonstration that a certain quantity, closely related to entropy, can only increase as a system moves forward in time. We have shown the same thing for the entropy in the Kac ring model. Have we derived the thermodynamic arrow of time?

We have not. With the understanding gained by studying the reversibility paradox, it is clear that all we have proved is that if you take an unusual state ('unusual' to be defined in a moment) and evolve away from it *in either time direction*, its entropy increases. This is illustrated in Fig. 2.1.3 where we took a state consisting of 92% white balls at time zero and evolved away from it in both time directions using the appropriate dynamical law.

What then was 'unusual' about the time zero state that we took? Simply this: its macroscopic characterization (δ) was such that there were relatively few microscopic states consistent with that characterization. In other words, it had low entropy. In this light, the version of the Boltzmann H-theorem that we proved begins to border on tautology. Things are more likely to look like what most things look like. You start a system somewhere in its universe (the 2^N ball sequences in our case) and let it wander. If it is not trapped where you started it, its macroscopic

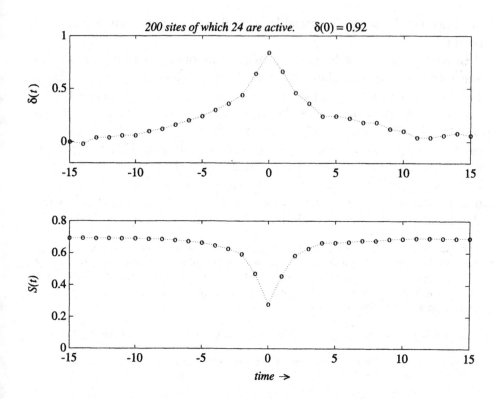

Fig. 2.1.3. Time symmetry. Moving in either time direction away from the unusual time-0 state (with over 90% white), the entropy increases. The plots show a computer simulation in which both active sites and time-0 ball color were randomly selected, appropriately biased to give the indicated values.

features will begin to change. And they are most likely to change to the macroscopic features possessed by most states.

But the past 150 years of statistical mechanics have not been wasted on tautologies. There are two substantive issues and one historical. The historical, or perhaps philosophical, point is not to underestimate the subtlety of a tautology. Realizing that a situation is the only one possible is significant progress. In a way, general relativity takes force out of gravity and makes it mere geometry. If the H-theorem is a tautology, it could be said to take the dynamics out of thermodynamics and make it mere counting.

The first substantive issue is the effectiveness of the dynamics at making the system 'wander' away from where you started it. This is what ergodic theory is all about, showing that the dynamical laws we believe obtain in Nature do indeed cause system states to sample significant fractions of phase space.

The second substantive issue returns to my question a few paragraphs back. Apparently, we have *not* derived the thermodynamic arrow of time. But the H-theorem has helped focus on what it would take to make such a derivation. You need to explain why the states we produce as *initial* conditions are the ones with low entropy, are the 'unusual' ones. I can take an ice cube and drop it in a glass of water, isolate this for an hour and find the ice cube melted. Its entropy has increased, because I was able to prepare something 'unusual.' What I cannot do is make something have an unusual *final* state. I cannot take a glass of water, force its microscopic state to be one of those relatively few that are the time reversals of a melted ice cube, isolate it for an hour and come back to find an ice cube spontaneously formed, floating in warmer water. So the statement that the entropy of isolated systems increases is the statement that **we** *prepare initial states* that are unusual. To understand this aspect of the thermodynamic arrow you must go back to our arrow as agents, and from ours to that of planet earth, and so forth.

Appendix for this section (at the end of the chapter):
2.1.A   Other ensembles and limitations of the ring model

2.2 Continuous dynamics: the cat map

Our next example is a caricature of classical mechanics. We still use a discrete time variable, but the 'motion' takes place in a continuum, the two-dimensional unit square. Strictly, it's a torus since we use periodic boundary conditions at opposite edges, but you should think of the space as classical phase space, for reasons I will explain later. The mapping sends the unit square into itself. For (x, y), with $0 \le x < 1, 0 \le y < 1$, take

$$
\begin{aligned}
x' &= x + y & \text{mod } 1 \\
y' &= x + 2y & \text{mod } 1
\end{aligned}
\tag{2.2.1}
$$

The transformation in Eq. (2.2.1) is known as the cat map because in the book of Arnold and Avez its effectiveness at inducing equilibration is illustrated with the image of a cat. It would be remiss of me not to provide this picture to the reader and in Fig. 2.2.1 will be found, in the upper left corner, the image of a cat; it consists of about 20000 individual points. To the right is shown what happens when each point is mapped by Eq. (2.2.1). The lower left and lower right illustrations continue the time evolution. After the first step you can still see where the eyes were, even pick out the bend for one of the ears. But by the third time step, little is recognizable. It is convenient to rewrite Eq. (2.2.1) in other forms:

$$\xi' = \varphi(\xi) = M\xi \qquad \text{mod } 1 \qquad (2.2.2)$$

with

$$\xi = \begin{pmatrix} x \\ y \end{pmatrix} \qquad \text{and} \qquad M = \begin{pmatrix} 1 & 1 \\ 1 & 2 \end{pmatrix}, \qquad (2.2.3)$$

which also defines φ.

To understand the mapping φ, it is helpful to show what it does to a square that is close to the origin so that the 'mod 1' operation is not invoked. This is shown in Fig. 2.2.2. The square is stretched along one axis, squeezed along the other. This also illustrates one of the most important features of φ: it is measure preserving. That is, when mapping one region into another, the area stays the same. In general, the area maps according to the Jacobian of the transformation, $\partial(x', y')/\partial(x, y)$.

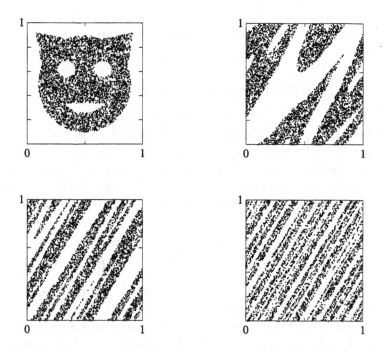

Fig. 2.2.1. Evolution of a cat under the transformation φ, known, by virtue of a similar illustration in the book of Arnold and Avez, as the 'cat map.' Top left is the initial configuration. Top right, the image after one time step. Bottom, next two steps.

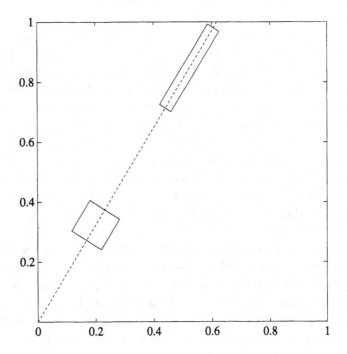

Fig. 2.2.2. Action of φ. The square is mapped into the rectangle. The square is close enough to the origin that the addition operations do not yield numbers greater than 1. The dashed line is along the direction of the eigenvector of the matrix M having the larger eigenvalue.

But that Jacobian is the matrix M of Eq. (2.2.3), whose determinant is 1. Now consider the eigenvalues and eigenvectors of M. They are

$$\lambda_\pm = \frac{3 \pm \sqrt{5}}{2}, \qquad v_\pm = \frac{1}{\sqrt{\lambda_\pm^2 + 1}} \begin{pmatrix} 1 \\ \pm\lambda_\pm \end{pmatrix} \qquad (2.2.4)$$

and the numerical values of $\lambda_\pm$ are about 2.618 and 0.382, or $\exp(\pm 0.962)$, with product unity because $\det(M) = 1$. Imagine a collection of points along a line from the origin in the direction of the eigenvector v_+. Applying M to this produces points along the same line that are (about) 2.6 times as far from the origin (before application of the mod 1 operation). This explains the long dimension of the rectangle in Fig. 2.2.2. The other eigenvector is orthogonal, and along it one has shrinking by a corresponding amount. This is the short dimension of the rectangle. The mutilation of the cat image is now understood as this distortion process

combined with the mod 1 operation (which brings stretched objects back
into the unit square).

The measure-preserving property of the transformation motivates the
analogy of this system to Hamiltonian dynamics in phase space. The
Liouville theorem of classical mechanics is the statement that volumes
in phase space are preserved under Hamiltonian dynamics. Furthermore,
like phase space, giving an initial point for the cat map is enough to
specify all future positions. Finally, the inverse of the cat map, which
exists everywhere, is also (necessarily) measure preserving, although its
functional form is slightly different from that of the cat map itself.

Let us look at simple equilibration properties for this map. In an
Appendix I will introduce the notions of mixing and ergodicity which
sharpen ideas along these lines, but here I prefer a more intuitive level.
A collection of points in the box can be thought of as gas molecules
in a container, perhaps imagining the x-axis of the unit square to be
coordinate space. As for real containers, if the gas starts in a small region
it will spread. We will define a kind of entropy, more ambitiously than
we did for the Kac ring, and watch it increase as the system approaches
equilibrium. It will also be convenient to study finite-size effects, which
have the interpretation of statistical fluctuations in the gas.

As an illustration of the equilibrating properties of the map, less dramatic perhaps than Fig. 2.2.1, we consider 250 points started in a small
0.2×0.1 rectangle in the lower left corner of the unit square. In Fig. 2.2.3,
we follow it for a few time steps, showing steps one and two to show the
initial stretching and squeezing, and step seven to show how soon the
perception of structure is lost.

Defining entropy

To deal with this 'perception' in a quantitative way and to connect as
well with thermodynamics, we turn to the definition of entropy. Because
cat map dynamics in the unit square is rich enough to have an association with classical phase space, I can model the definition of entropy
on an understanding of how entropy should be defined classically (which
also has a natural quantum version). However, despite the additional
structure the basic idea is the same as in Section 2.1. Entropy is the logarithm of the number of microscopic states consistent with a macroscopic
description. Again this highlights the problem of defining 'macroscopic.'
The intuitive idea is that if my ruler can measure a difference—including
good rulers like scanning tunneling microscopes—then the coordinate is
macroscopic. There are also devices for measuring velocity, and at any
particular stage of technology, laboratory and other devices are capable of

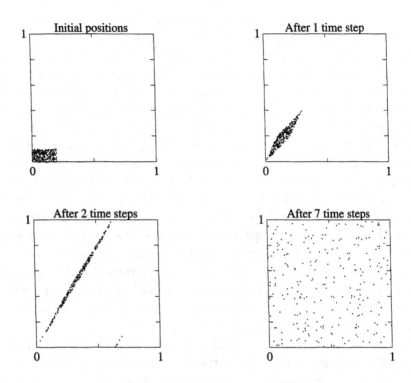

Fig. 2.2.3. Action of the cat map, φ, on 250 points initially in a small rectangle in the lower left corner. By time step 7 all trace of structure is lost to the eye.

defining different regions of phase space. Characterizing a measurement as defining a region of phase space also includes macroscopic observations like, 'the stone is green.' Color and other non-metric properties translate into statements about composition. In the full multidimensional phase space, having a piece of jade in a certain location with a ruby to its right is a different phase space region from reversing the stones. If your apparatus cannot distinguish a composition variable, despite good precision in position and velocity, your 'macroscopic' description will cut a wide swath in phase space. With this point of view, a macroscopic description becomes a coarse graining of phase space. The term 'coarse graining' suggests granularity, with the precision of the measurements defining the grains.[7] In multiparticle physical phase space, grains, even the best at our present level of technology, are large and intricate.

The foregoing discussion of 'macroscopic' and hence of entropy, introduces subjective or technology-dependent features into what is supposed to be fundamental physical law. There are various ways to deal with this problem. Perhaps you could justify an underlying minimal coarse graining.[8] If not, and my view is that there is no *a priori* coarse graining, the question can be dealt with at two levels. The first is to establish that the conclusions of the theory with a particular coarse graining are substantially independent of the precision.[9] (We consider the issue of grain-size independence in a Remark below.) At the next level there is the subtler issue of confronting the *subject* of the subjectivity. Are we the evolutionary products of Nature's intrinsically defined coarse grains? This would suggest that our choice of coarse graining reflects deeper properties (for example, spatial localization). Or could there be radically different coarse grains with correspondingly different measures of entropy? I believe these are valid questions, but since I don't have anything to say beyond asking them, I will leave the matter open.

Coarse graining then consists of a specification of regions in phase space. For the cat map we divide the unit square into small rectangles. If we have N points moving under cat-map dynamics, the microscopic state would consist of listing the positions of the points either in an idealized continuum sense or to the precision of a particular computer. The macroscopic state would consist of giving the number of points in each grain. In Fig. 2.2.4 we show a coarse graining consisting of 50 grains, each a rectangle of size 0.2×0.1. (In this case I have good justification for my grain size: it's the default grid dimension in my graphics program.)

---

[7] Our implementation of coarse graining involves an artificially sharp boundary between nearby, but practically indistinguishable, points (points just on opposite sides of a grain boundary). A more elaborate treatment would involve smearing functions and could lead to fuzzy logic.

[8] Taking grain size $\hbar^N$ for an N degree of freedom system would not be appropriate since this would be a *microscopic* description. In quantum theory this is the *most* you can say about a system, and saying that your microstate is in such a grain leaves you exactly one microscopic state consistent with the description.

[9] The 'theory' that should be independent of the coarse graining is the usual thermodynamic theory, steam engines and electrolytic cells. One can certainly expect new physics as contemporary measurements, for example, capture of single atoms in electromagnetic traps, push back the frontier of the micro/macro distinction. In fact this can be considered one of the possible routes to an experimental test of the quantum measurement theory developed later in this book. The 'special' states that I will speak of correspond to microscopic states that until now have been macroscopically indistinguishable from non-special ones. One of the experiments proposed below can be thought of as just such a shrinking of the coarse grains—an ability to affect special states in a way different from the non-special ones.

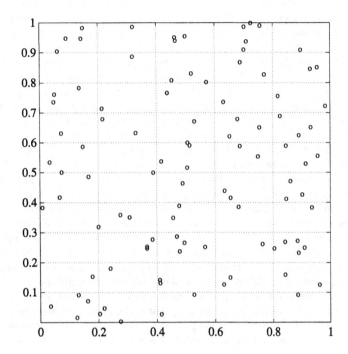

Fig. 2.2.4. Coarse graining of the unit square. Rectangles are the analogs of regions in phase space. A macroscopic observation tells how many points are in a rectangle, but not where in the rectangle they are. In the figure there are 50 grains and 100 points randomly distributed among them. Using our entropy definition, the entropy per point was calculated to be 3.7074. This is to be compared to log 50 ≈ 3.9120. See Exercise 2.2.2.

Remark: Physical insensitivity to coarse grain size. If you change your coarse grains do you change the physics? If the change reflects new apparatus and new phenomena revealed by that apparatus, then, yes; you may for example see fluctuations that were previously hidden beneath averages. However, here we wish to show that for *given* apparatus the physics is not sensitive to the 'arbitrary' grain size, provided that grain size is beyond the discriminating powers of the apparatus. Let us be specific about the coarse grain selection. Fix the system and divide phase space into subsets such that within each the equipment cannot distinguish between its constituent points.[10] Let the subset of smallest volume be Γ_0. Let its phase space volume be $\mu(\Gamma_0)$. Now divide the minimal-discrimination subsets into coarse grains of this size,[11] even though many grains will be smaller than needed; that is, the equipment is unable to tell in which of several

---

[10] For quantum mechanics replace 'subset of phase space' by subspace of Hilbert space, and volume by dimension.

[11] Incommensurate volumes can be handled by a finer cut and error estimates.

grains some points are. The entropy is $S = -\sum p_\alpha \log p_\alpha$, where $p_\alpha = \int_{\Gamma_\alpha} \rho$ and ρ is the phase space density. A microscopic state in Γ_0 has entropy 0, while one that might (indistinguishably) be in one of N grains, has entropy $\log N$. (The coarse graining makes the density into a normalized sum of the characteristic functions of the Γ_α in which the system might be found.) Now suppose a yet finer cut had been made: each Γ_α is the disjoint union of yet smaller sets $\gamma_{\alpha\sigma}$, say n of them, all of equal volume. Let the 'probability' $p_{\alpha\sigma}$ be defined with respect to $\gamma_{\alpha\sigma}$ as we defined p_α above. Let $s = -\sum_{\alpha\sigma} p_{\alpha\sigma} \log p_{\alpha\sigma}$. Writing $p_{\alpha\sigma} = (p_{\alpha\sigma}/p_\alpha)p_\alpha$ and $q_{\alpha\sigma} \equiv p_{\alpha\sigma}/p_\alpha$, it is immediate that $s = S + \sum_\alpha p_\alpha[-\sum_\sigma q_{\alpha\sigma} \log q_{\alpha\sigma}]$. Because the equipment cannot distinguish among the $\gamma_{\alpha\sigma}$, the sum $-\sum_\sigma q_{\alpha\sigma} \log q_{\alpha\sigma}$ is simply $\log n$. This is the same additive constant no matter what the original density, so long as it was constant on each of the coarse grains. The next question concerns transformations. Suppose there is a mapping ϕ of phase space. Under ϕ a phase space density ρ is mapped to another density, which is in turn replaced by its coarse graining; i.e., on each grain the density is replaced by its average on that grain. This can only lead to an increase of entropy (or no change). (For proof, see the next Remark.) Since by assumption our equipment is unable to differentiate states more finely than the original Γ_0, ϕ, which is a transformation available to our 'equipment,' will send the density of each $\gamma_{\alpha\sigma}$ uniformly to each image set (if the equipment could nonuniformly occupy image sets, then it could preferentially occupy preimage sets, contradicting its assumed resolution). A straightforward calculation shows that again the only effect of the finer coarse graining is the addition of a constant $\log n$ to the entropy. For comparisons, this constant is irrelevant. Note, by the way, that when quantum phenomena come into play one sometimes uses power-of-$\hbar$ fine graining. Thus a statement that entropy goes to zero as temperature goes to zero is essentially the assertion that the ground state is not exponentially (as a function of volume) degenerate. (That assertion is known to be false for some model systems.)

Remark: Insensitivity, cont. It's not obvious that with the scheme outlined above you always increase entropy (whether with the original coarse grains or with a finer set). If the original density is concentrated on a single grain, a transformation, ϕ, can only spread the density to several grains and surely entropy increases. But with an initial state supported on several grains one could imagine that density from different regions went to the same coarse grain and would lead to a decrease in S. Suppose that initially the probability distribution is $\{p_\alpha\}$, so that the density is $\rho = \sum_\alpha p_\alpha \chi_\alpha$, where χ_α is the characteristic function of the set Γ_α. Under the transformation, $\Gamma_\alpha \to \phi(\Gamma_\alpha)$, which may overlap with many grains Γ_β. For any particular Γ_β its total probability is the sum of the measures of all its intersections with $\phi(\Gamma_\alpha)$ that entered it. Note that within Γ_β these will be disjoint, by the uniqueness of the classical mechanics (or the unitarity of the quantum transformation). This means one can simply add probabilities. It follows that the density ρ' after transformation with ϕ is $\rho' = \sum_\beta p'_\beta \chi_\beta$ with

$$p'_\beta = \sum_\alpha p_\alpha \mu[\phi(\Gamma_\alpha) \cap \Gamma_\beta]$$

The new entropy is $S' = -\sum p' \log p'$, and we must show $S' \geq S$. Define $g(\alpha, \beta) \equiv \mu[\phi(\Gamma_\alpha) \cap \Gamma_\beta]$. Since the dynamics (and thus ϕ) are measure preserving, summing g over either α or β will give the measure of the elementary volume, $\mu(\Gamma_0)$. Hence, up to a factor, g is a doubly stochastic matrix.[12] So we need a theorem about doubly stochastic

---

[12] A stochastic matrix is a matrix of nonnegative elements whose columns add to unity. It is doubly stochastic if the rows also add to unity.

matrices and their relation to entropy. What we'll use is something known about general stochastic matrices. The quantity $S(p|q)$ is defined for two probability distributions p and q as $S = -\sum_x p_x \log(p_x/q_x)$ and is known as relative entropy (see Sections 2.4 and 5.2). If R_{xy} is a stochastic matrix ($\sum_x R_{xy} = 1$), then it is known that $S(Rp|Rq) \geq S(p|q)$ (references in Section 2.7). If you now take q_α to be constant, the doubly stochastic nature of g shows it to be unchanged when acted on by g. This gives the desired result. Note that for the quantum situation this derivation is relevant only if the grains are large enough for no interference effects to be important.

Aside: The word 'entropy' is related to the Greek for 'turning' or 'transformation.' Not only does entropy describe change but its meaning varies from discipline to discipline. In the book of Arnold and Avez you will find a definition of Kolmogorov, which is different from ours. It is the entropy of a *transformation*, not of a state.

Calculating the entropy

Having associated macroscopic variables with phase space regions called 'coarse grains,' we address the counting problem. We are given a sequence $n_1, n_2, \ldots, n_g$, where g is the number of coarse grains, n_j the number of points in coarse grain j, and $\sum n_j = N$. We want the number of detailed microstates consistent with that. Because our mapping has been defined on the continuum it is necessary to introduce a cutoff to give a finite result for entropy. Such a *small* dimension cutoff raises issues different from those encountered earlier. In particular, the cutoff only produces an overall, classically irrelevant constant entropy shift.[13] Another feature that influences the counting is whether or not the particles are distinguishable. For present purposes our 'N' gas points are more like a tensor product of N separate systems, so we will do the calculation for distinguishable points. In an Appendix we treat indistinguishable points.

Suppose then that each coarse grain is composed of G fine grains. G is taken to be enormous compared to anything else around. The number of ways (consistent with the sequence $n_1, n_2, \ldots, n_g$) to distribute the distinguishable points among the g coarse grains is

$$\frac{N!}{n_1! n_2! \ldots, n_g!}$$

Within grain k they can be placed in $G(G-1)\ldots(G - n_k + 1)$ ways. Therefore the number of microstates consistent with the given occupation numbers is

$$\exp \frac{S}{k} = N! \prod_{j=1}^{g} \binom{G}{n_j} \tag{2.2.5}$$

---

[13] In quantum mechanics there is a natural phase space minimum size scale, namely the physical, $\hbar$-related, quantum cell size.

Because G is large ($n \ll G$, even for $n = N$), the combinatorial coefficient simplifies: $\binom{G}{n} = \frac{G!}{n!(G-n)!} \approx \frac{G^n}{n!}$ and it follows that

$$\frac{1}{k}S = N \log G + \log N! - \sum_{j=1}^{g} \log(n_j!) \qquad (2.2.6)$$

Using Stirling's approximation this gives

$$\frac{1}{k}S = N \log G - N \sum \rho_j \log \rho_j \qquad (2.2.7)$$

where $\rho_j \equiv n_j/N$. In defining entropy we want the macrostate in which all points are in a single grain to represent maximal macroscopic information. This implies that the term $N \log G$ in Eq. (2.2.7) should be dropped.[14] Thus

$$\frac{1}{k}S = -N \sum \rho_j \log \rho_j \qquad \text{with} \qquad \rho_j = n_j/N \qquad (2.2.8)$$

With this definition, if the points are evenly distributed among the grains, we have $\rho_j = 1/g$ for all j and Eq. (2.2.8) gives $S/kN = \log g$. For the random (Poisson) distribution of points shown in Fig. 2.2.4, the entropy per point is slightly less than the maximum, $\log 50$.

Exercise 2.2.1.   Show that a continuum limit does not eliminate the logarithmic ambiguity due to fine grain size.

Relaxation and equilibrium

We now have the machinery to view the equilibration process. For the systems in Fig. 2.2.5, points were started from a single coarse grain and allowed to evolve, in one case 50 points, in the other 200. Entropy as a function of time is shown. This is a classic illustration of the effect of sample size on fluctuations.

It is amusing to embark on two exercises in statistical mechanics. We will estimate the time scale for the approach of S to its equilibrium value and the finite-size effect deviation—fluctuations—of S from its maximum, $\log g$.

For the approach to equilibrium it is the Lyapunov exponent that sets the time scale. That exponent is essentially the (logarithm of the) rate at which nearby points run away from each other and for the cat map is $\log \lambda_+ \approx \log 2.618 \approx 0.962$. Suppose that all the grains are very small and all the points are started in a single grain. Let φ be applied once. That set

---

[14] Consistent with continuity, we set $x \log x = 0$ for $x = 0$.

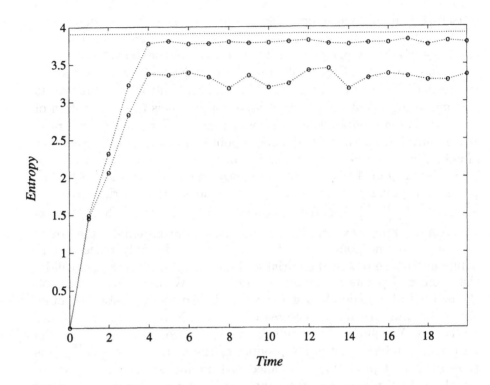

Fig. 2.2.5. Entropy per point as a function of time for two samples. For the upper curve 200 points were used, for the lower, 50. The number of grains was 50. In both cases, all points started in a single coarse grain. Note the larger fluctuations in the smaller sample as well as the greater deviation from the maximum, log 50 (the line near the top). For the 200 point sample the average deviation from log 50 for $t > 6$ is 0.118.

of points is now stretched and squeezed and (compare Fig. 2.2.2) spread among several other grains, on the order of 2.6 of them; that is, most of the points are in two of them, some in a third and possibly additional points in others due to 'mod 1' cuttings or figures that protrude slightly into other grains. For this first step the geometry can be complicated (and the entropy increase rapid), but as further steps are applied the image of the initial grain becomes thin and the number of grains occupied by that image is close to 2.6 times the number occupied by the preimage. This will remain true until individual grains begin to get contributions from more than one preimage, which obviously becomes likely as λ_+^t approaches g. By taking the grains sufficiently small the era during which the factor λ_+ growth obtains can be made long. Under this small-grain assumption, we estimate the evolving distribution of points by taking the number of grains effectively occupied at time step t to be λ_+^t. If the number of points is itself large each grain so occupied will have roughly N/λ_+^t points in it. Applying Eq. (2.2.8), we get $S \sim -N\log(1/\lambda_+^t) = Nt\log\lambda_+$. The rate of change of entropy per point is thus the Lyapunov exponent.

For the finite-size effect I will start the exercise and leave the rest for the reader. Many time steps have passed, the entropy is close to its maximum $(\log g)$ but fluctuates at lower values. It is the expectation of that fluctuation-induced defect that we calculate. For a particular grain, A, we want the probability that it has n points in it at time t. The number of points in A at time t is the number of points in its preimage at time $t-1$. Thinking of A as a set, this preimage can be written $\varphi^{-1}(A)$. Now φ^{-1} is a mapping very much like φ. In particular the matrix associated with φ^{-1}, $\begin{pmatrix} 2 & -1 \\ -1 & 1 \end{pmatrix}$, has the same eigenvalues as M, so that φ^{-1} has the same Lyapunov exponent as φ. Extending our argument, the points in A at t are the points in $\varphi^{-t}(A)$ at 0. But the only place to find points at time zero is in the initial grain, call it A_0. Therefore n will be the number of points in the set $A_0 \cap \varphi^{-t}(A)$. Within that set we took the points to be randomly distributed so the expected number of points should be proportional to the area of the set. Now, finally, we can use a theorem. We show in Appendix B to this section that as $t \to \infty$ the area of that intersection is the product of the areas of the sets.[15] This is an important property; it is called *mixing* and we will have more to say about it. Dynamical systems that are mixing equilibrate and we are now in a position to use that property. All grains have the same area, $1/g$. The area of the intersection, $A_0 \cap \varphi^{-t}(A)$, is therefore a fraction $1/g$ of the area of A_0. The original random process that filled A_0 dropped N points into it. The probability that any one of those points lands in the intersection $A_0 \cap \varphi^{-t}(A)$ is therefore $1/g$. This is a Poisson process with rate $r \equiv N/g$ and the probability that exactly n points land in $A_0 \cap \varphi^{-t}(A)$ is therefore

$$P_n = e^{-r}\frac{r^n}{n!} \tag{2.2.9}$$

We have thus found the probability distribution of the set A at time t. To calculate $\langle S \rangle$ you need to take the expectation value of $-\rho\log\rho$, that is, $N\log N - \sum P_n n \log n$.

Exercise 2.2.2.   Finish the calculation. The case of large r is straightforward.

Exercise 2.2.3.   As for the Kac ring, recurrence and reversibility paradoxes can be formulated for the cat map. Reversibility involves a slight variant of the forward motion, namely φ^{-1} with appropriate handling of the mod 1 operation. Recurrence follows from the measure-preserving property. The time scale for this phenomenon is g^N, but try

---

[15] Recall that the area of our universe is 1.

to get more information, namely, find—or estimate—the probability that a recurrence occurs on time step T for large T.

Exercise 2.2.4. The recurrence time depends on the dynamics. As an example, think of the points as gas atoms, and modify the dynamics so that they only traveled in pairs with particles in a pair never getting more than a distance 10^{-10} away from one another, a kind of molecule. This changes the recurrence time; for coarse grains larger than 10^{-10} on a side, the time scale would go from g^N to $g^{N/2}$. Consider the effect of other dynamical rules on the recurrence time. Could you modify the dynamics so as to make the recurrence time longer?

Exercise 2.2.5. Can you slow the cat? Show that for no choice of integers in the transformation of Eq. (2.2.1) is there a smaller Lyapunov exponent. What about a continuous time version? Show that using $M^{1/3}$ (instead of M) gives asymptotically non-uniform density. In studies of dynamical systems a variety of other transformations are considered, the baker's transformation, etc. Are some of them amenable to analytic calculation? For example are there coarse grainings for which the entropy increase could be computed analytically?

Appendices for this section (at the end of the chapter):

2.2.A   Entropy for cat-map dynamics with indistinguishable points
2.2.B   Mixing and ergodicity in the cat map
          The Poincaré recurrence theorem
          Ergodicity
          Mixing
          The cat map is mixing

2.3 Classical dynamics

For classical mechanics Goldstein's book (referred to below) is the canonical source. The measure preserving property to which we earlier referred holds on phase space ($6N$ dimensions for N particles moving in $\mathbb{R}^3$) and is the content of Liouville's theorem.

An important discovery of the past few decades is the existence of islands of regular motion in complex dynamical systems. These are called KAM tori, after Kolmogorov, Arnold and Moser, and they are called tori because, like the motion of a multidimensional harmonic oscillator, the orbit of the system is essentially a torus. References are given in Section 2.7. These tori are known to exist when the system has been perturbed slightly from regular motion. In contrast to ergodic motion, an orbit on a torus does not sample much of phase space. However, for the program of ergodic theory there are two mitigating factors. First, only part of phase space, presumably a small part, lies on tori. Second, with three or more degrees of freedom there is the phenomenon of Arnold diffusion allowing orbits to get around barriers.

I am loath to use these discoveries as the basis for conclusions about the foundations of statistical mechanics. First, there are still open questions in the study of dynamical systems. More important, the underlying physics is quantum mechanical. Although the classical mechanics is a correspondence limit of the quantum mechanics, the features that play a role in establishing the detailed picture of KAM tori and Arnold diffusion may not be robust. (My doubts are focused on macro/micro statistical mechanics issues. The implications for celestial mechanics are another matter.) The point is that the classical trajectory of a system of particles is only an approximation to the underlying wave propagation. It may be that the smearing of that trajectory is enough to invalidate the delicate conclusions of KAM theory.

2.4 Stochastic dynamics and the master equation

A principal tool in the study of non-equilibrium statistical mechanics is the master equation. This is one of a family of closely related equations, the Fokker-Planck equation, the Chapman-Kolmogorov equation, various transport equations. The framework is that of a Markov process. You have a state space and the configurations or states of the system are labeled by points of the state space, $\alpha, \beta, \ldots$. The system state undergoes transitions from one to the other of these with transition probability $W(\alpha, \beta)$. That is,

$$W(\alpha, \beta) = \Pr\left(\text{System state is } \alpha \text{ at } (t+1) \mid \text{System state is } \beta \text{ at } t\right)$$
$$(2.4.1)$$

The Markov property is the statement that this probability does not depend on the state of the system prior to t. To help remember what is going where,[16] it may be useful to write $W(\alpha \leftarrow \beta)$. Clearly $\sum_\alpha W(\alpha, \beta) = 1$, since starting from any β you must *somewhere*. In this context one studies $u(\alpha, t) \equiv \Pr\left(\text{System state is } \alpha \text{ at } t\right)$, which, from the definition of W, satisfies

$$u(\alpha, t+1) = \sum_\beta W(\alpha, \beta) u(\beta, t) \qquad (2.4.2)$$

The fact that W summed over its first index gives unity implies that probability is conserved, i.e., from $\sum u(\alpha, t) = 1$, you get $\sum u(\alpha, t+1) = 1$. For W close to the identity matrix, it is convenient to write its off-diagonal matrix elements as $w(\alpha, \beta)\Delta t \equiv W(\alpha, \beta)$, with Δt a fixed positive number for all α, β. The condition $\sum_\alpha W(\alpha, \beta) = 1$ now fixes the diagonal

---

[16] ... which can be difficult because W's indices are sometimes defined oppositely.

elements of W in terms of w and we get

$$W(\alpha, \beta) = \delta_{\alpha\beta} + \left[w(\alpha, \beta) - \sum_\gamma w(\gamma, \beta)\delta_{\alpha\beta}\right]\Delta t \qquad (2.4.3)$$

If Δt is small, this can be considered a short time approximation for an underlying continuous time process (so the ws are rates). For a continuous time process, the probability functions u satisfy

$$\frac{du(\alpha, t)}{dt} = \sum_\beta w(\alpha, \beta)u(\beta, t) - \sum_\gamma w(\gamma, \alpha)u(\alpha, t) \qquad (2.4.4)$$

Alternatively, one could consider w to be the given matrix and then derive a finite time transition matrix.[17] The matrix W is known as 'stochastic' and at this point we have imposed no conditions on it other than $0 \le W(\alpha, \beta)$ and having columns that sum to unity.

As part of a program to understand irreversibility, the use of Eq. (2.4.2) or Eq. (2.4.4) begs the question. Eq. (2.4.4) is first order in time and in general asymmetric. To show this I quote snippets from the extensive lore on the spectrum of stochastic matrices, some of which is immediate, some more subtle. W has an eigenvalue 1 and all other eigenvalues are of norm equal to or less than one. The eigenvector associated with the eigenvalue 1 has non-negative entries. Eq. (2.4.2) can be written in matrix notation as $u_t = W^t u_0$. Therefore for large t the largest eigenvalues of W dominate and in particular u works its way into the (generally) smaller space spanned by W's eigenvectors of largest eigenvalue. Often there is only one of these.[18] For large dimension there may be clustering near 1, a typical phenomenon where power law decay occurs.

Increase of entropy

Under the right circumstances you can go further with Eq. (2.4.4) and show that *entropy* increases. One such circumstance is that your system is isolated—you are in the microcanonical ensemble. Anticipating a result to be discussed below, time reversal invariance then implies that W (or w) is a symmetric matrix. In this context the state variables can provide the coarse graining and entropy is the missing information. Therefore

$$S(t) = -\sum_\alpha u(\alpha, t) \log u(\alpha, t) \qquad (2.4.5)$$

---

[17] W would be the propagator for Eq. (2.4.4).
[18] If for all α, β, there is an n such that $(W^n)_{\alpha\beta} > 0$, then the eigenvalue 1 is unique.

Taking the time derivative of Eq. (2.4.5), using Eq. (2.4.4), flipping
dummy indices, using the symmetry of w, and adding yields the result

$$2\frac{dS}{dt} = \sum_{\alpha,\beta} w(\alpha,\beta)[u(\beta,t) - u(\alpha,t)][\log u(\beta,t) - \log u(\alpha,t)] \qquad (2.4.6)$$

Since the logarithm is a monotonically increasing function and $w(\alpha,\beta)$
non-negative, $dS/dt \geq 0$. (If W is not symmetric but satisfies detailed
balance with a non-trivial function $\bar{p}(\alpha)$, then replacing $\log u(\alpha,t)$ by
$\log[u(\alpha,t)/\bar{p}(\alpha)]$ in the definition of S allows proof of increase for this
'relative entropy'—which is actually just another thermodynamic func-
tion. (See Sections 2.2 and 5.2.)

Exercise 2.4.1.   A *doubly stochastic* matrix is a non-negative matrix for which rows
and columns add to one. For a state space with N elements, would that condition be
sufficient to ensure that the entropy attain the value $k \log N$?

Exercise 2.4.2.   How could you produce (calculate, compute, whatever) W for a coarse
graining of the cat map?

As remarked, all this begs the question. Our goal is not to examine the
consequences of irreversibility but to find where it enters in an otherwise
time symmetric theory. The issue is where did we get Eq. (2.4.2) or
Eq. (2.4.4)?

Derivation of the master equation

The earliest derivations of the master equation from quantum mechanics,
are due to Pauli, and his arguments were extended by van Hove, Prigogine, Zwanzig and others. As you can imagine, the arguments involve
profound, difficult and at times obscure assumptions.

There are *two* subtleties that enter the argument, one having to do with
destroying coherence, which in this context is the essential requirement
for irreversibility, the other with a minimum time scale on which you can
have any hope of accomplishing the first goal. The minimum time scale
problem is in a way technical and microscopic, and not directly related
to statistical considerations.

To justify Eq. (2.4.4) one is naturally led to think of the probabilities
$u(\alpha,t)$ as $|\langle\alpha|\psi(t)\rangle|^2$, where $\psi(t)$ is the wave function of a system at time
t and α now signifies a Hilbert space basis vector. For the moment I
will take this approach, although it is inadequate and I will later argue
that it is only with coarse graining (or its equivalent) that the derivation
goes through. With the provisional identification $u(\alpha,t) = |\langle\alpha|\psi(t)\rangle|^2$,
it is clear that we will have to truncate somewhere. Quantum evolu-
tion depends on the phases of the wave function, while the master equation, employing only norms, drops them. This can be seen explicitly

by writing the time evolution of a probability. To avoid confusion with earlier material, the Hilbert space basis will be called $\{|n\rangle\}$ with n running over an index set. Let $U(t) \equiv \exp(-iHt/\hbar)$, let the wave function at time zero be $\psi^{(0)}$, let its matrix elements be $\psi_n^{(0)} \equiv \langle n|\psi^{(0)}\rangle$, and let $p_n(t) \equiv |\langle n|\psi(t)\rangle|^2 = |\langle n|U(t)|\psi^{(0)}\rangle|^2$. We use the notation $U(t)_{nm} = \langle n|U(t)|m\rangle$ and recall that $U(-t) = U(t)^{-1} = U(t)^\dagger$. It follows that

$$p_n(t) = \sum_m \sum_{m'} \psi_{m'}^{(0)*} \psi_m^{(0)} U(-t)_{m'n} U(-t)_{mn}^* \qquad (2.4.7)$$

For this to reduce to a master equation depending only on $\{p_m\}$, the double sum on the right would have to lose all but its diagonal terms. For a diagonal term, the product $U(-t)_{mn}U(-t)_{mn}^*$ is real and positive. The great and elusive objective of master equation derivations is to show that for all other terms the rapidly varying phases cause them to cancel against one another. This would be the random phase postulated by Pauli and invoked or justified in one form or another in every attempt to derive the master equation. In the justification I give below it is both the right selection of basis as well as the need for coarse graining that do the job. This 'job,' incidentally, is essentially the same as what is now called a calculation of 'decoherence.'

I will return to the randomizing of phases but first I want to discuss another problem (the 'technical' one mentioned above). It is this: no derivation can succeed if the time t is too small. Suppose, to the contrary, that even for small t a master equation depending only on p_n could be deduced. Obviously (cf. Eq. (2.4.7)) it could only involve the diagonal elements $U(-t)_{mn}U(-t)_{mn}^*$. If you expand that product for small t you find

$$U(-t)_{mn}U(-t)_{mn}^* = \delta_{nm}\left[1 - t^2\langle n|H^2|n\rangle\right] + t^2|\langle n|H|m\rangle|^2 + O(t^3) \quad (2.4.8)$$

The salient feature of this equation is that the product differs from δ_{nm} only by terms that are $O(t^2)$. If the time interval for the randomizing of phase were short enough for Eq. (2.4.8) to hold, then nothing would happen! For suppose we formed dp_n/dt by looking at $[p_n(t) - p_n(0)]/t$; from Eq. (2.4.7) and Eq. (2.4.8) the $t \to 0$ limit of this quantity is 0. If $dp_n/dt = 0$, then indeed, nothing happens. This reasoning is essentially the same as that for the phenomenon of dominated time evolution, also called the quantum Zeno effect. In the early stages of a quantum transition, although phases have begun to move, probabilities go more slowly. Only later does decay assume its characteristic exponential form. As far as I know, the time scale for this effect has only been investigated in the case of barrier penetration, but we here see that this 'Zeno time' also represents a minimal time below which the master equation cannot hold.

Suppose then that the time interval *is* long enough to avoid the problems just discussed. Then we must evaluate the diagonal products of the Us of Eq. (2.4.7), not using the expansion Eq. (2.4.8), but in another way. This other way is what one calculates by standard quantum mechanics using time-dependent perturbation theory and the result is the 'Golden Rule.' It gives the transition rate from one state to a different one as the square of the magnitude of the appropriate off-diagonal matrix element of the Hamiltonian, times phase space factors. In Section 2.5 we do such a calculation—the decay rate expression following Eq. (2.5.18), $\Gamma = 2\pi\rho(h)|\gamma(h)|^2$, is an example of the Golden Rule. In the present section we continue to use the matrix U, but the relation of 'UU' to the Hamiltonian is what provides (for a time-reversal invariant Hamiltonian) the symmetry of the coefficients w, mentioned above. It can also happen—and generally does—that because of the phase space factors, although H is symmetric, w and W are not. With the phase space factors, the property of w (or W) that is imposed by time-reversal invariance is known as detailed balance.[19] For the matrix W it takes the form[20] $W_{\alpha\beta}\bar{p}_\beta = W_{\beta\alpha}\bar{p}_\alpha$, where $\bar{p}_\alpha$ is the steady state under W.

We next consider the need for coarse graining in justifying Eq. (2.4.4), or more precisely the dropping of non-diagonal terms, $m \neq m'$, in Eq. (2.4.7). The particular coarse graining that I use is closely related to that given earlier for the classical entropy.

In speaking of a coarse graining, I have in mind that the variable 'n' for $p_n(t)$ is not a projection of the wave function ψ on a *single* dimension of Hilbert space, but on a much larger subspace.[21] Let the variable n refer to a subspace whose projection operator is

$$P_n = \sum_\gamma |n\gamma\rangle\langle n\gamma| \qquad (2.4.9)$$

and let the variable p_n correspond to the expectation of this larger object; that is,

$$p_n(t) = |\langle P_n \mid \psi(t)\rangle|^2 = \sum_\gamma |\langle n\gamma \mid U(t) \mid \psi^{(0)}\rangle|^2 \qquad (2.4.10)$$

The expansion of $\psi^{(0)}$ over intermediate states is now more complicated since the intermediate state labels, called m and m', are now coarse grain labels and have their own, additional, internal sums. Let the labels for the m and m' sums be δ and δ', respectively. Writing out Eq. (2.4.10) in

---

[19] Which was also mentioned above, in calculating entropy and generalized entropy increase.

[20] This relation need *not* be satisfied for a steady state of an arbitrary stochastic W.

[21] Recall that p_n was our candidate for $u(\alpha, t)$ with $n \leftrightarrow \alpha$.

terms of matrix elements of U, we obtain the extension of Eq. (2.4.7):

$$p_n(t) = \sum_{m\delta} \sum_{m'\delta'} \sum_\gamma \psi^{(0)*}_{m'\delta'} \psi^{(0)}_{m\delta} U(-t)_{(m'\delta')(n\gamma)} U(-t)^*_{(m\delta)(n\gamma)} \quad (2.4.11)$$

As before, our goal is to be left only with terms of the form p_m on the right-hand side, that is, $m \neq m'$ cross terms should drop out. That goal would be accomplished through the vanishing of $\sum_\gamma U(-t)_{(m'\delta')(n\gamma)} \times U(-t)^*_{(m\delta)(n\gamma)}$, with $m \neq m'$. The mechanism of that vanishing is now seen to be the sum over γ. Suppose the subspace associated with n is of dimension N_n, which would be the number of γs summed. On the diagonal (i.e., for $(m, \delta) = (m', \delta')$) there are N_n terms, all positive. Off the diagonal there is the same number of terms. Since macroscopically the various pairs (m, δ) and (m', δ') are similar to one another the magnitudes of the products 'UU' should be similar. (Such magnitudes are transition probabilities—if they differed significantly the states would not be in the same grain.) However, the phase is more sensitive to the details of the states, and need not add consistently. If indeed the phases do vary widely[22] the associated sum in Eq. (2.4.11) would be of order $\sqrt{N_n}$. If these coarse grains are like those envisioned in my earlier classical discussion, $N_n \gg 1$ and $(\sqrt{N_n}/N_n) \ll 1$. It follows that $\sum_\gamma U(-t)_{(m'\delta')(n\gamma)} U(-t)^*_{(m\delta)(n\gamma)} = N_n \delta_{mm'} \delta_{\delta\delta'} |U_{(m\delta)(n\gamma_0)}|^2$, with γ_0 in the 'n' grain (and with the assumption that there is little variation in the magnitude of $|U_{(m\delta)(n\gamma)}|$ as a function of γ). In Eq. (2.4.11) this leads to

$$p_n(t) = \sum_{m\delta} |\psi^{(0)}_{m\delta}|^2 N_n |U_{(m\delta)(n\gamma)}(-t)|^2 = \sum p_m^{(0)} W_{mn}$$

which identifies the transition matrix as an appropriate scaling[23] of the microscopic transition amplitude.[24] The coarse graining thus provides the statistical averaging needed for the phase cancellations and it is the coarse grained variables that appear in the entropy increase calculation above (Eq. (2.4.5)). Without this coarse graining the previous arguments about varying phases in Eq. (2.4.7) are wrong. Although as m' varies, the phase varies considerably, there would be a lot more off diagonal terms than diagonal ones. Such a large number, while reduced because of phase cancellation, need not be zero. In fact, for the sum of N random phase

---

[22] An example of an explicit justification of this phase randomizing is the stadium billiard, a system that is classically ergodic and whose quantum behavior has been extensively studied. See the notes below.

[23] The approximate equality of magnitude for the matrix elements of U from one particular grain to another also means that any state, γ, is equally likely to be reached. Hence the factor N_n in the above equation.

[24] So that, as mentioned above, W is given by matrix elements of the transition Hamiltonian times phase space factors.

objects one expects a sum of order $\sqrt{N}$, as is clear by thinking of the sum as a random walk in the complex plane.

Remark: Time-energy uncertainty. Further analysis of the phase cancellation in the foregoing argument leads to interplay of coarse grain definition and dynamics. First, it is reasonable to take grains to be Hilbert space subspaces. This is because if two microstates look the same their superposition should still look the same. Now suppose the states in a grain were all eigenstates of the Hamiltonian with the same energy. Then all would get the same phase (in the propagator U) and in fact there would *not* be phase cancellation. For the validity of the master equation we thus have a time-energy uncertainty relation, $\Delta E \Delta t > \hbar$, in the following sense: the spread in energies within a grain times the minimum time interval for validity of the master equation must be such that phase spread is much larger than 2π. A natural reason for such energy broadening is that the coarse grains may be defined in non-stationary terms, for example spatial locality or eigenspaces of only a part of the total Hamiltonian.

2.5 Quantum decay

Perhaps the simplest irreversible process one can study in quantum mechanics is the decay of an unstable state. We model the decay of an atom from an excited state to its ground state. Typically, a full quantum description is not simple and in atoms would require inclusion of the quantized electromagnetic field and the richness of the atom's multiparticle dynamics. Even a single particle trapped in a barrier has more richness than one might have expected and we begin with a discussion of this case.

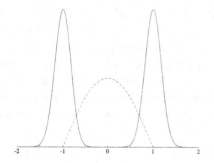

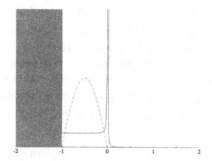

Fig. 2.5.1. A barrier to escape from. The potential has peaks at ± 1 and the initial wave function has support within those peaks.

Fig. 2.5.2. Situation for Eq. (2.5.1), with $a = 1$.

Escape from a barrier

A particle is trapped but not bound. For example, in one dimension, a potential of the form shown in Fig. 2.5.1 has no bound states, although the peaks at $x = \pm 1$ may hold a particle for a long time. If the wave function of the particle is initially within the barrier (as illustrated), then under time evolution amplitude will leak from the barrier, never to return. To avoid irrelevant complications we take the following Hamiltonian

$$H = \frac{1}{2}p^2 + h\Theta(-x) + \lambda\delta(x) + [\text{ Infinite barrier for } x < -a] \quad (2.5.1)$$

As suggested in Fig. 2.5.2, we again start a wave function entirely in the negative x region and ask about the rate with which it enters the right region. One way to do this is to expand the given initial wave function as a superposition of the eigenfunctions of the Hamiltonian and let this evolve in time, with each term in the expansion evolving with its appropriate $\exp(-iEt/\hbar)$. Because $V(x) \to 0$ for $x \to +\infty$ you need continuum wave functions. However, with the initial conditions given, no amplitude should be coming in from infinity and therefore you only use functions $\exp(+ikx)$; you do not use $\exp(-ikx)$. This is a boundary condition, a statement about what functions are acceptable solutions of Schrödinger's equation. With this prescription, solving Eq. (2.5.1) is an exercise. The wave function is of the form

$$\psi(x) = \begin{cases} A\sin k'(x+a) & x < 0 \\ B\exp(ikx) & x > 0 \end{cases} \quad (2.5.2)$$

with

$$\frac{1}{2}k'^2 + h = \frac{1}{2}k^2 = E, \qquad \psi(0^+) = \psi(0^-), \qquad \psi'(0^+) - \psi'(0^-) = 2\lambda\psi(0) \quad (2.5.3)$$

The boundary conditions at 0 imply

$$ik - \frac{k'}{\tan k'a} = 2\lambda \quad (2.5.4)$$

The most important feature of this equation, for our purposes, is the presence of an imaginary part. With k a function of k' from Eq. (2.5.3), we get a complex solution for k', hence for E. In itself this is no mystery. Although the operator H is formally symmetric, the outgoing wave boundary condition changes the rules of the game. Mathematically this leads to a certain amount of technical agony, but as far as we are concerned it's the best way to compute the evolution of the quasi-bound wave packet.[25] It also emphasizes that the inclusion of this boundary

---

[25] See Appendix B to Section 2.5 for mathematical comments.

condition is the place where we have said, 'This is the *initial* condition; *all* the amplitude is found in the well at $t = 0$ and we compute what will happen for $t > 0$.' That's why we took $\operatorname{Re} k > 0$, the outgoing wave boundary condition.

We turn next to solving Eq. (2.5.4). For large λ there is a family of quasi-bound states in the region on the left. These follow from Eq. (2.5.4) by letting $k'a \approx n\pi$ for integer n. To improve on this approximation, we set $k'a = n\pi + z$ and solve for z as an expansion in inverse powers of λ. We concentrate on the case $n = 1$, since it is the lowest eigenvalue of the corresponding infinite square well. It is also the state that our illustrated initial wave function resembles and is the slowest decay mode. The equation for z is

$$i\sqrt{h + \frac{(\pi + z)^2}{2a^2}} - \frac{\pi + z}{a\tan z} = 2\lambda \qquad (2.5.5)$$

The only subtlety in this equation arises because the first contribution to $\operatorname{Im} z$ is $O(\lambda^{-2})$, requiring a bit of care in the analysis. One obtains

$$\operatorname{Im} E = -\frac{\pi^2}{4a^2\lambda^2}\sqrt{h + \frac{\pi^2}{2a^3}} \qquad (2.5.6)$$

In this way the norm squared of the wave function in the well decreases by the factor $|\exp(-iEt/\hbar)|^2 = \exp(-2t\operatorname{Im} E/\hbar)$.

We have thus found irreversible behavior for a quantum system and we have seen where it and its time-direction came from. We set up an *initial* value problem and asked what happens later. Furthermore, the absence of recurrence results from the infinite size of the coordinate space, as we shall elaborate in Appendix A to this section, which presents the quantum recurrence theorem. In a sense it has been *too* easy to derive irreversibility here, and we turn to another system where the mechanism of irreversibility resembles the richer statistical mechanics phenomenon.

Exercise 2.5.1.   By a variation on the foregoing derivation it is possible to see that the unstable quantum state gives rise to a *pole in the S-matrix on the 'second sheet.'* For the wave function on $x > 0$ do not use Eq. (2.5.2), but rather $B\sin(kx + \delta)$, so that the phase shift δ is a function of k. The S-matrix is the operator which when applied to the incoming wave gives you the outgoing wave. A wave packet that bounces off this potential would go from positive x, to the region of 0, and to positive x again. Using this with the wave function $B\sin(kx + \delta)$ shows that the S-matrix is $S(k) = \exp[2i\delta(k) + i\pi]$. Solve the boundary value problem and show that the equation for the poles in $S(k)$ is precisely Eq. (2.5.4). Although Eq. (2.5.4) is specific to the Hamiltonian Eq. (2.5.1), show that the relation of the two ways of identifying the unstable state is general.

Aside: If you solved for the energy levels of Eq. (2.5.1) by putting another hard wall far off to the right, say at $x = b$ for large b, then the spectrum would be entirely real. Reconciliation with the complex spectrum found here can be made by means of dispersion relations. Consider the energy to be a function of the well depth parameter 'h.' With outgoing wave boundary conditions, $E(h)$ has a cut on the positive h axis. The discontinuity along this cut is $\operatorname{Im} E(h)$, given approximately in Eq. (2.5.6). With a hard wall at $x = b$, call the energy $E_b(h)$. It is true that $E_b(h)$ is real, but in fact it has many pairs of branch cuts near the real axis due to level crossings between approximate levels on both sides of the δ-function. Using a dispersion relation, $E(h)$ or $E_b(h)$ can be computed at h away from the singularities as integrals along its cuts, in the first case along the h-axis, in the second case around the paired branch cuts near that axis. For moderately large b the contributions turn out to be the same. This problem arose in thinking about singularities in the *free energy* in statistical mechanics, and whether a complex part could be associated with the lifetime of a metastable state. References are given in the notes below.

Decay of an excited state

An atom in an excited state often decays to its ground state through the emission of photons. If we use a basis of atomic eigenfunctions it is reasonable to neglect all but the initial excited state and the final ground state. The spatial coordinates of the atom disappear in this two-dimensional space. For the photons, however, it is not possible to neglect their spatial dependence completely since part of the process is that the photons, once created, leave the scene of their nascence at the speed of light. This plays a role in the irreversibility, but as we shall see, not the only role. We model this situation with the following Hamiltonian and wave function

$$H = \begin{pmatrix} h & C^\dagger \\ C & \Omega \end{pmatrix} \qquad \psi = \begin{pmatrix} x \\ Y \end{pmatrix} \qquad (2.5.7)$$

The upper component of ψ, the quantity x, is a complex number and if it is of norm unity, the system is in its excited state and there are no photons. The lower component of ψ, Y, is a complex N-vector. Amplitude in these components represents the atom in its ground state and the presence of one or more photons.[26] The total Hilbert space will be called $\mathcal{H}$, the subspace associated with x will be $\mathcal{H}_e$ (excited) and that associated with Y, $\mathcal{H}_d$ (decayed). Without loss of generality, Ω can be taken to be diagonal (if it isn't, diagonalize it and rearrange the components of C). Furthermore, although we have motivated the model in terms of an atom and photons, it could well be any decaying system with completely different decay channels, for example phonons.[27] An on-shell, energy conserving process in this language corresponds to going to an energy level of Ω with energy h. The atom has decayed, but its energy has been picked up by the other excitations.

---

[26] Note that a single component of Y can represent a state of several photons.
[27] With the right basis it could also model decay through a barrier. See Exercise 2.5.4.

The usual way of solving this problem is to take a Laplace transform of the equations of motion. Then you convince yourself that the inverse transform is dominated by a single complex pole and get exponential decay. (References are given below.) I will solve the time-dependent equations directly because it is easier to identify the roots of irreversibility in that framework.

Schrödinger's equation (with $\hbar = 1$) for the system Eq. (2.5.7) is

$$\dot{x} = -ihx - iC^\dagger Y, \qquad \dot{Y} = -iCx - i\Omega Y \qquad (2.5.8)$$

supplemented by the initial condition $x(0) = 1$, $Y(0) = 0$. The second equation is integrated to yield

$$Y(t) = -ie^{-i\Omega t} \int_0^t e^{i\Omega s} Cx(s)ds \qquad (2.5.9)$$

This is substituted into the first equation to give a renewal equation for x

$$\dot{x}(t) = -ihx(t) - C^\dagger e^{-i\Omega t} \int_0^t e^{i\Omega s} Cx(s)ds \qquad (2.5.10)$$

The essence of the process is further highlighted by defining

$$z(t) \equiv e^{iht} x(t) \qquad (2.5.11)$$

which produces a quantity that would be entirely stationary, but for the interaction. The equation satisfied by z is

$$\begin{aligned}
\dot{z}(t) &= -\int_0^t e^{iht} C^\dagger e^{-i\Omega(t-s)} C e^{-ihs} z(s)ds \\
&= -\int_0^t e^{iht} K(t-s) e^{-ihs} z(s)ds \qquad \text{with} \quad K(u) \equiv C^\dagger e^{-i\Omega u} C
\end{aligned}$$

$$(2.5.12)$$

Everything is still exact. The function, $K(u)$, appearing in Eq. (2.5.12) with argument $t - s$, is of central importance. It is the product of three operators. First C acts on x (cf. Eq. (2.5.8)) and brings amplitude into $\mathcal{H}_d$ (the subspace of decayed states). The amplitude in each mode then evolves for a time u, each mode with its own phase (from Ω). Next $C^\dagger$ sums the amplitude (the matrix product produces a *scalar*) and sends it back to $\mathcal{H}_e$. Now here's the point: to the extent that the modes in $\mathcal{H}_d$ get out of phase, the sum going back into $\mathcal{H}_e$ is reduced. In this way, K shrinks. The reduction will depend on two factors, u (larger u allows the modes to get *more* out of phase) and the range of Ω over which C is large. It is helpful to write $K(u)$ explicitly. The components of C will be denoted c_k and those of Ω (which is diagonal), w_k. K becomes

$$K(u) = \sum_k |c_k|^2 e^{-iw_k u} \qquad (2.5.13)$$

Obviously $|K|$ is symmetric in u. For small u the sum is largest and drops off as u increases. By going to a continuum limit, a familiar restatement of this dropoff can be made. The appropriate scaling for the continuum is $O(1/N)$ spacing for the levels of Ω, and transition matrix elements c_k of order $1/\sqrt{N}$. The independent variable is now taken to be ω, related to the previously used index k by $\omega = \omega_k$. The cs are written $\gamma(\omega) = c_k/\sqrt{N}$. One other object of importance enters in going from $\sum_k$ to $\int d\omega$. This is the density of states $\rho(\omega) = 1/[N(\omega_{k+1} - \omega_k)]$. In the limit, K becomes

$$K(u) = \int d\omega \rho(\omega)|\gamma(\omega)|^2 e^{-i\omega u} \qquad (2.5.14)$$

Define $f(\omega) \equiv \rho(\omega)|\gamma(\omega)|^2$. K is the Fourier transform of f. The usual uncertainty relation arguments apply, namely small spread in f means a big spread in K, and vice versa. Of course if a function is not smooth, both it and its Fourier transform can be widely spread, but the functions we deal with here are expected to be smooth. What interests us is the fact that $K(u)$ does drop off as u increases. This is the mathematical reflection of the arguments given earlier about modes getting out of phase.

I have dwelt on this point because it is the primary mechanism of irreversibility and the rationale for exponential decay. Further tricks will be needed later, but they take care of Poincaré recurrences and other exotic phenomena.

The consequence of this dropoff is that although Eq. (2.5.12) appears to use values of $z(s)$ all the way back to $s = 0$ in computing $\dot{z}(t)$, in fact only the recent history is relevant.

With this understanding of the structure of z's evolution equation, we rewrite it using the continuum form of K and with the change of variable $u = t - s$ within the time integral

$$\dot{z}(t) = -\int d\omega \rho(\omega)|\gamma(\omega)|^2 \int_0^t du\, e^{-i(h-\omega)u} z(t-u) \qquad (2.5.15)$$

As a consequence of our arguments concerning K, the range of the u integration does not extend far. Let us assume that z does not change much during this time interval. The validity of this assumption must be checked self consistently.[28] We thus remove z from the integral and perform the u integration, keeping only the contribution from the neighborhood of $u = 0$, again relying on the ω integral that forms K to kill

---

[28] It is not merely a matter of appealing to the validity of perturbation theory since the diagonal elements of our 'unperturbed' Hamiltonian (H of Eq. (2.5.7), without C and $C^\dagger$) do not set any scale. The self consistency will involve a comparison of the dropoff time of $K(u)$ and the lifetime, the latter, as we show below, depending on $\rho(\omega)|\gamma(\omega)|^2$ at a particular value of ω.

the large u contribution. This yields

$$\dot{z}(t) = -iz(t) \int d\omega \frac{\rho(\omega)|\gamma(\omega)|^2}{\omega - h} \qquad (2.5.16)$$

Now we have arrived at a problem that would have confronted us much sooner had we gone the route of Laplace transformations: the singular integral in ω. This is perhaps the most common singular integral in physics and is handled by the formula

$$\frac{1}{x \pm i\epsilon} = \mathcal{P}\frac{1}{x} \mp i\pi\delta(x) \qquad (2.5.17)$$

with $\mathcal{P}$ the principal value and it is understood that Eq. (2.5.17) is to be used[29] in an integral over x. The formula is fine, but which sign of ϵ should be used? If we had performed a Laplace transform in the beginning, now would be the time to invoke second sheets, physical sheets, unphysical sheets, the fact that Green's functions are defined only in certain half planes although they may be extendable elsewhere. I don't find that satisfactory. We began with a finite dimensional problem that was fully well defined and now we have an ambiguity to resolve. What does that have to do with sheets? Why does one need to invoke analytic continuation? Mathematically, the mischievous step was the continuum limit, but that observation does not in itself resolve the questions now raised. The advantage of working explicitly in the time domain, as we have done, is that we can see where the need for analytic continuation arose and how, as a consequence, to choose the sign of the imaginary part of ω.

The need arose when we accepted the argument that $K(u)$ goes to zero for large u. Our physical reasoning was that for large u the modes in $\mathcal{H}_d$ got out of phase and when their contributions were added would cancel one another—similar to the destructive interference that played a role in deriving the master equation. Now it happens that for finite dimensional $\mathcal{H}_d$ the phases *can* recombine constructively if you go to long enough times, while in the continuum limit that time is pushed to infinity— i.e., they never recombine constructively. Therefore for our continuum limit to be fully defined we should manhandle the Hamiltonian H just enough to eliminate recurrences. One way to do this is to add a small imaginary part to ω. That way, the norm of $\exp(i\omega t)$ (which is the sign with which ω appears above) will become $\exp(-t \operatorname{Im}\omega)$ and if ω has a

---

[29] This formula does not depend on analyticity of the integrand and is proved by multiplying numerator and denominator by $x \mp i\epsilon$ and separating real and imaginary parts.

positive imaginary part K will not recur.[30] This is closely analogous to what we did for the decay from a barrier. By taking outgoing wave boundary conditions we made sure that amplitude left, never to return.[31]

Taking $\text{Im}\,\omega > 0$ implies use of the upper sign in Eq. (2.5.17), which when applied to Eq. (2.5.16) yields

$$\dot{z}(t) = -iz(t)\left[-i\pi\rho(h)|\gamma(h)|^2 + \mathcal{P}\int d\omega \frac{\rho(\omega)|\gamma(\omega)|^2}{\omega - h}\right] \qquad (2.5.18)$$

$$\equiv z(t)[-\Gamma/2 - i\Delta E]$$

The time dependence of z is obviously $z(t) = \exp[t(-\Gamma/2 - i\Delta E)]$, with ΔE an energy shift arising from the real principal-value integral and Γ the decay rate (since $|z|^2$ drops like $\exp(-t\Gamma)$). The formula $\Gamma = 2\pi\rho(h)|\gamma(h)|^2$ is the familiar 'Golden Rule' applied to decay rates.

Exercise 2.5.2.   Interpret the solution with $\text{Im}\,\omega < 0$.

Exercise 2.5.3.   Derive exponential decay using the eigenstates of the full, finite-dimensional H of Eq. (2.5.7).

Exercise 2.5.4.   Find a basis in which the barrier problem of Eq. (2.5.1)—with a hard wall somewhere off to the right—can be written in the form Eq. (2.5.7). Your first tendency may be to take wave functions with support on one side or the other of the δ-function, but then how do you get C? (Like the previous exercise, this question is informed by joint work with B. Gaveau.)

Appendices for this section (at the end of the chapter):

2.5.A   Quantum recurrence theorem
2.5.B   More on decay through a barrier
2.5.C   Deviations from exponential decay

2.6 Different kinds of ignorance: a restatement of the second law

With an information-theory approach to entropy, the second law of thermodynamics takes the form of a sharp distinction between our knowledge of the future and of the past, between the way we estimate the later state of a system based on present data, and the way we estimate the earlier state based on present data. These estimates are called *prediction* and

---

[30] Built into this prescription is the fact that we are doing an *initial* value problem. It is for *positive t* that we demand good behavior.

[31] In a sense to be made precise in the quantum Poincaré recurrence theorem (Appendix 2.5.A), the finite Hamiltonian (with its recurrence) is like decay from a barrier in which there is a reflecting wall far off to the right. Note too that using a finite dimensional stationary state basis for the decay Hilbert space also makes less apparent the effect of escape of particles created in the decay.

*retro*diction. Consider the situation pictured in Fig. 2.6.1. We show three images of a glass of water with an ice cube in it. Suppose you are told that at 2 p.m. the water is at temperature T_2 and that the ice cube volume is V_2, as shown in the middle picture of Fig. 2.6.1. The system is completely isolated between 2 p.m. and 3 p.m. (the walls of the larger container are not shown) and you are asked to predict the water temperature and ice cube size at the later time.

If you are skilled in the ways of heat conductivity then your prediction will look like the right-hand illustration: a smaller ice cube and colder water, $V_3 < V_2$, and $T_3 < T_2$. Let us consider, at the level of principle, how you did your calculation. The situation in the middle picture is a *macroscopic* description and has associated with it a region of phase space,[32] call it Γ_2. Every microscopic state $\gamma \in \Gamma_2$ looks like the middle picture of Fig. 2.6.1 with the same temperature and cube size. To make your 3 p.m. prediction, you evolve each of those microscopic states forward for one hour. What do these microscopic states look like? To which macroscopic state do they correspond? The vast majority look like the right-hand picture in Fig. 2.6.1, and this is your predicted macrostate. This, in principle, is how the rules of heat conductivity are derived. For emphasis I want to make explicit something implicit in the foregoing steps. By identifying the final macrostate with the evolutes of the 'vast majority' of the 2 p.m. microstates, I am effectively averaging over the microscopic states of Γ_2, and, in particular, am giving them all the same weight. This procedure reflects a fundamental principle of statistical mechanics, an extension of what ergodicity is supposed to be doing for equilibrium

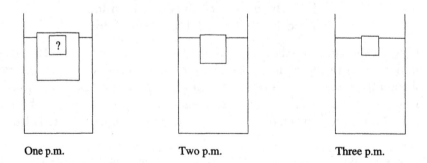

One p.m.   Two p.m.   Three p.m.

Fig. 2.6.1.   Three images of a glass of water with an ice cube in it.

---

[32] For convenience we adopt a classical description.

systems.[33] When a system is not in equilibrium, and our 2 p.m. glass of water and ice is not, it has other macroscopic features beside energy. The fundamental principle of statistical mechanics gives all microstates, *all those consistent with the given macroscopic conditions*, the same weight. An occupied coarse grain is taken to be uniformly occupied. The equal probability rule is implicit in ordinary statistical mechanics calculations. For example, in the above example, to make your 3 p.m. prediction you need the specific heat of water. You deduce that specific heat using the partition function. The partition function sums equally over all states, subject to various ensemble defining constraints.[34]

The foregoing procedure is also describable in phase space. In Fig. 2.6.2, the oval in the middle represents the phase space region Γ_2. It is from this region that all microstates begin at 2 p.m. Let them evolve for an hour. Let the operator that maps phase space points forward one

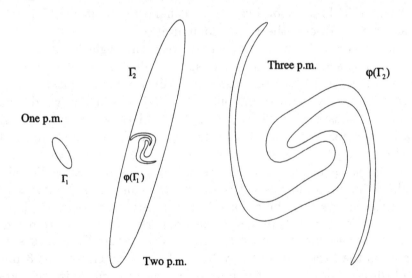

Fig. 2.6.2. Phase space. Schematic illustration of the phase space sets Γ_2, $\varphi(\Gamma_2)$ (the dynamical image of Γ_2 after 1 hour), Γ_1, and $\varphi(\Gamma_1)$. The degree of convolution of the dynamical images ($\varphi(\Gamma_1)$ and $\varphi(\Gamma_2)$) is severely understated.

[33] Recall that ergodicity asserts that in equilibrium all accessible microstates have the same weight. See Appendix B to Section 2.2. (Quantum mechanically this becomes the Trace operation on Hilbert space.)

[34] An ensemble may appear to have a non constant weight, for example the canonical ensemble $[\exp(-E/kT)]$, but such an ensemble does not describe a closed system. The weight is derived by taking equal weight for the larger combined system consisting of the original system and the reservoir with which it is in contact.

hour be called φ. The statement about equal *a priori* probabilities[35]
at 2 p.m. is the statement that our 2 p.m. phase space measure is the
characteristic function of Γ_2. By 3 p.m. the system has evolved to $\varphi(\Gamma_2)$.
(See the right-hand picture in Fig. 2.6.2.) This is no longer a macroscopic
state. If you think of macroscopic states as composed of coarse grains (as
discussed in Section 2.2), then, noting that the image of a coarse grain
need *not* be a coarse grain, the image of a macroscopic state is not in
general macroscopic. Compare Fig. 2.2.3, where although the points in
the lower left corner fall in a single grain (for the cross graining I used in
that section), their image does not. The typical situation is that $\varphi(\Gamma_2)$
falls into one or several coarse grains at the later time and this or these
are the predicted macrostate. Sometimes the image coarse grains will be
far apart, for example for the phase space description of a person throw-
ing dice. The point of a random dice throw is to make what looks like a
single macroscopic initial state go into several distinct macroscopic final
states. Another situation where they may be far apart is when amplified
quantum effects are important. Yet another feature of $\varphi(\Gamma_2)$ is that small
portions of it do not look like the right-hand figure at all. This will be
clear when we discuss the 1 p.m. configuration.

The macroscopic state associated with the right-hand picture in
Fig. 2.6.1 is some subset of phase space, call it Γ_3. (This is not pictured in Fig. 2.6.2.) The conclusion to be drawn from the last paragraph
is that, to a good approximation, $\varphi(\Gamma_2) \subset \Gamma_3$. What a 'good approximation' means is that $\mu(\varphi(\Gamma_2) - \Gamma_3) \ll \mu(\Gamma_2)$, where μ is the phase
space measure. So Γ_2 has found its way into Γ_3, but how much of Γ_3
does it occupy? The answer is, very little. The phase space volume Γ_2
has been stretched, whirled and tendriled so as to occupy Γ_3, all the
while maintaining its original phase space volume $\mu(\Gamma_2)$. But the filling is the way feathers fill a down jacket—there's more space in between
than there is filler. It is in making this remark precise that we reconnect
with the second law of thermodynamics. The entropy[36] of the system at
2 p.m. is the logarithm of the volume of Γ_2. The entropy at 3 p.m. is
the logarithm of $\mu(\Gamma_3)$. Therefore for a glass of normal size the ratio of
the phase space volumes will be on the order of $\exp(N_A)$, with N_A Avogadro's number (rather larger than for down jackets). So it is the fact
that a coarse grain at one time spreads into many coarse grains, or larger
coarse grains, at later times that corresponds to the second law of thermodynamics. Implicit in this is the assumption that the earlier grain was

---

[35] The phrase 'equal *a priori*' means that I take the probabilities equal because I have
no *a priori* knowledge about the system, except for its macroscopic description.

[36] ... up to base-of-logarithm and other ambiguities discussed earlier. These drop out
of the ratio I next mention.

uniformly occupied, otherwise you couldn't be sure that all the regions at the later time were also occupied. Recall too the results of Section 2.2 where grain spreading was directly related to entropy increase and both shown to depend on the Lyapunov exponent.

We turn next to a more subtle question. We are now given additional information: the system was not only isolated from 2 p.m. to 3 p.m., it had been untouched since 1 p.m. (and throughout the two hour period). What was the state of the system at 1 p.m.? One way to answer this would be to take the mapping φ^{-1}, apply it to Γ_2 and identify the coarse grains needed to contain most of $\varphi^{-1}(\Gamma_2)$. Since φ and φ^{-1} are the same or essentially the same, the 1 p.m. macrostate of the system would be the smaller cube shown on the left of Fig. 2.6.1 (along with colder water).

Of course it never happens that way. If you had started with a small cube at 1 p.m., you would have a *smaller* one at 2 p.m. Obviously there is a different rule for retrodicting. Guessing the 1 p.m. situation from 2 p.m. information is *not* done by propagating the set Γ_2 back to 1 p.m. How is it done?

There is no unique answer. Here is one way to do it. Make a guess, a hypothesis, about the 1 p.m. macrostate. This corresponds to some set Γ_1 in phase space (illustrated on the left of Fig. 2.6.2). Propagate it forward one hour to get $\varphi(\Gamma_1)$. If this lies substantially within Γ_2 (as it does in Fig. 2.6.2), if it *looks like* the middle picture of Fig. 2.6.1, then your hypothesis was acceptable. For example, the larger cube on the left of Fig. 2.6.1, might be a precursor to the middle picture. But other precursors are also possible. For example, you might have had a slightly smaller cube plus a little shaving of ice that melted entirely during the hour. The coarseness of your coarse grains may preclude your being able to discriminate between these cases. You may even have several hypotheses and want to use other information to weight your guesses. For example you may know that all available ice came from a machine whose maximum cube size is smaller than the large one pictured on the left, implying that there had to be some small chips at 1 p.m.[37]

One thing you do know though is that the combined phase space volume of all acceptable hypotheses is vastly smaller than $\mu(\Gamma_2)$. You know this because you know that most of the volume in Γ_2, when propagated back to 1 p.m., gave the single smaller ice cube, the one with the question mark in it in the figure. So if Γ_1 is now taken to represent the volume of all acceptable 1 p.m. phase space volumes, $\mu(\Gamma_1) \ll \mu(\Gamma_2)$.

---

[37] The non-uniqueness in retrodicting is related to the statistician's ambiguities in constructing estimator functions.

We summarize the rules for prediction and retrodiction. To predict, take—with equal weight—all microstates consistent with the macrostate and propagate them forward in time. Smear this set in accordance with the coarse graining to get your macroscopic prediction. To retrodict, go back to the earlier time and make macroscopic hypotheses. Use each such hypothetical macrostate to predict the present state. If it agrees, it was a possible precursor. In general, the phase space volume of all such precursors is less than or equal to that of the subsequent macrostate. In this way there is entropy non-decrease for an isolated system when moving forward in time. These different rules for prediction and retrodiction are an alternative statement of the second law of thermodynamics. Mathematically, the validity of this alternative statement depends on an interplay between the coarse graining and the dynamics: the evolute of a coarse grain is not a coarse grain. By virtue of the Liouville theorem (conservation of phase space volume) this mismatch of time evolution and coarse graining is the basic source of entropy increase (or information loss).[38]

For later purposes we make yet another statement of the past/future distinction. Suppose you know the 2 p.m. macrostate and are asked what you know about its microstates. If all you are concerned with is prediction, then you say, all microstates are equally likely. But if you happen to know that the system had been isolated since 1 p.m., then the vast majority of those microstates are rejected and all that are allowed are those in the set $\varphi(\Gamma_1)$, with Γ_1 your collection of acceptable hypotheses. In other words, for an *initial* state you treat all microstates equally, but if something is a *final* state the vast majority of its microstates are rejected.

The foregoing observation allows us to notice that the assumptions of statistical mechanics are even stronger than what we have already stated. One way to make a prediction is to give equal *a priori* probability to all initial microstates. But that's not the only way. For our system isolated from 1 p.m. to 3 p.m., the set of 2 p.m. microstates is now known *not* to be Γ_2, but $\varphi(\Gamma_1)$, a much smaller set. That means that our 3 p.m. macrostate should not be calculated from $\varphi(\Gamma_2)$, but from $\varphi(\varphi(\Gamma_1))$. Does anyone think the 3 p.m. result will be any different? All our experience says otherwise. It does not matter whether you use all of Γ_2 or only the part that results from a state that had been isolated earlier. Presumably, the set $\varphi(\Gamma_1)$ fills Γ_2 in a pseudo-random way, so that its points are representative of what all of Γ_2 is doing. Perhaps, recalling the rationale and goals of the ergodic hypothesis, I should turn

---

[38] Compare the Remark at the end of Section 2.4.

this around. We assume that taking *all* of Γ_2 gives us a good idea of what any particular microstate will do, and restricting to $\varphi(\Gamma_1)$ doesn't change that.

The issue touched on in the foregoing paragraph will arise later. We will propose a restriction on which microstates are permissible as *initial* conditions and then consider whether this could lead to observable differences. Another point for future reference concerns exceptional microstates. As remarked, most microstates of Γ_2 when propagated back to 1 p.m. give smaller cubes. But those of $\varphi(\Gamma_1)$ do not—more or less by definition. This exceptional set occupies the same sort of minuscule fraction of Γ_2 that we have come to expect, $\exp(-N_A)$, because there is a macroscopic entropy change involved. A similar minuscule fraction of Γ_2 states, when propagated forward to 3 p.m., will give larger ice cubes or no ice cubes or ice statues of polar bears. My point is that although these unlikely outcomes represent unimaginably small fractions of the phase space, they still have a lot of phase space themselves. Thus $\mu(\Gamma_1)$ is still a big number (in appropriate units, for example $\hbar^N$). There is a lot of room in phase space and losing a few million powers of 10 doesn't make a dent.

Exercise 2.6.1. The qualitative discussion of this section can be carried out in quantitative detail using the Kac ring model. One can consider the state of the ring at times $T = 100$, 200 and 300, given that $\delta(200) = 0.4$. Take the ratio of active sites to all sites (what we called μ) to be 0.00345. Assume the active sites to be distributed randomly and that the number of sites is much larger than 200 (so the problems associated with correlations are not present). The space Γ_2 would be the set of ball configurations for which $\delta = 0.4$ and after 100 time steps these would constitute a very small subset of all possible sequences with $\delta \approx (1 - 2\mu)^{100} \approx 0.2$. Just how small is easily calculable. Similarly one can check the sizes of the relevant sequence spaces for retrodiction.

Remark: We earlier commented on the non-uniqueness of retrodiction schemes. In practice what we do when we consider various hypotheses is to take the 'most reasonable' ones. Effectively we have *a priori* probabilities for a collection of possible earlier states, say $\bar{p}_\alpha$, with α labelling states. Then, using Bayes's rule, we use present information (say, that the system is in a state β) to estimate the relative probabilities of the precursors: $\Pr(\alpha) \propto \Pr(\beta|\alpha)\bar{p}_\alpha$. The quantities $\Pr(\beta|\alpha)$ are those calculated from the microscopic dynamics. For *prediction* too it may be convenient to speak in these terms. Thus, for a given coarse graining, one may have less information than is ideally available. Predictions would then also rely on *a priori* notions of likelihood. For example, for our melting ice cube one may not have checked whether in the middle of the ice cube there is a chunk of glass, stopping the melting when it is exposed. True, with available optical equipment the index of refraction discontinuity could have been discovered (so the presence of the glass is a coarse grained feature), but we simply didn't look. In this case, retrodiction and prediction are not all that different.

Appendix for this section (at the end of the chapter):
2.6.A Past/future reliability: *not* a restatement of the second law

2.7 Notes and sources

Section 2.0. Throughout the chapter we made repeated reference to the Boltzmann H-theorem—but in practice we only dealt with simplifications that captured the essence of its irreversible behavior. For the record, we here include a
brief statement of the theorem and its context.

The physical quantity studied is $f(r, v, t)$, the density of particles in a volume $d^3r\,d^3v$ of position and velocity space. This is a projection from the $6N$-
dimensional space in which all molecular positions and velocities are individually
specified. By a series of physical and mathematical arguments one finds that f
satisfies the 'Boltzmann transport equation'

$$\left(\frac{\partial}{\partial t} + v_1 \cdot \nabla_r + \frac{F}{m} \cdot \nabla_{v_1}\right) f_1 = \int d\Omega \int d^3v_2 \sigma(\Omega) |v_1 - v_2| (f_2' f_1' - f_2 f_1) \quad (2.7.1)$$

where F is an external force, m the mass of the molecules, $f_1 \equiv f(r, v_1, t)$,
$f_1' \equiv f(r, v_1', t)$, $f_2 \equiv f(r, v_2, t)$, $f_2' \equiv f(r, v_2', t)$, and $\sigma(\Omega)$ is the center-of-mass
scattering cross section for a pair of molecules whose relative velocity goes from
$v_2 - v_1$ to $v_2' - v_1'$. The left side of this equation represents the change of the
function in the absence of collisions between molecules. The right side is derived
by assuming that the relative velocities of molecules *are not correlated with their
positions*. This is the assumption of molecular chaos, the *Stosszahlansatz*. Suppose $F = 0$. Then it is consistent to take f independent of r, and we drop r
from the notation. One now defines a quantity $H(t)$, dependent on f, which is
essentially the negative of an entropy:

$$H(t) \equiv \int d^3v f(v, t) \log f(v, t)$$

An immediate consequence of Eq. (2.7.1) is that $\frac{dH}{dt} \leq 0$. This is the H-theorem.

Section 2.1. In 1956 Kac gave a series of colloquia in Dallas at the research
offices of the Socony Mobil Oil Company, [KAC 56]. Besides the ring model he
spoke about the telegrapher equation and about the path integral, topics that
turned out to be intimately related. In our present era of shrinking industrial
budgets for basic research one looks back on Socony Mobil's program with nos-
talgia. The lectures were a prelude to a book, [KAC 59], but the lectures convey
more fully Kac's personal charm.

Besides those who have exploited the Kac ring's pedagogical virtues (e.g.,
[THOMPSON]), some authors have used it to study other phenomena. An example
is Exercise 5.1.3, where the model's tractability allows solution of a two time
boundary value problem. Other projects (e.g., quantization) can be found in
[COOPERSMITH] and [TAVERNIER].

An excellent source for the information theory approach to statistical mechanics is [KATZ], where the work of E. T. Jaynes is mentioned. To some extent this
approach was stimulated by confronting 'Maxwell's demon,' a creature whose discriminating power allows reduction of entropy. Articles on this topic have been
collected in [LEFF], along with commentary. The negative temperature work

mentioned in the text can be found in [RAMSEY] and the spin echo experiments are described in [BREWER].

Section 2.2. A classic book for the study of ergodicity, distinguished by its use of models, with its even more classic illustration of the cat map, is [ARNOLD]. My wisecrack is not without justification: their stretched cat illustration has been reproduced, line for line, in at least half a dozen books.

Coarse graining of phase space, along the lines described here, is done in [SCHULMAN 75]. My remark about fuzzy logic can be understood from the relation that the paper just cited finds between coarse graining of phase space and the concept of implication. The proof of the increase of relative entropy under stochastic evolution (referred to in a Remark) can be found in [COVER] or [GAVEAU 96]. A model system with non-zero entropy at zero temperature is given in [AVRON].

Discussion of ergodicity and its implications for physics can be found in [WIGHTMAN] and [THOMPSON]. Hilbert space approaches to classical mechanics were pioneered by B. Koopman. See [ABRAHAM].

A pair of parallel line segments capped at both ends by semicircles creates a figure that looks like a stadium ($\bigcirc$). A particle moving freely within the stadium and bouncing specularly off the walls is known to be ergodic. The quantum mechanics of this system has provided rich insights into quantum chaos and has even led to reverse modeling in which experimentalists confine particles in stadium-like regions. See [HELLER] and [MARCUS].

Section 2.3. *The* reference for classical mechanics is [GOLDSTEIN]. Recent works going beyond the traditional topics are [LASOTA], [TABOR] and [LICHTENBERG].

Section 2.4. The master equation has a long history and often appears in more elaborate guise than that which we presented. It was called *master* by Pauli because it did not deal merely with a projection to a single coordinate and velocity space, as in the Boltzmann equation, but incorporated more of the system's multiparticle degrees of freedom. Background on the derivation of the master equation can be found in [CHESTER]. See also [PRIGOGINE]. [KATZ] has a derivation of the master equation in which the randomizing of phases gets done by information/ignorance arguments—a kind of coarse graining. More application oriented is the book by [OPPENHEIM], although it also contains a reprint of a van Hove paper on deriving the master equation. Less specialized books that include material on the master equation are [DICKE, KITTEL, ROBERTSON] and [VAN KAMPEN].

Properties of stochastic matrices can be found in [BERMAN] and [GANT-MACHER].

References for increase of generalized or 'relative' entropy were given above. The same quantity is used by [VAN KAMPEN], Chap. 6. In information theory it plays a prominent role and is variously known as conditional uncertainty ([ASH]), the Kullback-Liebler difference or simply as relative entropy ([COVER]).

Slow, non-exponential short time decay is often called the quantum Zeno effect. See the notes on Section 2.5 for further information. The 'Golden Rule' for transition and decay rates was so named by [FERMI]. Phase randomizing of the stadium billiard leads to an extension of the correspondence principle for chaotic systems. See [SCHULMAN 94a].

In the last few years a 'post-Everett' approach to quantum measurement theory has been put forth. Everett's ideas are often called the 'many worlds interpretation.' The post-Everett work takes the same point of view but talks about decoherence and consistent histories, managing thereby to avoid the more garish images inspired by the earlier terminology. The claims about decoherence lead to the need for calculations that are the same sort that appear in justifying the master equation. In both cases one needs a loss of phase coherence so as to allow the use of classical or collective variables for the description of the system. See [GELL-MANN, GRIFFITHS, ZUREK] and several articles in the book [COLEMAN].

Section 2.5. The earliest derivations of the exponential decay law are associated with the names Wigner, Weisskopf, Gamow and several others. A cogent presentation of the Wigner-Weisskopf method can be found in [KABIR].[39] The first to realize the limitations of the method was Khalfin, pointing out problems both for the long-time behavior ([KHALFIN 61]), and for short times as well ([KHALFIN 57]). He relates that when he first told his mentor, Fock, about the breakdown in long time behavior, Fock did not believe him. Later, Fock called him back and said, yes, it was remarkable but unassailable. With Landau, Khalfin only got to the first stage. There has recently been experimental confirmation of the short-time effect [ITANO].

In the text we showed that an unstable quantum state gives rise to an imaginary part in the energy. The reconciliation of this with the *reality* of the energy when walls are used, is in [SCHULMAN 78a]. See also [MCCRAW] for the relation of level-crossing and lifetimes in metastable statistical mechanics systems.

In the 1970s there was renewed interest in this field, perhaps because of concern that the anomalous decay of the K meson might not be due to *CP* violation, but to a breakdown in the exponential decay law. Careful studies were made and two worthwhile reviews are [FONDA] and [TERENT'EV].

As mentioned earlier, the short time departure from exponential decay has been given the name 'quantum Zeno effect' which I consider unsuitable for reasons given in [SCHULMAN 94b]. A more accurate designation is 'dominated time evolution;' however, I don't expect my complaints to dissuade physicists from the opportunity to make a learned classical reference. A mark of this realization is my own definition of the 'Zeno' time in Appendix C to Section 2.5. A recent study of anomalous decay behavior is [GAVEAU 95].

The proof of quantum recurrence follows the paper [SCHULMAN 78b] and earlier work on quantum recurrence is found in [BOCCHIERI].

Section 2.6. Much of the material of this section appeared in [SCHULMAN 75] and [SCHULMAN 91]. The work of Gold on arrows of time (about which more later) is [GOLD].

---

[39] Kabir's concern is the two K meson formalism, but the treatment is convenient and accessible even if all you want is the single mode case.

References

ABRAHAM 78	R. Abraham & J. E. Marsden (1978) *Foundations of Mechanics*, 2nd ed., Addison-Wesley, Reading, Mass.
ARNOLD 68	V. I. Arnold & A. Avez (1968) *Ergodic Problems of Classical Mechanics*, Benjamin, New York.
ASH 90	R. B. Ash (1990) *Information Theory*, Dover, New York. (Orig. pub., Wiley-Interscience, 1965.)
AVRON 81	J. E. Avron, G. Roepstorff & L. S. Schulman (1981) Ground State Degeneracy and Ferromagnetism in a Spin Glass, *J. Stat. Phys.* **26**, 25.
BERMAN 89	A. Berman, M. Neumann & R. J. Stern (1989) *Nonnegative Matrices in Dynamic Systems*, Wiley, New York.
BOCCHIERI 57	P. Bocchieri & A. Loinger (1957) Quantum Recurrence Theorem, *Phys. Rev.* **107**, 337.
BREWER 84	R. G. Brewer & E. L. Hahn (1984) Atomic Memory, *Sci. Amer.* **251** (June), 50.
CHESTER 63	G. V. Chester (1963) Theory of Irreversible Processes, *Rep. Prog. Phys.* **26**, 411.
COLEMAN 91	S. Coleman, J. B. Hartle, T. Piran & S. Weinberg (1991) *Quantum Cosmology and Baby Universes*, World Scientific, Singapore.
COOPERSMITH 74	M. Coopersmith & G. Mandeville (1974) Irreversible Behavior of Interacting Systems. I. The Approach to Equilibrium, *J. Stat. Phys.* **10**, 391.
COVER 91	T. M. Cover & J. A. Thomas (1991) *Elements of Information Theory*, Wiley, New York.
DE ALFARO 65	V. De Alfaro & T. Regge (1965) *Potential Scattering*, North-Holland, Amsterdam.
DICKE 60	R. H. Dicke & J. P. Wittke (1960) *Introduction to Quantum Mechanics*, Addison-Wesley, Reading, Mass.
FERMI 50	E. Fermi (1950) *Nuclear Physics*, Univ. Chicago Press, Chicago.
FONDA 78	L. Fonda, G. C. Ghirardi & A. Rimini (1978) Decay theory of unstable quantum systems, *Rep. Prog. Phys.* **41**, 587.
GANTMACHER 59	F. R. Gantmacher (1959) *The Theory of Matrices*, Chelsea, New York. See vol. II, Chap. 13.
GAVEAU 95	B. Gaveau & L. S. Schulman (1995) Limited quantum decay, *J. Phys. A* **28**, 7359.
GAVEAU 96	B. Gaveau & L. S. Schulman (1996) Master equation based formulation of non-equilibrium statistical mechanics, *J. Math. Phys.* **37**, 3987.
GELL-MANN 90	M. Gell-Mann & J. B. Hartle (1990) Quantum mechanics in the light of quantum cosmology, in *Complexity, Entropy, and the Physics of Information*, W. H. Zurek ed., Addison-Wesley, Reading, Mass.
GOLD 62	T. Gold (1962) The Arrow of Time, *Am. J. Phys.* **30**, 403.
GOLDSTEIN 80	H. Goldstein (1980) *Classical Mechanics*, 2nd ed., Addison-Wesley, Reading, Mass.
GRIFFITHS 84	R. B. Griffiths (1984) Consistent Histories and the Interpretation of Quantum Mechanics, *J. Stat. Phys.* **36**, 219.
HELLER 93	E. J. Heller & S. Tomsovic (1993) Postmodern Quantum Mechanics, *Phys. Today* **46**, July, 38.
ITANO 90	W. M. Itano, D. J. Heinzen, J. J. Bollinger & D. J. Wineland (1990) Quantum Zeno Effect, *Phys. Rev.* A **41**, 2295; Reply to "Comment on 'Quantum Zeno Effect,'" (1991) *Phys. Rev.* A **43**, 5168.

KABIR   P. K. Kabir (1968) *The CP Puzzle: Strange Decays of the Neutral Kaon,*
68   Academic Press, New York.

KAC   M. Kac (1956) Some Stochastic Problems in Physics and Mathematics,
56   Lectures at the Field Research Lab. of the Socony Mobil Oil Co.

KAC   M. Kac (1959) *Probability and Related Topics in Physical Sciences,* In-
59   terscience, New York.

KATZ   A. Katz (1967) *Principles of Statistical Mechanics: The Information
67   Theory Approach,* Freeman, San Francisco.

KHALFIN   L. A. Khalfin (1957) *Doklady Akad. Nauk USSR,* **115**, 277; (1957) *Zh.
57   Eksp. Teor. Fiz.* **33**, 1371; Eng. trans. (1958) *Sov. Phys.–JETP* **6**, 1053.

KHALFIN   L. A. Khalfin (1961) *DAN USSR* **141**, 599; (1968) *Pis'ma Zh. Eksp.
61   Teor. Fiz.* **8**, 106; English trans. (1968) *JETP Lett.* **8**, 65.

KITTEL   C. Kittel (1958) *Elementary Statistical Physics,* Wiley, New York.
58

LASOTA   A. Lasota & M. C. Mackey (1994) *Chaos, Fractals, and Noise: Stochastic
94   Aspects of Dynamics,* 2nd ed. Springer, New York.

LEFF   H. S. Leff & A. F. Rex (1990) *Maxwell's Demon: Entropy, Information,
90   Computing,* Princeton Univ. Press, Princeton.

LICHTENBERG   A. J. Lichtenberg & M. A. Lieberman (1983) *Regular and Stochastic
83   Motion,* Springer, New York.

MARCUS   C. M. Marcus, A. J. Rimberg, R. M. Westervelt, P. F. Hopkins & A.
92   C. Gossard (1992) Conductance fluctuations and chaotic scattering in
              ballistic microstructures, *Phys. Rev. Lett.* **69**, 506.

MCCRAW   R. J. McCraw & L. S. Schulman (1978) Metastability in the Two Dimen-
78   sional Ising Model, *J. Stat. Phys.* **18**, 293.

NEWTON   R. G. Newton (1982) *Scattering Theory of Waves and Particles,* 2nd ed.,
82   Springer, New York.

OPPENHEIM   I. Oppenheim, K. E. Shuler & G. H. Weiss (1977) *Stochastic Processes
77   in Chemical Physics: The Master Equation,* MIT Press, Cambridge.

PRIGOGINE   I. Prigogine (1962) *Non-Equilibrium Statistical Mechanics,* Interscience,
62   New York.

RAMSEY   N. F. Ramsey (1956) Thermodynamics and Statistical Mechanics at Neg-
56   ative Absolute Temperatures, *Phys. Rev.* **103**, 20.

ROBERTSON   H. S. Robertson (1993) *Statistical Thermophysics,* Prentice-Hall, Engle-
93   wood Cliffs.

SCHULMAN   L. S. Schulman, R. G. Newton & R. Shtokhamer (1975) Model of Impli-
75   cation in Statistical Mechanics, *Phil. Sci.* **42**, 503.

SCHULMAN   L. S. Schulman, M. Stone & D. J. Wallace (1978) Metastable Lifetimes,
78a   Level Crossings and Vertical Cuts in Quantum Mechanics, *J. Phys.* A
              11, 1933.

SCHULMAN   L. S. Schulman (1978) Note on the Quantum Recurrence Theorem, *Phys.
78b   Rev.* A **18**, 2379.

SCHULMAN   L. S. Schulman (1991) Definite Quantum Measurements, *Ann. Phys.*
91   **212**, 315.

SCHULMAN   L. S. Schulman (1994) Accuracy of the semiclassical approximation for
94a   the time dependent propagator, *J. Phys.* A **27**, 1703.

SCHULMAN   L. S. Schulman, A. Ranfagni & D. Mugnai (1994) Characteristic scales
94b   for dominated time evolution, *Phys. Scripta* **49**, 536.

TABOR   M. Tabor (1989) *Chaos and Integrability in Nonlinear Dynamics: an
89   Introduction,* Wiley, New York.

TAVERNIER 76   J. Tavernier (1976) Classical and Quantum Descriptions of the Kac Ring Model, *J. Stat. Phys.* **14**, 101.

TERENT'EV 72   M. V. Terent'ev (1972) On the Exponential Decay Law of Nonstable Particles, *Ann. Phys.* **74**, 1.

THOMPSON 72   C. J. Thompson (1972) *Mathematical Statistical Mechanics*, Macmillan, New York.

VAN KAMPEN 81   N. G. van Kampen (1981) *Stochastic Processes in Physics and Chemistry*, North-Holland, Amsterdam.

WIGHTMAN 81   A. S. Wightman (1981) The Mechanisms of Stochasticity in Classical Dynamical Systems, in *Perspectives in Statistical Physics*, H. J. Raveche ed., North-Holland, Amsterdam.

ZUREK 91   W. H. Zurek (1991) Decoherence and the transition from quantum to classical, *Phys. Today*, October, 36.

Notes on the exercises

Exercise 2.0.1 Wheeler's answer (according to the story) was, 'I heard them strike each other.'

Exercise 2.1.2 Let the color changing sites lie between the ball locations.

Exercise 2.2.2 For large r, the per point deviation from $\log g$ is $-1/2r$.

Exercise 2.4.1 No.

Exercise 2.4.2 A numerical way would be to go through the grains, one at a time, putting points in each grain and counting occupancies on the next time step.

Exercise 2.5.1 The expression for $\exp(2i\delta)$ is $J(k)^*/J(k)$ with $J(k) = k' + 2\lambda \tan(k'a) - ik \tan(k'a)$. To see that the two formulations agree in identifying the unstable state, use the definition of $S(k)$ and write the $x > 0$ wave function $B \sin(kx + \delta)$ (up to factors) as $\exp(+ikx) - \exp(-ikx)/S(k)$. The function $J(k)$ is essentially the Jost function. See [DE ALFARO] or [NEWTON].

Exercise 2.5.3 See [GAVEAU 95], discussion following Eq. (4.2). Beware of misprints; also we are stingy with details, although the idea is stated.

Appendix 2.1.A: Other ensembles and limitations of the ring model

There is another kind of ensemble averaging that comes to mind, namely to average over initial ball states. One could take an initial condition with the number of white balls, W, less than N, and average over the $\binom{N}{W}$ configurations consistent with this. For the Kac ring model this doesn't work.

To see that randomizing the initial conditions of the balls is not a sure way to equilibrate, consider a contrived realization of the Kac ring. Let the active sites always come in pairs. That is, take A even, suppose that the active sites cluster, and that they cluster so that every cluster has an even number of active sites. I will now show that the equilibrium value of δ depends on the initial configuration, and for the case of an initial all white state is $(1 - \mu)^2$, not zero. For the active-site configurations just described (even clusters only) it is possible to divide the active sites into two classes, odd-active and even-active, as follows. As you move counterclockwise call the first site of an active cluster odd. The next is even, and so on. With no further calculation it is already possible to see that something is amiss. Consider an initial configuration in which there are white balls on all inactive sites, white balls in all odd-active sites, and black balls in all even-active sites. See Fig. 2.1.A.1. A moment's reflection shows that this ball state does not change at all in time.

Let us go back to general $\mathcal{A}$. Start a white ball in site 1 and let it go. Let its color when it reaches site p be γ_p $(= \pm 1)$. Take as an initial ball state precisely that sequence. Then, again, nothing happens! This history, taken as an initial ball state, is invariant. Obviously, $-\gamma_p$ also provides an invariant state. For the first invariant state $\delta(t)$ is a constant and is given by $\delta_1 = \frac{1}{N} \sum_p \gamma_p$, which may or may not be zero. For the other invariant state, $\delta(t) = -\delta_1$. Now consider an arbitrary initial state. Some number of its balls, call it N_1, take the values associated with history 1, some number, $N_2 = N - N_1$, take the values associated with history 2. Our present discussion is aimed at long time behavior, and the time period we take is a full cycle around the ring, N steps. In averaging δ every member of the history 1 set gives the same contribution, δ_1; similarly for history 2. Therefore $\bar{\delta} = (1/N)[N_1 - N_2]\delta_1$. This can be different from zero, although in uncontrived situations it has two reasons to be small, namely $N_1 \approx N_2$ and $\delta_1 \approx 0$.

We return to the even-active-clusters sets. One invariant state has already been identified and its value of δ is seen to be $(1/N)\{[(N - A) + A/2] - [A/2]\} = (1 - \mu)$. For the all white initial state, the number of balls having this history is $N - A + A/2$ and the number with the other history is $A/2$. By the formula in the last paragraph, we find $\bar{\delta} = (1 - \mu)^2$, as asserted.

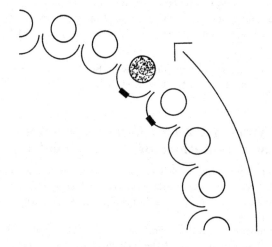

Fig. 2.1.A.1. Invariant ball state for an active site configuration with even clusters only. The even-active site contains a black ball. Other sites contain white balls. On the next time step the black ball turns white and the white ball entering the even-active site turns black. All other balls are remain white.

Ball state averaging does not force equilibrium because the dynamics are not rich enough. For these uncoupled color-switchers, only a tiny fraction of the 2^N states are traversed. A further clue is the short recurrence time, polynomial in N, rather than exponential. There is yet another way to see the simplicity of the dynamics. Define a new dynamical variable:

$$\xi_p(t) \equiv \gamma_p \eta_p(t) \qquad \text{with} \quad \gamma_p \equiv \prod_1^{p-1} \epsilon_p \qquad (2.1.A.1)$$

(This γ_p is the same as that above.) The 'equation of motion' for $\xi_p(t)$ is trivial:

$$\xi_p(t) = \epsilon_{p-1}\gamma_{p-1}\epsilon_{p-1}\eta_{p-1}(t-1) = \xi_{p-1}(t-1) = \ldots = \xi_{p-t}(0) \qquad (2.1.A.2)$$

The change of coordinates, Eq. (2.1.A.1), is a kind of gauge transformation and shows that in this theory all the dynamics can be gauged away.

Remark: A point of relevance to the 'cryptic constraint' thesis of this book: despite this 'non-ergodicity' in the dynamics, for the vast majority of $\mathcal{A}$ s equilibration seems normal. For a reasonable coarse grain definition the flaw will usually be microscopic.

Appendix 2.2.A: Entropy for cat-map dynamics with indistinguishable points

We do the same entropy calculation as in Section 2.2, but letting the N-point cat map have indistinguishable points. This would be a reasonable choice if one wanted to think of the points as atoms of an ideal quantum gas.

As in the main text, N points undergo cat-map dynamics in the unit square, to be thought of as phase space. The square is again divided into g coarse grains with G fine grains within each coarse grain. We have the following macroscopic state: grain k has n_k points ($k = 1, \ldots, g$). For each grain the number of ways to distribute n indistinguishable points among G sites is $\binom{G}{n}$. The entropy is the logarithm of the number of possible distributions

$$\exp \frac{S}{k} = \prod_{j=1}^{g} \binom{G}{n_j} \qquad (2.2.A.1)$$

This differs from Eq. (2.2.5) by a factor $N!$. By the usual approximations

$$\frac{1}{k}S = N \log G - N(\log N - 1) - N \sum \rho_j \log \rho_j \qquad (2.2.A.2)$$

where again $\rho_j \equiv n_j/N$. As before we drop the entropy contribution associated with the fine grained cutoff to yield

$$\frac{1}{k}S = -N(\log N - 1) - N \sum \rho_j \log \rho_j \qquad (2.2.A.3)$$

An interesting question is extensivity. It is the introduction of the $1/N!$ due to indistinguishability that ordinarily provides extensivity for single species substances. For the cat map you would like to double the volume and the number of particles and find the entropy doubled. But there isn't an obvious way to do this since the 'volume' is a part of the dynamics.

To pose the extensivity problem with a doubled volume I use the approach suggested by the comoving coordinate system discussed in Chapter 4. The dynamics are unchanged, but the coarse graining distance scale is halved. This means twice as many spatial coarse grains can be fitted along the x-axis. We now have a doubled number of particles and a doubled number of grains. The number of points in each grain is thus the same (assuming the statistical distribution is the same, which must be the case if you want extensivity) and the density that enters the entropy calculation halved (because the denominator for ρ is now $2N$). Before the doubling, the entropy was given by

$$\frac{1}{k}S_1 = -N(\log N - 1) - N\sum_{j=1}^{g} \rho_j \log \rho_j \qquad (2.2.\text{A}.4)$$

Afterwards

$$\frac{1}{k}S_2 = -2N(\log(2N) - 1) - 2N\sum_{j=1}^{2g} \frac{\rho_j}{2}\log\frac{\rho_j}{2} \qquad (2.2.\text{A}.5)$$

Note that the ρ_js in Eq. (2.2.A.5) are the same as those in Eq. (2.2.A.4), there are just twice as many of them. They therefore add to 2 (not 1). Rewrite Eq. (2.2.A.5) as

$$\frac{1}{k}S_2 = 2\left\{ -N(\log N - 1) - N\log 2 - \frac{N}{2}\sum_{j=1}^{2g}\rho_j\log\rho_j + \frac{N}{2}\sum_{j=1}^{2g}\rho_j\log 2 \right\} \qquad (2.2.\text{A}.6)$$

Cancellation of the $N\log 2$ term is the elimination of the entropy of mixing. Using the fact that when summed to $2g$ the ρs add to 2, we get $S_2 = 2S_1$.

Appendix 2.2.B: Mixing and ergodicity in the cat map

In this Appendix, we turn to more abstract considerations. As a warmup we prove the Poincaré recurrence theorem. We then define and discuss ergodicity and mixing and show that the cat map is indeed mixing and therefore also ergodic.

The Poincaré recurrence theorem

The Poincaré recurrence theorem is frugal in its assumptions, immediate in its proof and far reaching in its consequences. All you need is a measure-preserving transformation that is one-to-one and maps a set of positive, finite measure into itself under the action of the transformation. The cat map satisfies this, and more importantly, so does the time evolution of a confined system in classical mechanics.

Because common dynamical systems possess small subsets for which atypical behavior can occur (for example, $(0,0)$ in the cat map, which goes nowhere, or $(0.4, 0.2)$, which is periodic) it is necessary to formulate the recurrence theorem using the language of measure theory. If you are unfamiliar with Lebesgue measure, just think of it as ordinary area or volume, generalized so that you can talk about the area or volume of sets defined through limit processes. Countable sets or sets of lower dimension (lines in the plane, planes in 3-space) have measure zero. A useful notion once you have Lebesgue measure is 'almost always' or 'almost everywhere,' referring to statements that are true except possibly on a set of measure zero.

The theorem states that almost all points return arbitrarily closely to their initial positions. More precisely, let the transformation be called φ_t with the label t taking discrete or continuous values and let $\varphi_t\varphi_s = \varphi_{t+s}$. Let φ_t map a space Ω of finite measure into itself in a measure preserving way. That is, if μ is the measure and A a subset of Ω, $\mu(\varphi_t(A)) = \mu(A)$ for all t. In addition $\varphi(\Omega) \subset \Omega$ and $\mu(\Omega) < \infty$. We also assume that φ_t has an inverse.[40] The theorem states that if $A \subset \Omega$, then for almost all $\omega \in A$, there are arbitrarily large values of t such that $\varphi_t(\omega) \in A$.

To prove this, let B be the subset of A that does not recur after a time τ, i.e., for all $t \geq \tau$, $\varphi_t(B) \cap A = \emptyset$. Then certainly for all $t \geq \tau$, $\varphi_t(B) \cap B = \emptyset$. We show that for every τ, $\mu(B) = 0$. Let $\Phi \equiv \varphi_\tau$; it is sufficient to restrict discussion to powers of Φ. The proof examines the orbit of B, i.e., the set of points $\Phi^n(B)$ for $n = 0, 1, \ldots$. By the hypothetical non-recurring property of B it follows (I'll show this in a moment) that the orbit does not intersect itself, so all those images of B are disjoint. But if they are disjoint their measure adds and if the measure of B were anything but 0 the orbit would have infinite measure, which is impossible since that orbit is a subset of the finite-measure set, Ω. The non-intersecting statement for the orbit is the following: for every $n, m \geq 0$ and $n \neq m$, $\Phi^n(B) \cap \Phi^m(B) = \emptyset$. To see this, suppose it to be false and that the intersection is a non-empty set C. Let $n < m$. Apply Φ^{-n} to both sides to get $B \cap \Phi^{m-n}(B) = \Phi^{-n}(C)$. Because $[(C \neq \emptyset) \Rightarrow (\Phi^{-n}(C) \neq \emptyset)]$, this would give us a point in B that recurs after time $(m - n)\tau$, contrary to hypothesis.

Ergodicity

To predict specific heat, or pressure, or just about anything that the tools of statistical mechanics allow you to predict, you average over phase space (or Hilbert space). Most quantities are derivable from a partition function and a partition function is a sum over all states consistent with ensemble-defining constraints. Historically, ergodicity addressed the following question: if you measure the specific heat of a particular sample, why does the result equal that obtained by averaging over phase space? The answer was supposed to be that your measurement took a finite time. During this time the system covered a lot of ground in phase space, so that your physical measurement was effectively a time average. If the ground covered by your system didn't leave out any important regions in phase space, then this time average would be the same as the phase space average. Under the well meaning attentions of mathematicians, the time average became the limit as the interval for averaging went to infinity. In this Appendix we accept that formulation, although it ignores the physical time scale of typical measurements.

The equality of phase space averages and time averages is equivalent to having only one constant of the motion survive in equilibrium.[41] We prove this in the following context. There is a dynamical law that moves points around in phase space. It is one-to-one (almost everywhere) and measure preserving. We confine attention to a space Ω, which for classical mechanics is a surface of constant energy, a subset of phase space.[42] For the

---

[40] If the mapping fails to have an inverse on a set of measure zero, discard the offending set, followed by the necessarily zero-measure sets now rendered without image or preimage by the excision of the discarded set. Continue this process. On what remains the mapping is one-to-one, while the discarded sets have total measure zero, since only countably many removals were performed.

[41] Some definitions of equilibrium make this statement a tautology. What I mean is that you take an isolated system and let it sit until it settles down. Experimentally you never know how long is long enough.

[42] On phase space, defined by ordinary position and momentum, the measure is Lebesgue measure generated in a straightforward way from ordinary volume in the coordinates and the momenta. On the energy surface, it is the measure induced by the constraint $H(p, q) = E$; in particular there is an additional factor in the integrand, $1/|\nabla H|$. This factor arises from standard operations on integrals. For example, in the plane, let $f(x, y) = c$ define a curve (as $H = E$ defines the energy surface). Consider the integral $J \equiv \int dx\, dy\, g(x, y)\delta(f(x, y) - c)$. Then J can be writ-

cat map, Ω is the unit square. The dynamical law is a mapping φ_t with t taking discrete or continuous values. Let f be a real valued function on Ω. Its time average depends on the point that you start from, and therefore is another function[43]

$$\mathcal{T}(f;x) \equiv \lim_{T \to \infty} \frac{1}{T} \int_0^T f(\varphi_t(x)) \qquad (2.2.\text{B}.1)$$

The phase space average is denoted

$$\mathcal{P}(f) \equiv \int_\Omega f(x) \qquad (2.2.\text{B}.2)$$

Ergodicity means that for almost all $x \in \Omega$

$$\mathcal{T}(f;x) = \mathcal{P}(f) \qquad (2.2.\text{B}.3)$$

I will now show that for classical mechanics this is equivalent to having only one constant of the motion, the energy, survive the macroscopic limit. Since we are already on the energy surface, this means that there can be no other constants. For isolated macroscopic systems, this implies that in equilibrium they all look the same. If there were another constant of the motion, you might start one of the systems in a region of phase space with one value and another in another region. Then they would remain different forever.[44]

The mathematical formulation of this assertion is that there is no function, other than a constant (almost everywhere), that is invariant under φ_t (for non-zero t). We now prove that this assertion is equivalent to ergodicity; that is, Eq. (2.2.B.3) is equivalent to the absence of non-constant invariant functions.

Suppose $f(x)$ is invariant under φ_t but is not a constant on Ω. Then there is a number f_0 such that on one set of non-zero measure $f(x) > f_0$, and on another set of non-zero measure $f(x) \le f_0$. Call the first set A and the second set B. Without loss of generality take $f_0 = 0$. It follows that $\int_A f > 0$.

Aside: This is an opportunity to see what measure theory can buy you. You might have raised the following objection to the assertion about the strict positivity of the integral. Although $f(x) > 0$ on A, it need not be bounded away from zero. The possibility then arises that the difference is positive but inconsequential. That is, for any $\epsilon > 0$, $\int_{A(\epsilon)} f(x) = 0$ where $A(\epsilon)$ is the set of points where $f(x) > \epsilon$. However, I can write A as the disjoint union[45] $\cup_n [A(1/(n+1)) - A(1/n)]$. The measure of A is the sum of the measures of the sets in that union. If every one of them has zero measure, so does A.

It follows that there are orbits that start in A for which the time average is larger than 0. (Average $\mathcal{T}(f;x)$ over A and use the invariance of A under φ.) Any orbit starting in B must yield a time average equal to or less than 0. Therefore the time averages are unequal and cannot both equal the phase space average. A non-constant invariant function therefore implies no ergodicity.

---

ten as an integral over x alone using the usual transformation rules for δ-functions:
$J = \int dx\, g(x,y)/|\partial f/\partial y|$ with y evaluated at a point on the curve fixed by x.

[43] My notation is clumsy here because of a desire for $\mathcal{T}$ and $\mathcal{P}$ to be similar objects. The *function* $\mathcal{T}(f;x)$ is a mapping of x to the reals. To emphasize its functional character one could also write it as $\mathcal{T}(f;\cdot)$.

[44] The system is assumed 'pinned to the table.' Otherwise other additive constants of the motion, e.g., angular momentum, can survive.

[45] For sets C and D the difference $C - D$ is the set of points in C that are not in D

The converse is the statement that the absence of non-constant invariant functions implies ergodicity. Consider the function $F(x) \equiv \mathcal{T}(f; x)$. It is invariant since the difference between $(1/T) \int_t^{T+t} f(\varphi_{t'}(x)) dt'$ and $(1/T) \int_0^T f(\varphi_{t'}(x)) dt'$ vanishes for $T \rightarrow \infty$. An absence of non-constant invariant functions then implies that $F(x)$ is a constant. Averaging F over all x shows that the constant must be $\mathcal{P}(f)$.

From the foregoing discussion you should expect that ergodicity helps explain equilibration. If a system is ergodic its points get around and the entropy of the system should increase. Unfortunately, our definition permits silly exceptions to that increase. Systems in one space dimension are always ergodic, but points on the energy surface do not separate from one another, stymieing equilibration. For more general systems, when you can actually show them to be ergodic it is because nearby orbits on the energy surface separate from one another rapidly. This is true on spaces of constant negative curvature (on which the dynamics are ergodic) and is the essence of Sinai's demonstration of ergodicity for certain hard disk dynamical systems.

Although ergodicity has never been shown to hold for any non-trivial natural system, we go on to define a yet stronger notion, one that brooks no exceptions to its equilibrating power, and which is in fact a property of the cat map.

Mixing

The stronger concept is 'mixing' and to define this I will continue to use the same notational conventions, except that now it is convenient to normalize the measure of the total space Ω to 1. We introduce the definition: a transformation is mixing if $\lim_{t \rightarrow \infty} \mu(\varphi_t(A) \cap B) = \mu(A)\mu(B)$. Intuitively this means that φ_t spreads A around so much (for large t) that its image is as likely to be in B as out of it (proportional to the size of B). Compare this formula to that for conditional[46] probability: $\Pr(C|D) = \Pr(C \cap D)/\Pr(D)$. If $\Pr(C \cap D) = \Pr(C)\Pr(D)$ it would follow that $\Pr(C|D) = \Pr(C)$; that is, conditioning on D implies nothing, the two events are independent. Mixing thus means that for long enough times you lose any correlation[47] between B and $\varphi_t(A)$.

It is easy to show that mixing implies ergodicity. All we need establish is that there cannot be a non-constant invariant function on Ω. Suppose there were such a function. As above we call it f; again there is a number f_0 for which there are sets of non-zero measure on both sides of f_0. Again we define the set A to be those points for which $f(x) > f_0$ and the non-constancy of f implies that $0 < \mu(A) < 1$. Since f is assumed invariant, $\varphi_t(A) = A$. Consider $J \equiv \mu(\varphi_t(A) \cap A)$. Since $\varphi_t(A) = A$, J must be $\mu(A)$. On the other hand, if φ_t is mixing we also have that for $t \rightarrow \infty$ J approaches $\mu(\varphi_t(A))\mu(A)$. By the measure preserving property of φ_t this implies $\mu(A) \cdot \mu(A) = \mu(A)$ which implies $\mu(A) = 0$ or 1, a contradiction.

The cat map is mixing

We complete this Appendix by showing that the cat map is mixing. To give this proof we are drawn into what might appear to be irrelevant formalism, Fourier analysis that is useful for studying the cat map and seemingly not much else. Later in this book (Section 4.3 and its Appendix), the cat map will reappear; it is simple enough to allow

---

[46] $\Pr(C|D)$ is the probability of the event C, given the event D.
[47] To be more explicit, identify probability (Pr) and measure (μ). Then if φ is mixing
$\Pr(\varphi_t(\omega) \in B | \omega \in A) = \mu(\varphi_t(A) \cap B)/\varphi(A) \rightarrow \mu(B)$.

solution of two-time boundary value problems, and the Fourier analysis we now develop
will be useful for that too.

Recall that the characteristic function of a set is the function that is one for points
in the set, zero otherwise. For a set A we call its characteristic function χ_A. With this
notation, the measure of A can be written

$$\mu(A) = \int \chi_A(x) = \int \chi_A \qquad (2.2.B.4)$$

If you are uncomfortable without any differential-like object inside the integral, you could
insert a $d\mu(x)$ in Eq. (2.2.B.4). Note that $\chi_{A \cap B} = \chi_A \chi_B$. With characteristic functions,
the mixing property takes the form

$$\lim_{t \to \infty} \int \chi_{\varphi_t(A)} \chi_B = \int \chi_A \int \chi_B \qquad (2.2.B.5)$$

Specializing φ to be the cat map, the behavior of functions under φ can be neatly
expressed using Fourier analysis. For the unit square, the following form a complete set
of functions

$$e_{pq}(x,y) \equiv \exp[2\pi i(px + qy)] \qquad \text{with} \qquad p, q = \text{integers} \qquad (2.2.B.6)$$

Notice that for points in the square we sometimes use the notation (x, y) and sometimes
ξ, as in Eq. (2.2.2). For the pair (p, q) of Fourier coefficients we will also use the notation
ν. Any function f on the unit square, $\mathcal{I}$, can be written as a sum over functions e_{pq} in
the usual way:

$$f(x, y) = \sum_{p,q} \hat{f}_{pq} e_{pq}(x, y) = \sum_\nu \hat{f}_\nu e_\nu(\xi) \qquad (2.2.B.7a)$$

with the Fourier coefficients given by

$$\hat{f}_{pq} = \int_{\mathcal{I}} dx dy f(x, y) e_{pq}^*(x, y) \qquad (2.2.B.7b)$$

When a transformation acts on a space it induces a natural mapping on the functions
on that space. If ψ is a transformation, it defines a new function $U_\psi(f)$ by

$$[U_\psi(f)](\xi) \equiv f(\psi(\xi)) \qquad (2.2.B.8)$$

When ψ is measure preserving, the map U is unitary and this is the basis of certain
Hilbert space approaches to classical mechanics. The cat map is measure preserving
and its action on functions, written in terms of the action on the Fourier coefficients, is
beautifully simple. Obviously

$$f(\varphi(\xi)) = \sum_\nu e_\nu(\varphi(\xi)) \hat{f}_\nu \qquad (2.2.B.9)$$

On the functions e_ν the action of φ can be transferred to ν

$$e_{pq}(\varphi(x, y)) = \exp\{2\pi i[(p(x + y) + q(x + 2y)]\} = \exp\{2\pi i[(p + q)x + (p + 2q)y]\}$$

This implies that $e_\nu(\varphi(\xi)) = e_{\varphi(\nu)}(\xi)$. Eq. (2.2.B.9) becomes

$$f(\varphi(\xi)) = \sum_\nu e_{\varphi(\nu)}(\xi) \hat{f}_\nu = \sum_\nu e_\nu(\xi) \hat{f}_{\varphi^{-1}(\nu)} \qquad (2.2.B.10)$$

Therefore the mapping on function space induced by the cat map moves the Fourier
coefficients around, using the inverse of the matrix M (given earlier), without even the
need for mod 1 complications. Now the action of the matrix M^{-1} on integer pairs (p, q)

is as dramatic as the action of M on points (x, y). After a few (say n) iterations, the image $M^{-n}(p, q)$ grows rapidly, picking up the factor λ_+ in its distance from the origin with each additional step. In a way the action is *more* dramatic because it allows no exceptions. In the plane, if two points are separated precisely along the eigenvector of M having the eigenvalue λ_-, they will come together, rather than separate. The catch is that their orientation must be precise and a slight error will introduce a piece of the other eigenvector, which will cause separation for large enough n. For the integer pairs (p, q) you *cannot* produce the eigenvector of M^{-1} of smaller eigenvalue, because it involves the irrational number $\sqrt{5}$ (cf. Eq. (2.2.4)) and p and q are integers.

We prove mixing by showing that Eq. (2.2.B.5) holds for any two characteristic functions. For sets A and B we call the Fourier coefficients of their characteristic functions a_ν and b_ν. Thus

$$\chi_A(x) = \sum_\nu a_\nu e_\nu(x) \tag{2.2.B.11}$$

As usual, we know that for given $\epsilon > 0$ you can find an R such you can discard terms (p, q) in the sum with $\sqrt{p^2 + q^2} > R$ with error (when the truncated expansion is used in an integral) less than ϵ. For the particular coefficient $(p, q) = (0, 0)$ (to be called $\nu = 0$) we have $a_0 = \int \chi_A = \mu(A)$, and similarly for B. Therefore, to prove mixing we must only show that the norm of[48]

$$\Delta_n \equiv a_0 b_0 - \int \chi_{\varphi^n(A)} \chi_B \tag{2.2.B.12}$$

can be made as small as we wish, where the time parameter has been called n. Take the ϵ for the aforementioned truncations to be half of whatever it was you wished for $|\Delta_n|$ and define $\widetilde{\Delta}_n$ in terms of the Fourier transforms of χ_A and χ_B:

$$\widetilde{\Delta}_n \equiv a_0 b_0 - \int {\sum_\nu}' a_\nu e_{\varphi^n(\nu)} {\sum_\mu}' b_\mu e_\mu \tag{2.2.B.13}$$

The sums have been given primes to indicate that they have been truncated to have both ν and μ within the radius, R, determined by ϵ. $\widetilde{\Delta}_n$ therefore differs from Δ_n by some fraction of ϵ. In the integral in Eq. (2.2.B.13) there is at least one term that survives, namely $\nu = \mu = 0$. This cancels the term $a_0 b_0$, as it should. It remains to show that what's left vanishes. For a product from the double sum to survive the integration, we would need a pair, μ and ν, with $\mu + \varphi^n(\nu) = 0$, by the orthogonality of the expansion functions. But every μ is within the circle defined by R. On the other hand, the fact that the νs are themselves from a finite set implies that for $\nu \neq 0$ and for sufficiently large n every $\varphi^n(\nu)$ can be sent outside that circle. The remaining case, $\nu = 0$, also leads to zero integrals, except for $\mu = 0$, which has already been dealt with.

Appendix 2.5.A: Quantum recurrence theorem

Consider a system confined to a box and let it have wave function $\psi(0)$ at time 0. The quantum recurrence theorem states that the time-evolved wave function gets arbitrarily close to $\psi(0)$ at arbitrarily large times. That is, for any ϵ and any T_0, there exists a T such that $\|\psi(T) - \psi(0)\| < \epsilon$, with $T > T_0$.

---

[48] Notation: because the time-n transformation is the n-th power of φ, φ_n and φ^n mean the same thing.

As for the classical Poincaré recurrence theorem, the proof depends on the system's being confined to a finite region. (A free particle just keeps going.) For the classical case, the finiteness was expressed by the requirement that the phase space volume 'Ω' was finite. For quantum mechanics the discreteness of the spectrum does the job.

We will prove the theorem with a trick—appeal to the classical theorem. The idea is that when you truncate the modes of a confined quantum system, its dynamics become those of a finite collection of classical oscillators. Then use the classical Poincaré recurrence theorem for those oscillators.

Let the given initial wave function be $\psi(0)$, call the Hamiltonian H, let its eigenvalues be E_n and the corresponding eigenfunctions u_n. Then at time t the wave function is given by

$$\psi(t) = \sum_n u_n r_n \exp(-iE_n t) \tag{2.5.A.1}$$

with $r_n = \langle u_n | \psi(0) \rangle$. Without loss of generality, we define the phases of the u_n so that the r_n are real. We must prove that for any $\epsilon > 0$ and T_0 we can find $T > T_0$ such that

$$\|\psi(T) - \psi(0)\|^2 = 2 \sum_n r_n^2 (1 - \cos E_n t) < \epsilon \tag{2.5.A.2}$$

Since $\sum_n r_n^2 = 1$, we can find N such that

$$2 \sum_{n=N+1}^{\infty} r_n^2 (1 - \cos E_n t) \leq 4 \sum_{n=N+1}^{\infty} r_n^2 < \frac{\epsilon}{4} \tag{2.5.A.3}$$

The foregoing holds uniformly in t. It remains to show that there is a $T > T_0$ such that

$$2 \sum_{n=1}^{N} r_n^2 (1 - \cos E_n T) < \frac{3\epsilon}{4} \tag{2.5.A.4}$$

This part of the proof is usually accomplished by quoting theorems about almost periodic functions. However, the sum in Eq. (2.5.A.4) can immediately be related to the motion of N harmonic oscillators—classical ones—and then we will be able to use the *classical* recurrence theorem. The idea is simple, but technicalities will engulf us in a plague of ϵs.

Consider a collection of N harmonic oscillators with frequencies $(E_1, \ldots, E_N)$. Let the action-angle coordinates be $\gamma \equiv (I_1, \phi_1, \ldots, I_N, \phi_N)$. The solution to the equation of motion is $I_k = $const, $\phi_k = \phi_k(0) + E_k t$, $k = 1, \ldots, N$. Because the angle is defined modulo 2π this is motion on a multidimensional torus with radii given by the $\{I_k\}$. If the oscillators have initial values $I_k = r_k$, $\phi_k(0) = 0$, for $k = 1, \ldots, N$, then the collection of numbers $\{r_k \exp(-iE_k t)\}$ stand in one-to-one correspondence with the motion $\gamma(t)$ on the torus. The Poincaré recurrence theorem says that the oscillators almost always get arbitrarily close to the initial point, which implies that the sum in Eq. (2.5.A.4) will get close to zero.

With this preamble, there's nothing to be done but bring on the 'ϵ's. For $\gamma = (I_1, \phi_1, \ldots)$, define a volume in phase space

$$\Gamma(\gamma, \alpha) \equiv \left\{ \gamma' = (I_1', \phi_1', \ldots) \,\middle|\, \sum_{n=1}^{N} |I_n' e^{i\phi_n'} - I_n e^{i\phi_n}| < \alpha \right\} \tag{2.5.A.5}$$

The quantum recurrence of $\psi(0)$ will happen when there is a classical recurrence in the neighborhood $\Gamma(\rho, \epsilon/4)$ of ρ. Pick a time T_0; we seek $T > T_0$ for which there is a quantum recurrence. The classical recurrence theorem does not guarantee that there is a T for which ρ recurs (because it allows measure zero exceptions), but it does guarantee that there are points in its neighborhood that *do* recur (for some time $T > T_0$, and for a neighborhood of size 'α' $= \epsilon/4$). Pick one of them and call it $\rho' = (r_1', \phi_1', \ldots)$. By virtue of the fact that ρ' recurs we have

$$\sum_{n=1}^{N} |r_n' e^{i\phi_n'} - r_n' e^{i(\phi_n' + iE_n T)}| < \epsilon/4 \qquad (2.5.A.6)$$

Since $\rho' \in \Gamma(\rho, \epsilon/4)$ we also have

$$\sum_{n=1}^{N} |r_n' e^{i\phi_n'} - r_n| < \epsilon/4 \qquad (2.5.A.7)$$

By the triangle inequality it then follows that

$$\sum_{n=1}^{N} |r_n e^{-iE_n T} - r_n| = \sum_{n=1}^{N} |r_n e^{+iE_n T} - r_n|$$

$$\leq \sum_{n=1}^{N} \left[|[r_n - r_n' e^{i\phi_n'}]e^{iE_n T}| + |r_n' e^{i(\phi_n' + iE_n T)} - r_n' e^{i\phi_n'}| + |r_n' e^{i\phi_n'} - r_n| \right] < \frac{3\epsilon}{4}$$

$$(2.5.A.8)$$

This would finish the proof, except that we defined the neighborhood $\Gamma(\gamma, \alpha)$ in terms of absolute value, not its square (whereas Eq. (2.5.A.2) uses the square). However, since the right hand side of Eq. (2.5.A.8) can be taken to be less than one, every term on the left must be less than one. Therefore for each n, $|r_n e^{-iE_n t} - r_n|^2 < |r_n e^{-iE_n t} - r_n|$ and the theorem is proved.

A few comments are in order. First, quantum recurrence allows no exceptions, not even as infrequent as the measure zero initial conditions of classical mechanics. Second, note where the finite volume restriction entered. It was essential to Eq. (2.5.A.4); discreteness of the spectrum allowed truncation to a finite sum. In our discussion of quantum decay through a barrier, the system did not recur, because we did not put a wall on the far right. Third, it would be interesting to see whether things could be turned round and the Poincaré recurrence theorem used to establish results in the theory of almost periodic functions.

Appendix 2.5.B: More on decay through a barrier

The 'eigenfunction' that we get from the complex k' solution of Eq. (2.5.4) is not in the Hilbert space. That also happens for continuum 'eigenfunctions' but for them the spectrum is still real. The function obtained from the complex k' solution corresponds to a resonance, and in the terminology of scattering theory would be a pole in the 'second sheet.' Note that while the sign of Im E is such as to give decay of the amplitude in the well to the left of 0, it also gives k an imaginary part of the same sign (k and k' are related by Eq. (2.5.3)). This implies that the function $\exp(ikx)$ has *growing* magnitude as $x \to +\infty$, not a good candidate for Hilbert space, even the generalized sort that accepts continuum wave functions. As remarked, this induces a level of mathematical agony, but makes sense physically, from the following considerations. The equation we solved with the outgoing wave boundary conditions (which required Re $k > 0$) was the *time-independent* Schrödinger equation. The time independence, however, is a myth, since we set up a certain initial condition and evolved forward in time from that. However, the mathematical formalism takes the myth seriously. The steady state has amplitude in the well steadily draining out, but it also assumes that this has been happening forever. Therefore if you go to large positive x you will find the amplitude that drained out a long time ago. But a long time ago there was a lot more amplitude in the well, so what you find at large positive x is proportional to the early well occupancy. There is nothing mysterious about this solution to Schrödinger's equation. It's just that squeezing it into Hilbert space is tricky. Incidentally, the outgoing wave solution is closely related to the Jost function used in scattering theory.

These observations on our complex energy wave function are relevant when one considers how it is supposed to help solve the initial value problem. It's true that our wave function resembles the complex energy solution for negative x and even for moderate positive x, but for large x they differ considerably. To deal with this we observe that Schrödinger evolution is *local*. If two wave functions coincide in a region and only differ far away, then, in the given region, they will evolve in the same way, at least for a while. A quantitative demonstration can be made as follows. Let the complex energy solution we found earlier be called $\psi_c(x,t)$. For $x > 0$ it is of the form $\exp(ikx - iEt)$ (dropping $\hbar$s), and for any fixed t blows up for $x \to \infty$. Let $\ell > 0$ and let $\sigma(x)$ be a smooth function that is identically 0 for $x > \ell$ and identically 1 for $x < -\ell$. Consider the wave function $\psi(x,t) = \psi_c(x,t)\sigma[x - L - (\operatorname{Re} k)t]$, where L is another positive number. There are well known ways of making σ smooth, even C^∞. Now apply $H - i\partial/\partial t$ to $\psi(x,t)$. Our function ψ *almost* satisfies the time-dependent Schrödinger equation. The only place it fails is near the truncation. If you do the arithmetic you find

$$\left[H - i\frac{\partial}{\partial t} \right] \psi(x,t) = (\operatorname{Im} k)\psi_c\sigma' - \frac{1}{2}\psi_c\sigma''$$

The right-hand side is non-zero only in the neighborhood of the truncation. For a long time it will have little effect on the decay out of the unstable state. This can be seen using the time-dependent propagator (the thing you evaluate in a path integral), which is the solution to $(H - i\partial/\partial t)G(x,t) = -i\delta(x)\delta(t)$. By integration over the inhomogeneous portion of the above equation (its right-hand side), the propagator gives its effect everywhere (see the exercise, below). If the right-hand side is smooth, small and distant, then the wave emanating from it will be unimportant in the neighborhood of the origin for a long time. Making it smooth, small and distant involves optimizing the parameters L and ℓ. It should also be clear that achieving this goal depends on the decay rate not being too large, and for this our large λ assumption is important. This justifies our conclusions drawn from the complex energy solution, in particular the use of $2 \operatorname{Im} E$ as the decay rate.

Exercise 2.5.B.1.   Suppose $(H - i\partial_t)\psi = \phi$, where ϕ is a function of x and t. We seek a $t > 0$ solution to the partial differential equation with $\psi(x,0) = f(x)$. Let $\psi^{(0)}$ be the solution of this equation for $\phi = 0$, namely $\psi^{(0)}(x,t) = \int dy G(x,t;y)f(y)$ with G the propagator, as mentioned above. Show that
$\psi(x,t) - \psi^{(0)}(x,t) = i\int_0^t ds \int dy G(x,t-s;y)\phi(y,s)$.

Appendix 2.5.C: Deviations from exponential decay

Since we have built up a lot of machinery on quantum decay, I will digress and use it to demonstrate well known deviations from the exponential decay law.

Our argument in Section 2.5 depended on having a time beyond which the function $K(u)$ became small, which in turn depended on mode cancellation among the ωs. For times shorter than this a deviation from exponential decay is expected. Let us calculate

$\dot z$ in the extreme small-t limit. Expand Eq. (2.5.15) in small u and integrate to get

$$\dot z(t) = -z(t) \int d\omega \rho(\omega) |\gamma(\omega)|^2 \left[t - i(h-\omega)t^2/2 + \ldots \right] \qquad (2.5.C.1)$$

Keeping only the lowest non-vanishing term gives $z(t) = 1 - \frac{1}{2}t^2 \int d\omega \rho(\omega) |\gamma(\omega)|^2$. Define a 'Zeno' time, τ_z, to be

$$\tau_z \equiv \sqrt{\frac{1}{\int d\omega \rho(\omega) |\gamma(\omega)|^2}} \qquad (2.5.C.2)$$

then $z(t) \approx 1 - t^2/(2\tau_z^2)$ and two things are clear. For $t < \tau_z$ the decay is not exponential. And if τ_z is not much less than the lifetime $(1/\Gamma)$, then the entire scheme is questionable. This is the consistency check alluded to earlier. Further material on dominated time evolution is cited in Section 2.7, including references to a more general form for τ_z.

The long time breakdown in exponential decay arises because the function $\rho(\omega)|\gamma(\omega)|^2$ that appears in K vanishes for sufficiently negative ω; that is, the function is not just small but it is zero. This is because energy is bounded from below. Let us suppose we have taken the continuum limit. It is known in Fourier analysis (Paley-Wiener theory) that the transform of a function that vanishes on a half line cannot vanish exponentially; that is, $|K(t)|$ must go to zero more slowly than $\exp(-\alpha t)$ for any positive α. This means (cf. Eq. (2.5.12)) that even for large times, K is taking pieces from $z(0)$ with larger than exponentially small size. But if $\dot z$ is not dying exponentially, neither is z.

Appendix 2.6.A: Past/future reliability: *not* a restatement of the second law

The feeling that the past has occurred, is known and unchangeable, while the future is open, is a psychological arrow of time that we discuss in Chapter 3. Here I distinguish this arrow from the notion of reliability. The relative accuracy of prediction and retrodiction— reliability—is another issue and, in contrast to the psychological arrow, can be stated in the quantifiable terms used in Section 2.6.

This point was made by Gold in 1962. He quotes a Mr. Levinger who gave the example of a Russian satellite. In this case prediction is more effective than retrodiction. A satellite in orbit allows accurate prediction of its free trajectory. However, it is difficult to deduce where it was launched. (In those Cold War days, such information was often secret.)

We use the notation of Section 2.6, Γ_2, etc., although now the system is not a glass of water, but something else, like a roulette wheel. To say that the system is difficult to predict means that the set $\varphi(\Gamma_2)$ has significant weight in more than one macrostate, i.e., coarse grain.[49] To state this as an equation, let Δ_k be grains of the relevant phase space region at time T_3 (k runs over an index set). In this language, uncertainty means that $\Pr\left(\Delta_k | \varphi(\Gamma_2)\right) = \mu\left(\Delta_k \cap \phi(\Gamma_2)\right) / \mu(\Gamma_2)$ is well away from zero for several k values. In terms of a roulette wheel, from the initial ball and wheel condition, you cannot tell into which trap the ball will fall. To display a symmetric example of an uncertain retrodiction imagine a variation of the conventional roulette wheel. Prior to time T_1 the ball is at rest in one of the traps. At T_1 the wheel is spun vigorously. It leaves the trap it is in, bounces

---

[49] Again we speak of three times, T_1, T_2 and T_3, which in the glass of water fable were 1 p.m., 2 p.m., and 3 p.m. The operator φ is the phase space transformation that moves the system forward one time unit. The measure μ is phase space volume.

around for a while as the wheel spins, and then (at T_3) is caught by another trap. Now you look at the time T_2 position, with $T_1 < T_2 < T_3$. You don't know where it came from and you don't know where it is going. In our formal language, we could call the coarse grains of the earliest time δ_k. With this notation, your various time T_1 hypotheses (i.e., coarse grains δ_k) would be realized with probability proportional to $\mu\big[\Gamma_2 \cap \phi(\delta_k)\big]$, and your uncertainty would correspond to this being relatively large for several values of k. In this way we see that uncertainty of retrodiction (respectively, prediction) depends on the numbers $\mu[\Gamma_2 \cap \varphi(\delta_k)]$ (respectively, $\mu[\varphi(\Gamma_2) \cap \Delta_k]$) having several comparable non-zero values. By applying φ^{-1} to the second expression we see that having large competing contributions is essentially the same for either time direction. What will be different about past and future is the states (and entropies) of interacting systems, for example the number of chips in your pile which results from one or another ball position, or the increase in air temperature due to the spinning of the wheel. But these have little quantitative relevance for the computation of the probabilities discussed above.

All this can also be studied in the context of the cat map.

Remark: Why then when we *remember* something does it seem more certain than when we predict? This is because we are not retrodicting from the system itself, but from an auxiliary system, our brain, on which the earlier state of the system caused modification of neural coarse grains, i.e., wrote to memory. The fact that coarse grains in the brain have relatively few precursors is a property you want for any recording system.

3

Arrows of time

In our experience, time is *not* symmetric. From cradle to grave things happen that cannot be undone. We remember the past, predict the future. These arrows, while not always—or even now—deemed suitable for scientific investigation, have been recognized since the dawn of thought. Technology and statistical mechanics give us a precise characterization of the thermodynamic arrow. That's what the previous chapter was about. But the biological arrow (memory, etc.) is elusive. Then we come to arrows that only recently have been recognized. The greatest of these is the fact that the universe is expanding, not contracting. This is the cosmological arrow. Related, perhaps a consequence, is the radiative arrow. Roughly, this is the fact that one uses outgoing wave boundary conditions for electromagnetic radiation, that retarded Green's functions should be used for ordinary calculations, that radiation reaction has a certain sign, that more radiation escapes to the cosmos than comes in. Yet more recently, the phenomenon of *CP* violation was discovered in the decay of K mesons. As a consequence of *CPT* invariance, and some say by independent deduction, there is violation of *T*, time reversal invariance. This *CP* arrow could be called the *strange* arrow of time, not only because it was discovered by means of 'strange' particles, but because its rationale and consequences remain obscure. There is another phenomenon often associated with an arrow, the change in a quantum system resulting from a measurement. It is designated the quantum arrow of time. In my opinion there is no such arrow. In the second part of this book I present a theory of quantum measurement in which 'measurement' is no different from ordinary quantum evolution ($\psi \rightarrow \exp(-iHt/\hbar)\psi$) and no new arrow results. For other theories of quantum measurement it is also questionable whether there is an arrow, but I will leave it to the advocates of those theories to sort this out.

In this chapter I present various temporally asymmetric phenomena
with as much detail as is required for later chapters. More information
can be found in the references in Section 3.6.

3.1 The thermodynamic arrow of time

The thermodynamic arrow of time was the subject of Chapter 2. Here
are the versions developed there.

1) The entropy of an isolated system increases. 'Increase' means with respect to our conventional clock time, that for which Newton preceded
   Einstein. This arrow can also be phrased as the impossibility of converting heat to work in a closed cycle. Recall that we did not find the
   definition of entropy to be trivial. Ultimately it depended on a coarse
   graining, a division of phase space into volumes or of Hilbert space into
   subspaces, such that macroscopically we are unable to distinguish the
   different microstates in the same coarse grain.

2) When predicting the development of an isolated macroscopic state,
   take all microscopic states consistent with the macrostate and evolve
   them forward in time. Taking equal weight for the initial microstates,
   final macrostates with substantial weight in them are the possible
   later macrostates. When there are several competitors, the relative
   weight in each represents the probability of the particular outcome.
   In retrodicting, deducing an earlier state of an isolated system, make
   hypotheses about its earlier state and use them for prediction, as just
   described. Those that yield the present macrostate are possible precursors. The assignment of relative weights to competing hypotheses
   is not uniquely determined and is related to the statistician's ambiguities in constructing estimator functions. This statement of the second
   law can be summarized by 'Equal *a priori* microscopic weights for *initial* conditions, but not for final conditions.' In fact, final microstates
   are selected with a rejection rate on the order of $1 - \exp(-N_A)$, with
   N_A Avogadro's number.

Creative dissipation

The emphasis on dissipation and loss of information in the second law of
thermodynamics belies a creative richness that exists in *non*-equilibrium
systems. When a system is open, not isolated, an influx of negentropy can
create order. The preeminent example is planet earth. A flux of photons
at 6000 K rains down upon us. We in turn discharge photons (radiate) at
our black body temperature, about 300 K. The energy balance is approximately maintained, but it is the entropy difference that drives the highly

structured, far from equilibrium, phenomena on the planet. And preeminent among these phenomena is life itself. In Section 3.5 we elaborate, and in Section 3.6 provide references.

3.2 The cosmological arrow of time

The central observation, first made in the early twentieth century, is that most galaxies are red shifted; identifiable absorption lines are shifted to longer wavelengths. The systematic correlation of these red shifts with distances measured by other means was made by Hubble in 1929. The attribution of the red shift to a Doppler shift means that most galaxies are receding from us. Taken in conjunction with results from general relativity about possible cosmological structures, this implies that the universe is expanding. It is this expansion that is called the cosmological arrow of time. The rate of expansion is measured in terms of the Hubble parameter, H_0, which is the slope of the velocity versus distance curve. Because of uncertainties in the distance scale there are big error bars on H_0 and the best contemporary guess[1] is $H_0 = 100h_0$ km s^{-1} Mpc^{-1} with $0.4 \leq h_0 \leq 1$. At the time of writing, results from the Hubble space telescope seem to be pushing this number to larger values (implying a younger universe—see comments below). One can also define a *deceleration* parameter, essentially the next derivative of that curve. However, observation does not pin this down and in particular it is not known well enough to discriminate among cosmological models.

The next significant datum is the presence of photons with a wavelength of about 2 mm in every direction in the heavens that you look. This radiation has a thermal spectrum with a temperature of about 2.7 K. The interpretation of this radiation enjoys wide acceptance and it is considered to be a remnant, expanded gas of photons dating back to the significant decoupling of matter and radiation in the early history of the universe. There are recent observations showing that this radiation is not uniform, that it possesses large-scale spatial structures, and this too has significant bearing on cosmological models.

The distribution of luminous matter also enters when one tries to determine the global cosmology. It appears that galaxies are distributed with a fractal structure as far out as anyone can check. In particular, the two-point function, the probability that a given galaxy has another galaxy a distance r from it, drops off like $1/r^{1.7}$, approximately. This is an observation (modulo assumptions about the interpretation of red shifts). The theory for assimilating this observation lags, although in recent years a

---

[1] The value is given in 'Mpc,' Megaparsecs. A parsec is 3.26 light years.

correct treatment of the fractal structure has been introduced. However, fractals have the disturbing property that as the volume considered increases, the average density goes to zero. If we don't see an end to fractal structure, then what right have we to assume a non-zero average density? Contemporary (general relativity based) cosmological models assume a homogeneous matter distribution with a positive average density. If the fractal structure extends over cosmologically significant distances, is a homogeneous matter distribution a reasonable basis for proposing cosmological solutions? Another feature of fractal structures is the wide variation in mass density (aside from the asymptotic zero average) associated with voids and clustering. There have been studies (cited below) which address this problem at a perturbative level.

Yet another problem, which may obviate the one just raised, is known as 'missing mass.' As we will see, space-times with relatively little mass can expand forever, while with large mass they recollapse. Call the current density for which our universe would be at the boundary of those two behaviors, ρ_c. Based on Friedmann-Robertson-Walker cosmologies (to be outlined below) ρ_c is given by

$$\rho_c = \frac{3H_0^2}{8\pi G} \tag{3.2.1}$$

with G the gravitational constant. Our estimate of the value of ρ_c is subject to the same uncertainties as that of H_0 and it is roughly 10^{-29} gm cm^{-3}. When you add all the matter whose existence you can account for observationally, the resulting density is a small fraction of ρ_c. On the other hand, the relatively high velocities of stars within galaxies (and a bit of Newtonian dynamics) suggest more matter than we see. Furthermore, contemporary theories of the early universe, in particular those involving a period of rapid inflation, are strongly prejudiced in favor of the density being close to ρ_c. The difference between what we see and what we believe is the 'missing mass.' Besides many exotic proposals, from time to time there are observations that do provide a basis for believing that there is more matter out there. Personally, and for my own reasons, I am sympathetic to this assertion. I also find it a confirmation of the humility that the human race ought to show. As remarked earlier, there was little understanding of the thermodynamic arrow of time for at least the first 10000 years of homo sapiens' existence as city dwellers. The cosmological arrow, fully as important, escaped the attention of these same sapient creatures well past their invention of steam engines, gunpowder and telescopes. And now it seems that perhaps 90% of the matter in the universe is still hidden from us. Note that if there is missing mass, it need not follow the distribution of luminous matter, and the fractal-related issues raised in the previous paragraph may not arise.

A third kind of information that significantly constrains cosmological models is the abundance of light elements. Making use of this information is relatively dependent on astrophysical theory. Nevertheless it represents data whose validity will survive current fashions.

There is yet another result, also dependent on astrophysical theory, that is not directly concerned with cosmology, but is (at the time of writing) presenting a puzzle in this connection. This is the age of the oldest stars. There seem to be objects aged 14 to 16 billion (10^9) years. But if the Hubble constant is pushed (as suggested by recent observations) to 80 km s^{-1} Mpc^{-1}, they would be older than the age of the universe (in the usual models). This means that there are still big gaps in our understanding of the universe (surprise!), but for the purposes of the forthcoming discussion I will assume that the broad outlines will survive the resolution of this problem.

Cosmological models

General relativity is a geometrical theory of gravity. No longer does one speak of a 'force' that draws us to earth, but rather there is a modification of space-time geometry. The geodesics of this 4-geometry—which we freely follow—provide the effect of gravity. But the dynamical rules for the geometrical structures give new predictions of their own. The most dramatic of these describe the entire cosmos.

I will now present a few highlights of the well-known cosmological models. They all assume the 'Cosmological Principal,' which is to say that our spatial location in the universe is not special. This presentation is only intended as a refresher; in Appendix 3.2.A is a review of definitions and notation. Textbooks are listed in the notes.

If the universe is homogeneous and isotropic on a sufficiently large scale, then the global space-time metric is of the form

$$ds^2 = g_{\mu\nu}dx^\mu dx^\nu = dt^2 - R^2(t)d\sigma^2 \tag{3.2.2}$$

The quantity $R(t)$ is a so-far undetermined function of time and $d\sigma^2$ is the metric for the 3-dimensional spatial geometry

$$d\sigma^2 = \frac{dr^2}{1 - kr^2} + r^2(d\theta^2 + \sin^2\theta d\phi^2) \tag{3.2.3}$$

Here, (r, θ, ϕ) parametrize the 3-space in the same way that the usual polar coordinates do. The parameter k takes the values $0, \pm 1$ and gives the sign of the curvature of the space. Eq. (3.2.2) is known as the Robertson-Walker metric and represents a space-time of maximal spatial symmetry and spatially constant curvature.

For $k = 1$, the volume of the 3-space associated with the metric Eq. (3.2.2) is finite (actually $2\pi R^3(t)$), and $R(t)$ is a kind of radius of the universe. For other values of k, $R(t)$ plays the role of a scale factor. The time dependence of $R(t)$ will be obtained below from the Einstein equations relating the geometry to the matter.

Assuming our universe bears an approximate resemblance to Eq. (3.2.2), it is the time dependence of $R(t)$ that constitutes our primary cosmological observation. Suppose an observer moves freely in this space (i.e., follows a space-time geodesic). Then for a slowly increasing $R(t)$, there are two immediate consequences. 1. Other mass points (at rest in the global frame) move away, moving more rapidly the farther they are. 2. Photons emitted from those receding mass points are red shifted (this is just a general relativistic Doppler shift). I will not derive these effects, but will quote the distance-red shift relation. Call the observer's position the origin. Let D be the coordinate distance of a mass point from the observer (in the spatial portion of the metric Eq. (3.2.2)) and let a photon be emitted from that mass point. Let z be the usual red shift parameter, i.e., if the original wavelength is λ_0 and the observed wavelength λ, then $z = (\lambda - \lambda_0)/\lambda_0$. It can be shown that

$$z = \frac{R(t)}{R(t_0)} - 1 = HD + \frac{q+1}{2}H^2 D^2 + \ldots \qquad (3.2.4)$$

where t_0 is the emission time of the photon, and H and q are Hubble and deceleration parameters at time t. In terms of $R(t)$ these are

$$H = \frac{\dot{R}}{R}, \qquad q = -\frac{\ddot{R}R}{\dot{R}^2} \qquad (3.2.5)$$

with the dot the usual time derivative. H and q are the quantities referred to in our earlier discussion of observational results. Thus if one has a collection of values of z and corresponding distances D (measured in a red shift independent way), then Eq. (3.2.4) can give information about H and perhaps q.

The Einstein equations relate the metric of a space-time to the matter within it. Their general form is

$$R_{\mu\nu} - \frac{1}{2}g_{\mu\nu}R - \Lambda g_{\mu\nu} = -8\pi G T_{\mu\nu} \qquad (3.2.6)$$

where $R_{\mu\nu}$ is the Ricci tensor, R the curvature, $T_{\mu\nu}$ the energy momentum tensor, and Λ a parameter known as the cosmological constant. When the homogeneity and isotropy assumptions used for the metric are applied to $T_{\mu\nu}$, it takes the form

$$T_{\mu\nu} = -P g_{\mu\nu} + (P + \rho)u_\mu u_\nu \qquad (3.2.7)$$

where u_μ is the 4-velocity of the matter, namely $(1, 0, 0, 0)$, and P and ρ are respectively the pressure and energy density of the matter.

Because of the considerable symmetry, it is straightforward to obtain (from Eq. (3.2.6)) an ordinary differential equation for R. It must satisfy

$$\dot{R}^2 = \frac{8\pi}{3}G\rho R^2 + \frac{1}{3}\Lambda R^2 - k \qquad (3.2.8)$$

subject to the condition

$$\frac{d}{dR}\left(\rho R^3\right) = -3PR^2 \qquad (3.2.9)$$

The only further input required is the 'equation of state,' namely the relation between ρ and P, as well some notion of what number should be taken for the 'cosmological constant,' Λ. Cosmological models within the framework we have so far developed are known as Friedmann-Robertson-Walker (FRW) models.

Even within the context of FRW models there is an elaborate zoology of possibilities. The infamous cosmological constant was introduced by Einstein to allow a static universe—but he sorely regretted this introduction once Hubble's work showed that things were not static. Nevertheless, it *is* a possibility, irrespective of its history. There are observational upper bounds on Λ, but the only tool for its complete excision seems to be Occam's razor. About the equation of state one can say a good deal. Depending on whether one is dealing with matter-dominated or radiation-dominated epochs in the history of the universe, different, but well-understood, equations of state are appropriate.

For the present era, pressure can be neglected, and Eq. (3.2.9) implies that $M \equiv \rho R(t)^3$ is constant. If the cosmological constant is assumed zero, Eq. (3.2.8) becomes a simple differential equation which can be solved by introducing a new variable $\chi \equiv \int dt/R$. In terms of χ, the time (t) and scale factor (R) are given by

$k = 0$:

$$R(\chi) = \frac{2\pi GM}{3}\chi^2, \qquad t(\chi) - t_0 = \frac{2\pi GM}{9}\chi^3 \qquad (3.2.10)$$

$k = +1$:

$$R(\chi) = \frac{4\pi GM}{3}(1 - \cos\chi), \qquad t(\chi) - t_0 = \frac{4\pi GM}{3}(\chi - \sin\chi)$$
$$(3.2.11)$$

$k = -1$:

$$R(\chi) = \frac{4\pi GM}{3}(\cosh\chi - 1), \qquad t(\chi) - t_0 = \frac{4\pi GM}{3}(\sinh\chi - \chi)$$
$$(3.2.12)$$

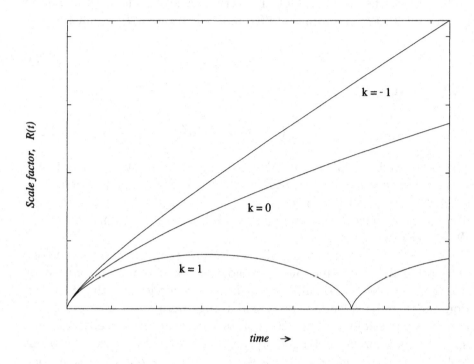

Fig. 3.2.1. Time dependence of the scale factor in Friedmann universes. (See Eqs. (3.2.10–3.2.12)).

In all cases, t_0 is a time at which $R = 0$. The time (t) dependence of R for the various cases is shown in Fig. 3.2.1. Presumably the zero of $R(t)$ at the left of the figure represents the physical big bang. The same presumption suggests that for the $k = 1$ cosmology there will be a future recollapse, an event referred to as the 'big crunch.'

If you want to estimate which picture describes our world, you don't measure 'k' directly. To see what you can measure, go back to Eq. (3.2.8) and divide by R^2. Let us assume that $\Lambda = 0$. Recalling that $H = \dot{R}/R$, Eq. (3.2.8) is rewritten as

$$\frac{k}{R^2} = \frac{8\pi}{3}G\rho - H^2 = \frac{8\pi}{3}G(\rho - \rho_c(t)) \qquad (3.2.13)$$

The quantity $\rho_c(t)$ in this equation is the same as that which appears in Eq. (3.2.1) except that now it is given an explicit time argument (the object in Eq. (3.2.1) is the current value of ρ_c). The left-hand side of Eq. (3.2.13) is positive, zero, or negative, depending on k. Specializing to the current time, this means that our universe has positive, zero, or negative k, as a direct consequence of ρ being greater than, equal to, or less than ρ_c. This is the importance of estimates of the average mass density of the universe.

Is the FRW universe qualitatively correct?

The Robertson-Walker metric assumes an apparently false homogeneity for the universe, and it is not clear whether the ever larger distance scales on which we observe the inhomogeneity will finally force us to abandon this metric. But there is more trouble. In the course of time evolution we expect there to develop black holes. There is now an accepted body of literature dealing with singularities in space-time structure. It is believed that if you start with the Robertson-Walker metric things will get crinklier and crinklier. A more precise version of this belief is phrased in terms of the Weyl tensor (see Appendix A to this section). It is zero for the Robertson-Walker metric but as black holes and other strains in the metric develop it becomes non-zero. It has been conjectured that some scalar built out of the Weyl tensor is a monotonic function of time. If that is the case, then the collapsing phase of the $k = 1$ FRW universe could not have the symmetric features suggested by Fig. 3.2.1. In the opinion of some thinkers, this conjectured arrow is the master arrow of time, driving the others.

My own view, is that such an asymmetry in the *solutions* of the equations of general relativity need not exist in order to explain those arrows we experience. There is a long history of expectations of asymmetry—the Boltzmann H-theorem was such a hope—and I am frankly skeptical that the formally time-symmetric general relativity will lack physically plausible time-symmetric solutions. When I later discuss two-time boundary value problems I will go into detail. In any case, those who believe the master arrow to come from a fundamental asymmetry in solutions to the Einstein equations have expressed their views in numerous books and articles, and reference is made in the notes below.

For the purposes of discussion we assume the $k = 1$ picture in Fig. 3.2.1 is a qualitative possibility. It may be that from the beginning the Weyl tensor was non-zero. It may be that strains in the metric develop and then undo themselves in the reverse phase. Arguments against the latter scenario are sometimes based on its *unlikelihood*. It would be as if the ice cube *grew larger* between 3 p.m. and 4 p.m. in the example of Section 2.6. Similar reasoning was used by early workers to show entropy increase in 'oscillating' universes. There is a severe flaw in these uses of probability. A probability calculation is only as good as your assumptions, and one of the goals of our exposition on irreversibility was to sensitize the reader to such assumptions. In particular the use of equal *a priori* probability for *initial* conditions is a *consequence* (or alternative formulation) of the second law of thermodynamics. So at the cosmological level, where the second law is something to be derived, not assumed, such probabilistic assumptions are inappropriate. This point will be treated in Chapter 4.

Quantum cosmology

A major thrust in the study of cosmology is known as *quantum cosmology*. It is an outgrowth of theories of quantum gravity and represents the time-honored practice of taking the theories you have and pushing them all the way. In the present state of human knowledge, neither quantum mechanics nor general relativity has any limitation—in the large or in the small. So there is no choice but to see what you get when they are combined.

Actually there is a choice—in the selection of problems. For the questions that I wish to confront, namely, the explanation of the thermo-dynamic arrow of time and an understanding of quantum measurement theory, I don't believe the quantization of gravity need be invoked. My claim, to be developed below, is that the distance scale for the puzzles of the quantum measurement problem are at the 'mesoscopic' level (perhaps 10^{-6} cm), a regime where one also confronts the foundational problems of statistical mechanics. The thermodynamic arrow may well involve cos-mological issues, but here too one can begin by studying them for epochs where we have an understanding of the dominant physical laws (e.g., beginning a few minutes after the big bang). On the other hand, the distance scale that appears to be important for quantum gravity can be estimated by combining various physical parameters. The characteristic constant for gravity is 'G' ($\sim 6.7 \times 10^{-8}$cm^3/g s^2) and for quantum me-chanics $\hbar$ ($\sim 1 \times 10^{-27}$g cm^2/s). To get a quantity with the dimensions of length, use c ($\sim 3 \times 10^{10}$cm/s). The result is known as the Planck length

$$L^* = \sqrt{\frac{\hbar G}{c^3}} \approx 1.6 \times 10^{-33}\text{cm} \qquad (3.2.14)$$

This dimensional analysis is a guide to where the interplay of the two the-ories may be expected to have physical significance. L^* is about 18 orders of magnitude away from the characteristic lengths in nuclear and parti-cle physics. As such I do not take seriously the predictions of quantum gravity (and by extension arrow-of-time arguments based on quantum cosmology). There is nevertheless a role for this work, namely a search for consequences at more accessible distance scales with the possibility of seeing one or the other of these theories contradicted. But *studying con-sequences* and *believing extrapolated conclusions* are two different things. Anyone who confidently relies on this theory and assumes that there isn't a major paradigm change for the next 18 orders of magnitude is in my opinion a victim of either pipedreams or hubris. I am chastened by the 1920s scientific revolution inherent in the small step from 10^{-4} cm (which people could then see with a microscope) to 10^{-8} cm, the atomic level. In only four orders of magnitude one suddenly needed quantum mechanics. And if one only had information from the larger (10^{-4} cm) level, could anyone have had the vaguest guess as to what quantum mechanics would bring? A related example is the prediction of nineteenth-century physics that the sun's fires ought to be extinguished in a few thousand years because no mechanism for their continuation was known.

The foregoing diatribe is only intended as an explanation of how I decide to spend my own time. Some of my best friends quantize gravity. In the notes I give references for those who may disdain my advice.

Appendix for this section (at the end of the chapter):
3.2.A General relativity on one foot

3.3 The radiative arrow of time

In opening this chapter, I described the radiative arrow as 'the fact that one uses outgoing wave boundary conditions for electromagnetic radiation, that retarded Green's functions should be used for ordinary calculations, that radiation reaction has a certain sign, that more radiation escapes to the cosmos than comes in.' By contrast, the cosmological arrow was defined in a single phrase: the universe is expanding. It does not take a cynic to realize four vague definitions are used because there isn't a single firm one. The fact is that Maxwell's equations are linear and symmetric in time and anything you can describe with retarded Green's functions, you can describe with advanced Green's functions as well.[2] All that changes are the (homogeneous, i.e., sourceless) boundary conditions. Now it may be true that the boundary conditions are simpler to express when one kind of Green's function is used rather than the other, but this line of thinking begins to sound like the thermodynamic arrow. Another aspect of the radiative arrow that was mentioned above is the fact that the universe is a sink for radiation, rather than a source. Light departs from identifiable sources and goes to oblivion. But is this a feature of radiation or of the universe? Yet another attempt to define the radiative arrow might be based on the fact that an accelerated particle *loses* energy, rather than gains it, the particle radiates *away* energy. But with the

---

[2] The *coupled* equations for particles and fields are not linear, but are still time symmetric.

well-developed theory that same outgoing radiation can be considered to be coming from the future and accelerating the particle.

It is nevertheless possible to introduce a degree of precision by starting with a time-symmetric theory and seeing how it can become one or another apparently asymmetric theory. In this way the connection of this arrow to other arrows can be explored. In the remainder of the present section we undertake this project. Although mathematical difficulties prevent a firm conclusion, I think the case is good that radiation is not an independent player and whatever arrows we see for it can be related to the thermodynamic and cosmological arrows. In future chapters we will deal with the connection between *those* two. Another reason for going into detail on the time-symmetric form of electrodynamics is that it provides experience for treating things time symmetrically. Such thinking will play a role later in this book, and indeed for me, the theory now to be presented was a training ground.

Note: *The remainder Section 3.3 is superfluous if you are prepared to accept the following: the dynamics of electromagnetic fields coupled to particles form, like Newton's equations for particles, just another time-symmetric theory. As such, to the extent that an arrow (of time) is evinced, that arrow must arise from another source. If your characterization of the radiative arrow is the fact that the universe is a sink for radiation, then you are relating it to the cosmological arrow. If your characterization is the observation of radiation reaction for accelerating particles, then your underlying arrow is the thermodynamic arrow.*

Action-at-a-distance electrodynamics

About a century ago—and on several later occasions—there was proposed a theory of particles in which the electromagnetic field was subsumed into an action-at-a-distance variational principle for the particles alone. Using the Minkowski metric (Appendix A to Section 3.2) the action for a collection of N particles is written

$$S = \sum_{k=1}^{N} \int d\tau_k m_k \sqrt{\dot{x}_{k\mu}\dot{x}_k{}^{\mu}} + \sum_k \sum_{\ell \neq k} q_k q_\ell \int d\tau_k d\tau_\ell \dot{x}_{k\mu}\dot{x}_\ell{}^{\mu} \delta((x_k - x_\ell)^2)$$

$$(3.3.1)$$

The parameters $\{\tau_k\}$ are proper time. The quantities m_k and q_k are properties of the particles, mass and charge, respectively. The dot is derivative with respect to the appropriate τ. The first term in Eq. (3.3.1) is the usual free particle relativistic action. The interaction term contributes only when one particle finds itself on the forward or backward light cone of another. This interaction is the electromagnetic interaction between particles—except that instead of particles reacting only to each other's past, they respond to both past and future, with half the usual strength for each. This field-free form of electrodynamics eliminates self-interaction for a charged elementary particle.

The variational procedure for the above action is almost standard. You vary the x_ks, integrate by parts, and arrive at the following (Euler-Lagrange) equations of motion for each of the particles:

$$m_k \ddot{x}_k{}^\mu = q_k F_\nu^\mu(x_k) \dot{x}_k{}^\nu \tag{3.3.2}$$

where

$$F_{\nu\mu}(x) \equiv \frac{\partial A_\nu(x)}{\partial x^\mu} - \frac{\partial A_\mu(x)}{\partial x^\nu} \tag{3.3.3a}$$

and

$$A^\mu(x_k) \equiv \sum_{\ell \neq k} q_\ell \int \delta((x_k - x_\ell)^2) \dot{x}_\ell{}^\mu d\tau_\ell \tag{3.3.3b}$$

These are the usual equations, with 'A' playing the role of vector potential. The non-standard part of the variational principle is the non-locality in time, and it is manifested in the manipulations needed to arrive at the Euler-Lagrange equations. As we shall see, eliminating the terms generated by those manipulations introduces the issue of data specification.

This brings us to a significant problem for action-at-a-distance theories. What data should one give to get a solution? This question recalls the collegial tension that exists between practitioners of theoretical and mathematical physics. Generally, if one's goals are physical one should not do *more* mathematics than is necessary. However, one must not do less either. We will soon discuss the Wheeler-Feynman absorber theory, and if one attempts to follow some of the arguments in this subject the absence of a well-defined problem—what data for what solution—is sorely felt.

To get a feel for action-at-a-distance as a theory in its own right, consider the simpler variational problem for the real scalar function $x(t)$ with the action

$$S = \frac{1}{2} \int dt \left\{ [Dx(t)]^2 - \omega^2 x(t)^2 - \alpha x(t + \tau/2) x(t - \tau/2) \right\} \tag{3.3.4}$$

where $D \equiv d/dt$. Formal variation of S yields the Euler-Lagrange equation

$$D^2 x(t) + \omega^2 x(t) + \tfrac{1}{2}\alpha[x(t + \tau) + x(t - \tau)] = 0 \tag{3.3.5}$$

This is the analog of Eq. (3.3.2). To discover the natural boundary value specification for this action we keep all terms generated by the variation.

Let the range of integration in Eq. (3.3.4) be the interval $[a - \frac{1}{2}\tau, b + \frac{1}{2}\tau]$; vary $x(t)$ with no endpoint stipulations and compute $\delta S \equiv S(x + \delta x) - S(x)$ using integration by parts and translation of the time-integration variable:

$$
\begin{aligned}
\delta S = &\int_a^b dt\delta x(t) \left\{ -D^2 x(t) - \omega^2 x(t) - \frac{1}{2}\alpha[x(t+\tau) + x(t-\tau)] \right\} \\
&+ Dx(t)\delta x(t) \Big|_{a-\tau/2}^{b+\tau/2} \\
&- \left(\int_{a-\tau/2}^a + \int_b^{b+\tau/2} \right) dt\delta x(t)(D^2 x + \omega^2 x) \\
&- \frac{\alpha}{2} \int_{a-\tau}^a dt\delta x(t)x(t+\tau) - \frac{\alpha}{2} \int_b^{b+\tau} dt\delta x(t)x(t-\tau)
\end{aligned}
$$

$$(3.3.6)$$

To sort out the terms above, note that the first integral has as its integrand $\delta x(t)$ times the Euler-Lagrange expression, so that $\delta S = 0$ implies Eq. (3.3.5). Guaranteeing that the rest of Eq. (3.3.6) vanish is the job of the boundary conditions. Recall that in ordinary classical mechanics one gives $x(t)$ at the endpoints (say t_1 and t_2): this is dictated by the requirement that $Dx\delta x|_{t_1}^{t_2}$ vanish. Such a term appears above and indeed we require it to vanish. But that is just a beginning. From the last line of Eq. (3.3.6), it is clear that $\delta x(t)$ must vanish in the ranges $[a - \tau, a]$ and $[b, b + \tau]$. Vanishing of $\delta x(t)$ is the same as saying that all paths in the class of functions over which we vary have the same values in that interval; i.e., those values are part of the data specification for the variational problem, hence an appropriate set of data for the Euler-Lagrange equation. (Note that with $x(t)$ specified on $[a - \tau, a]$ and $[b, b + \tau]$ the other terms arising from the integration by parts also vanish.) To summarize: the data for this variational problem are the values of the function on two intervals, each of length τ.

For action-at-a-distance electrodynamics, the generalization is obvious. If you seek solutions between two space-like surfaces of Minkowski space, you should give enough boundary data so that integrations by parts drop out. See Appendix A to this section for elaboration.

We return to Eqs. (3.3.2–3.3.3) to see in detail their relation to the usual matter-field formulation of electrodynamics. The essential step in recovering the familiar expressions is the integration of the δ-function in Eq. (3.3.3b). The relation, $\delta[f(x)] = \sum_{\text{zeros of } f} \delta(x)/|f'(x)|$, implies

$$
\delta(x_\mu x^\mu) = \frac{1}{2r} [\delta(t-r) + \delta(t+r)] \tag{3.3.7}
$$

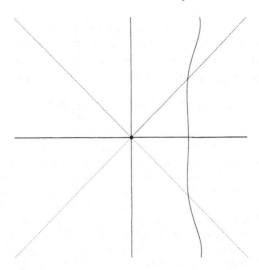

Fig. 3.3.1. Future and past interaction events. The light cones from the origin intersect the particle trajectory $(\dot{x}_\ell{}^\mu(\tau_\ell)$, the wavy line).

where $x = (t, \vec{x})$ and $r = +\sqrt{\vec{x} \cdot \vec{x}}$. Consider a single term from the sum in Eq. (3.3.3b), which with hindsight I call the vector potential at x due to the charge ℓ:

$$A_\ell^\mu(x) = q_\ell \int \delta((x - x_\ell)^2) \dot{x}_\ell{}^\mu d\tau_\ell \qquad (3.3.8)$$

The δ-function with 4-vector-squared argument is a Lorentz scalar, as is $d\tau_\ell$ (and $d/d\tau_\ell$). This makes A^μ a Lorentz (contravariant) 4-vector. From Eq. (3.3.7), there will generally be two points along the time-like trajectory $x_\ell(\tau_\ell)$ where the argument of the δ-function vanishes. Using the time t_ℓ the result can be written

$$A_\ell^\mu(x) = \frac{1}{2} q_\ell \sum \frac{\dot{x}_\ell{}^\mu}{r \, dt_\ell/d\tau_\ell} \qquad (3.3.9)$$

with the sum running over the two points on the light cones out of x that intersect the path $x_\ell(\tau_\ell)$. See Fig. 3.3.1. Now $dt_\ell/d\tau_\ell$ is the fourth component of $\dot{x}_\ell{}^\nu$, while the factor r in Eq. (3.3.9), which is the space-like distance from x to x_ℓ, is equal to plus or minus the time interval t as well, by the δ-function condition. Therefore the denominator in Eq. (3.3.9) can be written $\pm(x - x_\ell)_\nu \dot{x}_\ell{}^\nu$. Since this is a Lorentz scalar, the covariant form of Eq. (3.3.9) is

$$A_\ell^\mu(x) = \frac{1}{2} q_\ell \left\{ \frac{\dot{x}_\ell{}^\mu}{(x - x_\ell)_\nu \dot{x}_\ell{}^\nu} \bigg|_{t=r} - \frac{\dot{x}_\ell{}^\mu}{(x - x_\ell)_\nu \dot{x}_\ell{}^\nu} \bigg|_{t=-r} \right\} \qquad (3.3.10)$$

The conditions $t = \pm r$ are covariant by virtue of the distinction between past and future along a light cone.[3] The expressions in Eq. (3.3.10) are recognized as the retarded and advanced Green's functions appearing in the usual formulation of electrodynamics, except for the factor $1/2$.

Exercise 3.3.1.   Show that $A^\mu(x)$ and $F_{\mu\nu}(x)$ satisfy Maxwell's equations. What is the current? How is gauge invariance expressed here?

Remark: Quite a bit more can be said about this formulation of electrodynamics. One straightforward matter concerns conservation laws. We are used to the idea that the electromagnetic field carries energy, momentum and angular momentum. The functions A^μ do the job here. To see this apply Noether's theorem to the action Eq. (3.3.1). This yields constants of the motion (for the Poincaré group symmetries) that contain integrals over A^μ and its derivatives. As to the possibility of a quantum framework, it was the desire to quantize this Lagrangian-formulated theory that led Feynman to develop the path integral. The hope was that with neither self-interaction ($k = \ell$ is excluded in Eq. (3.3.1)) nor an independent electromagnetic field, one would avoid divergences. In his thesis Feynman studied the action Eq. (3.3.4) in his attempts to quantize Eq. (3.3.1). In the event, the quantization of the electromagnetic field was accomplished otherwise, but the gauche form taken by the successful theory suggests that Feynman's original hope may yet be realized.

Remark: The possibility that this formulation of electrodynamics is of fundamental significance is clouded by several problems. First, it lacks a quantized form (but see the notes below for recent progress). It also runs into trouble in the general relativistic con-

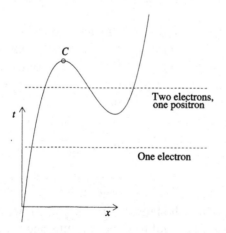

Fig. 3.3.2.   Space-time diagram of a particle whose trajectory is not monotonic in time. For the earlier timelike surface (the lower dashed line) there is one electron, but later there can be considered to be two electrons and a positron.

---

[3] Trouble could develop in the case of particle coincidence, but covariance would hardly be the only problem for solutions in which this occurred.

text because the δ-function in the action no longer need work.[4] Finally, there is the way multiparticle and particle annihilation/creation issues are handled. This is the bane of one-particle quantum theories, but here there are puzzles at the classical level. Feynman and Wheeler developed the beautiful idea that a Minkowski-space path for an electron that doubles back on itself in time would look like two electrons plus a positron (see Fig. 3.3.2). This idea has made its way into modern quantum field theory. But what happens now to the sum in Eq. (3.3.1), '$\sum_{\ell \neq k}$'? Does the doubled-back particle interact with itself? Presumably it should, but then you must have a way to describe the dynamics during the reversal (for example, at the point C in the figure).

Remark: If, despite the drawbacks just discussed, one does take the action-at-a-distance theory seriously, a different view of space-time is suggested. For practical purposes, our image of the world is electromagnetic. What we *see* is obviously conveyed by electromagnetic radiation. But electromagnetic dominance is more pervasive. Sound waves wave because of electromagnetic forces between atoms and molecules. The same forces initiate the vibrations in our sensory organs and those organs send their message electromagnetically through a network and to a brain whose workings are governed by electromagnetic forces. In other words, weak and gravitational forces play no role in our functioning, notwithstanding their importance in establishing the context within which we function. In a similar vein, the difficulties mentioned in the previous remark refer mainly to whether action-at-a-distance electrodynamics can be a fundamental theory. For describing our sensory processes it is adequate.

The image of the world that we have created with our electromagnetic senses is that of three spatial dimensions and one temporal dimension. But if you look at the action Eq. (3.3.1), the electromagnetic interaction doesn't deal with distance or time; it asks, does a given pair of events have zero 4-distance, are they on a light cone relative to one another? If they are, it draws on additional information, an indefinite inner product of their tangent vectors. In stating this, I use an underlying Euclidean 4-manifold, because that's the only way I know. But one can propose that *the light cone relation is fundamental.* For every pair of points in a so far unspecified set, they either do or do not bear the relation that (we know of as) they are on a light cone with respect to one another. Furthermore, this set has enough structure to define curves, tangents to curves, and an (indefinite) inner product of those tangents when the associated points have light-like separation.

The proposal then is this. You start with a set and a relation; call the relation *contiguity*. You want appropriate definitions and axioms for both set and relation such that: (1,) the set corresponds to the 4-dimensional space-time manifold we consider ourselves to live in, and (2,) the set of points contiguous to a given point corresponds to what we ordinarily call its light cone. Thus one begins with the primitive notion of contiguity and *defines* the 'cone' of a point A to be the set of points contiguous to A. Now recall a property of the usual Minkowski space: for two points of Minkowski space that are not light-like separated we know that the light cones intersect either in a sphere, if the points are time-like separated, or a hyperboloid, if they are space-like separated. Going back to our proposed axiomatic scheme, if we were to assume the initial set to be a topological space, we could define a concept of 'space-like' for a pair of points to mean

---

[4] What lies behind the success of the action Eq. (3.3.1) is the fact that the Green's function for a massless particle is itself a δ-function, that is $\partial_\mu \partial^\mu \delta(x^2) = 4\pi\delta(x)$. For a general Riemannian space this is not true and a pulse does not remain confined to the wave-front. This is a breakdown in Huygens' principle.

that the intersection of their respective cones is an open set. 'Time-like' would then be defined as having a closed set intersection. I'm not keen on that definition, as it seems too global, but it is an example of what I have in mind. One would also include in the axioms of contiguity other features that one 'secretly' knows about the usual relation of being light-like, for example the property of being symmetric but not transitive.

At this embryonic stage, it is not clear whether the tangent space structure implicit in Eq. (3.3.1) should be an assumption or a conclusion. In the most optimistic scenario one would derive topology and fiber bundle structure from contiguity alone and the indefinite metric would be a consequence. The justification for our program (the human image of space-time is based only on Eq. (3.3.1)) does not by itself support this optimism. In a more modest scenario, besides 'contiguity' one would introduce enough structure to have curves and tangents. An (indefinite) inner product of tangents would then be postulated (or perhaps derived) that would be defined for contiguous points (but not necessarily for all pairs of points).

This kind of axiomitizing is no longer popular, although with this proposal in mind I've noticed discussions with a similar flavor (see the notes). Speculations along these lines are daunting for other reasons as well. Implicit in a successful construction would be the possibility of rewriting other physical theories in terms of this characterization of space-time.

Absorber theory

We take up the matter of making Eqs. (3.3.2–3.3.3) look like ordinary physics. From Eq. (3.3.10) it follows that the vector potential, A_ℓ^μ, can be broken into advanced and retarded parts, corresponding respectively to the contributions from $t = r$ and $t = -r$. These will be distinguished with the subscripts 'A' and 'R.' The same notation will be applied to the fields F_ν^μ derived from A^μ. We can therefore write the field seen by particle #k at position x as

$$F^{(k)}(x) = F^{(k)} = \frac{1}{2} \sum_{j \neq k} [F_{Aj} + F_{Rj}] \qquad (3.3.11)$$

where the x is dropped from the arguments on the right, the tensor indices μ and ν are suppressed, and the index with parentheses in the superscript refers to the particle at whose position the field is evaluated. As in Eq. (3.3.10), the subscripts j, k and ℓ refer to the *source* of the field. By adding and subtracting $\frac{1}{2}F_{Ak} + \frac{1}{2}\sum_{\text{all } i} F_{Ri}$, Eq. (3.3.11) can be identically rewritten as

$$F^{(k)} = \sum_{j \neq k} F_{Rj} + \frac{1}{2} \sum_{\text{all } j} [F_{Aj} - F_{Rj}] - \frac{1}{2} [F_{Ak} - F_{Rk}] \qquad (3.3.12)$$

Let us examine Eq. (3.3.12) term by term. The first quantity (on the right), $\sum_{j \neq k} F_{Rj}$, is the retarded field—on particle k—that we ordinarily associate with a collection of charges acting on k. The middle term in Eq. (3.3.12), $F_{\text{free}} \equiv \frac{1}{2} \sum_{\text{all } j} [F_{Aj} - F_{Rj}]$, satisfies the *homogeneous* wave equation. To see this, apply $\partial_\mu \partial^\mu$ (using Exercise 3.3.1) and note that the source terms associated with the advanced and retarded parts are the same and cancel. We will return to a discussion of F_{free}, but for now let us go on to the third term on the right in Eq. (3.3.12). This quantity, $-\frac{1}{2}[F_{Ak} - F_{Rk}]$, is the difference between k's own advanced and retarded fields. Even before the work of Wheeler and Feynman, Dirac had examined this expression, and by deft and subtle manipulations had found that it worked out to the same quantity that had been used in radiation reaction. In summary, particle k's equation of motion—beginning from Eq. (3.3.2) above, brought to the form Eq. (3.3.12), with the dropping of F_{free}, and with Dirac's calculation—becomes

$$m_k \ddot{x}_k{}^\mu = q_k \sum_{j \neq k} F_{Rj\nu}{}^\mu (x_k) \dot{x}_k{}^\nu + \frac{2}{3} q_k^2 \left(\dddot{x}_k{}^\mu + \dot{x}_k{}^\mu (\ddot{x}_k)^2 \right) \qquad (3.3.13)$$

The additional 'radiation reaction' terms depend on first to third (time) derivatives and are *not* symmetric in time. The third derivative allows a runaway mode, but once this is fixed one recovers ordinary radiation damping and Eq. (3.3.13) thus represents the recovery of 'ordinary' physics. (The 'fixing' involves a future boundary condition, with preacceleration effects on the time scale $2e^2/3mc^3 \approx 6 \times 10^{-24}$ s.)

We return to F_{free}. The essence of the 'absorber theory' is the claim that this sourceless field vanishes. Why should it? Wheeler and Feynman gave four 'proofs.' Over the years they have been examined and in one way or another found to be problematic. For example, I earlier complained that the arguments based on changes in particle trajectories failed to specify what were the given data and what was changed while those data were held fixed. Not only do we not know how to prove the thesis, we don't even know how to state it.

Nevertheless, there is evidence that seems to bear on the problem. There is little question that there was an early era in which the universe was opaque and functioned in the desired way as an absorber. Furthermore, in our present era, any radiation that we know of seems to have an identifiable charged particle source,[5] again suggesting the absence of free fields. However, the expansion of the universe means that we don't know that all this radiation eventually hits a charged particle.[6] This highlights a reason for skepticism over this entire line of inquiry. Absorber theory arguments ultimately draw cosmology into the argument, whereas time-

---

[5] Even the cosmological background radiation is not viewed as Nature's electromagnetic field creation, but rather the result of annihilation radiation of charged particles.

[6] In discussing the issue of whether there are free fields we adopt conventional electrodynamics. In this context, a time-symmetric theory is a particular kind of field boundary condition.

symmetric electrodynamics is only valid on a flat, infinite space. The action of Eq. (3.3.1) does not work on a general Riemannian manifold. But irrespective of the possibility of generalizing this action, the mere fact of expansion means that although at one era it looks like there is almost no free field, at a later stage there can be fields that are never absorbed. In a similar vein, one of the Wheeler-Feynman arguments required a region outside all matter, thereby allowing arguments for the separate vanishing of outgoing and incoming waves. Contemporary ideas on cosmology do not favor such a region.

Even if one could overcome these theoretical issues, it would still be true that the vanishing of F_{free} depends on the properties of the world. If F_{free} does not vanish, then maybe the action-at-a-distance theory is wrong. Or, perhaps it is not wrong, but the rearranging of terms for Eq. (3.3.12) is less useful than was hoped, which would mean that radiation reaction is not what we took it to be. This is the origin of an experiment (referenced below) designed to determine whether the power required to run an antenna depended on where its output was being sent. That is, was there a difference between having the radiation directed into an absorbing material or into the night sky? The experiment showed that pointing the antenna at the sky did not change measured radiation reaction.

Implications of time-symmetry for the absorber theory

Any purported demonstration of the equivalence of the time-symmetric and the usual electrodynamics must contend with a simple observation about the steps from Eq. (3.3.11) to Eq. (3.3.12). With a minor change in those steps we could have obtained

$$F^{(k)} = \sum_{j \neq k} F_{Aj} - \frac{1}{2} \sum_{\text{all } j} [F_{Aj} - F_{Rj}] + \frac{1}{2} [F_{Ak} - F_{Rk}] \qquad (3.3.14)$$

The term F_{free} appears as before, but with a minus sign. If it vanishes we have a description in which particles respond to advanced fields and have an oppositely directed radiation reaction. Why should this description be inferior to the usual one? From the standpoint of electrodynamics I see no preference. The reason for using Eq. (3.3.12) rather than Eq. (3.3.14) is that it's easier, that the accelerated charges giving rise to $\sum F_R$ are easier to identify than those giving rise to $\sum F_A$. In effect, you can do calculations for Eq. (3.3.12) using coarse grains for fields and matter. For Eq. (3.3.14) coarse grains represent insufficient precision. Thus the selection of Eq. (3.3.12) over Eq. (3.3.14) is a consequence of the thermodynamic arrow of time. If this selection is the way to define the radiative arrow, then we have [thermodynamic arrow $\Rightarrow$ radiative arrow].

However, one should not underestimate the potential for confusion inherent in vague definition. If you want your radiative arrow to mean that more radiation goes off into the universe than comes back, then for *that* arrow things look better the other way round. In the next chapter I will present an argument purporting to show [radiative arrow $\Rightarrow$ thermodynamic arrow].

Remark: In the so-called standard model, electromagnetic forces are part of a larger dynamical scheme that includes the weak interactions. Whether it is possible to cook up a classical action-at-a-distance theory for this, I don't know. However, as indicated earlier, we will not ultimately be using absorber theory or action at a distance electrodynamics—it's just good practice for thinking about time symmetric theories.

Appendix for this section (at the end of the chapter):
3.3.A   Temporally non-local differential equations

3.4 The *CP* arrow of time

The 1964 observation of *CP* violation was a surprise, even after the discovery a few years earlier that the world was not reflection symmetric (i.e., *P* is violated).[7] The phases for the states of the neutral K meson and its antiparticle can be taken so that

$$CP|K^0\rangle = |\overline{K}^0\rangle \quad \text{and} \quad CP|\overline{K}^0\rangle = |K^0\rangle \qquad (3.4.1)$$

The kaon can decay into two or three pi mesons and into various leptonic modes as well. For the two-pi mode, the decay state is an eigenstate of *CP* with eigenvalue $+1$. For some of the three-pi modes $CP = +1$ is also possible, although for the most common (including angular momentum zero) the value is $CP = -1$.

The states $|K_1\rangle \equiv [|K^0\rangle + |\overline{K}^0\rangle]/\sqrt{2}$ and $|K_2\rangle \equiv [|K^0\rangle - |\overline{K}^0\rangle]/\sqrt{2}$ are eigenstates of *CP* with eigenvalues $+1$ and -1, respectively. If we lived in a *CP* conserving world the decay of K_1 and K_2 could be studied separately; each would have its own lifetime, and in particular the K_2, with *CP* eigenvalue -1, would never decay to two pions. Moreover, the K_2 would be expected to have a much longer lifetime because only 3-particle modes (with smaller phase space) are available to it. Now consider what would happen when (say) a K^0 meson is produced by strong interactions. At birth there is no admixture of $\overline{K}^0$ because the strong interactions are strangeness conserving. But if the $|K^0\rangle$ is written as a superposition of the K_1 and K_2, it is seen that in time—as the

---

[7] The reader is assumed to be familiar with the operators C (charge conjugation), P (parity), and T (time reversal). See Section 3.6 for references.

decay progresses and one eigenstate decays more rapidly than the other—
a certain amount of $|\overline{K}^0\rangle$ is produced and ultimately the state becomes
almost all K_2. Therefore after many K_1 lifetimes you should not see the
two-pi decay mode. But it was seen! Thus the eigenstates of Nature's
true Hamiltonian are not CP eigenstates; hence the true Hamiltonian
does not commute with CP.

Without the assumption of CP conservation, one must do a more careful analysis since the state associated with the long lived particle will
contain an admixture of K_1. One must also formulate in what sense the
unstable state is an eigenstate of the Hamiltonian. That sense turns out
to be that it has a complex eigenvalue, with the imaginary part providing the decay rate. For the physical Hamiltonian one concludes that the
strength of the CP-violating term is on the order of 10^{-3}.

The reason this has potential significance for our study of time's arrow
is that there is a more comprehensive operation that can be applied to
quantum states, CPT, which does seem to be conserved. The evidence
for this conservation is of two sorts. First there are proofs within the
context of quantum field theory. What such a 'proof' means is that no
one can imagine a reasonable theory in which CPT would be violated.
Second, there are physical predictions, such as the equality of masses and
lifetimes of particles and antiparticles, that are obeyed, to the limits of
experimental precision. If one does accept CPT invariance, the violation
of CP implies the violation of T, the time reflection operation. The
definition of this operation is more subtle than that of C and P, and in
a sense it is not truly a reflection. However, these matters do not seem
to affect the general considerations of the present book and I will not go
into them here (references are given below).

Another violation of T invariance would be a non-zero electric dipole
moment for an elementary particle. Intense searches, focusing on the
neutron, have not produced any observation inconsistent with zero.

Remark: The smallness of the CP violation does not imply a lack of cosmological significance. One scenario for the history of the world has an enormous number of baryons
and antibaryons at an early, hot stage. Because CP violation is small they are approximately, but not exactly, equal in number. With the cooling of the universe most of the
baryon-antibaryon pairs annihilate and the material content of our world is the leftover
small difference.

3.5 The biological arrow of time

There are several, possibly inequivalent, biological arrows of time. The
most prominent and the most difficult to tackle with the tools of physics
is our different conception of past and future, the one complete, realized,
unchangeable, the other without concreteness, only a potentiality. In a

similar, but not identical vein, we *remember* the past but only *predict* the future. A more tractable kind of arrow (for the physicist) is death, that is, aging and dissipation in life's processes. Finally, and providing a paradoxical counterpoint to mortality, is biological evolution, with the emergence of ever more complex structures.

Of these concepts, death and dissipation are straightforwardly related to the thermodynamic arrow. In my opinion, the emergence of complexity is also a consequence of that arrow, and will ultimately prove to be deducible from the statistical mechanics of non-equilibrium systems; this will be discussed below. I also believe that the remembering/predicting distinction will yield to this kind of analysis. As to the 'most prominent' arrow, a better understanding of consciousness and of the brain, its structure and evolution, is needed before subjective distinctions between the actual and the possible can be treated mechanistically. One also begins to encroach on philosophy and religion, although I expect the frontier between them and science to shift (as usual) with advances in the latter. Why should these matters be taken up in this book? First, they certainly belong in a catalog of time's arrows. Second, current research activity on complexity deserves to be mentioned. Third, we will later propose that the thermodynamic arrow is not nearly so unidirectional as it appears. To reconcile that with our perceptions one should know what kind of arrow is involved in those perceptions. There are also quantum measurement issues that impinge. The following statement has had its advocates: 'The collapse of the wave function requires human observation, perhaps even *my* observation.' Although I do not attach validity, nor even meaning to this claim, a resolution of the questions addressed by its proponents does demand attention.

Our discussion has covered both being alive and being conscious. Let's concentrate on being alive. Life processes are dissipative. Recall the 1-2-3 o'clock water and ice cube example of Section 2.6. Enlarge the box and establish that at 2 p.m. it contains a well-fed, pregnant cat. At 3 o'clock you'll have kittens, a big mess, digested food and higher entropy. As usual, the system is isolated from 1 p.m. to 3 p.m. The 2 p.m. status is known by automatic measurements, photographs, etc, recorded at that time. As in our earlier discussion, we ask, 'What was the 1 p.m. situation?' The answer is the same as for the ice cube. You won't have kittens, but you might have a mouse with a dismal future— and the entropy is lower. (The 2 p.m. cat may also 'remember' eating the mouse.) There is no question that life in the small is dissipative. We take negative entropy from the sun to grow wheat, to make sandwiches, to maintain our metabolism, to return about the same amount of energy in the form of high entropy heat and waste products for the environment. Without this, an isolated living system soon dissipates its way to death.

This assertion, that life is in step with thermodynamics, is straightforward: digestion is no different from any other chemical reaction. However, some reactions are more interesting than others and in particular some of them give rise to patterns and structure. This is where studies of complexity begin. Why should *any* structure emerge? Why should those structures become ever more elaborate, ever more intertwined with one another? At the time of writing I have not seen any solid answer for the more ambitious questions, but the first stages of the program are becoming an important part of non-equilibrium statistical mechanics. In the next paragraphs I will give the flavor of things that are known about the formation of structure and also discuss the sort of ideas that would need better definition for solid answers to the tougher questions. Throughout it is important to emphasize an essential point: I am now talking about open systems. This could be a chamber whose walls are maintained at different temperatures or a planet bathed in sunlight. For the selected subsystem—chamber or planet—entropy is below the equilibrium value and may grow lower. But for the larger system, including for example the thermal control systems for the chamber, total entropy increases. The second law is not challenged, but in its simplest form has little to say about the subsystem.

A famous example of a far from equilibrium *steady state* in an open system is Rayleigh-Bénard flow. A fluid is held in a rectangular box $h \times \ell \times L$ with L sufficient large to be considered infinite. The height of the box is h. On its top and bottom the box is maintained at temperatures T_{top} and T_{bot}, and it is $T_{top} \neq T_{bot}$ that prevents the system from having a true equilibrium in which microscopic states have their *a priori* probability assigned by a Boltzmann factor. For sufficiently small $\Delta T \equiv T_{bot} - T_{top} > 0$, heat is *conducted* from bottom to top in a gradual, diffusive way, by diffusion of phonons if you like. However, for larger ΔT the conductive mode becomes unstable and gross motion of the fluid begins, providing *convective* heat transfer. In particular, when a mass of fluid on the bottom is heated its density decreases and it rises relative to the cooler fluid above it. Similarly the fluid at the top, cooled by contact with the upper plate, will sink. An additional feature is that the geometry forces a symmetry breaking. *All* the fluid on the bottom tends to rise; all that on top tends to sink. But they get in each other's way. What actually happens is that some hot fluid breaks through to the top and vice versa. This settles into an orderly process and a set of cylindrical rolls is formed. See Fig. 3.5.1. In practice this neat pattern can be difficult to maintain if ℓ is too large, but for small ℓ and other experimental precautions (e.g., good heat conductivity for the upper and lower surfaces) the illustrated pattern will emerge. A well defined onset threshold also requires care in establishing horizontal homogeneity of the temperature, otherwise rolls will form no matter how small the top-to-bottom temperature gradient.

There are many lessons to be learned from this system. For ΔT small, the near-equilibrium system is reasonably well understood, but it is still worth examining the distinction between this and equilibrium, checking for example that the entropy deviation in the system is matched by what it takes for an outside force to maintain the two temperatures. But the rich phenomenon is the creation of structure *far* from equilibrium. The structure can be maintained because this is an open system, and negentropy (*negative* entropy) is being pumped in from outside: the upper and lower walls would not retain their different temperatures without external support. Presumably, life on earth is a grandiose generalization of this. We receive radiation from the sun at 6000 K. On earth this radiation powers complex non-equilibrium processes, notably photosynthesis. Eventually, this energy is degraded, i.e., lost to heat degrees of freedom. But the earth is warmer than outer space, and so it reradiates the energy at its black body temperature, about 300 K. The structures on earth are more complex than those in Fig. 3.5.1, but the principle is the same. The planet Venus also gets its radiation at 6000 K, reradiating at more than 300 K, partly because of its largely CO_2 atmosphere (which humans may yet emulate), and it too has non-equilibrium structures. We can see some of these in the form of wind patterns on the planet.

That an open system can convert negentropy to structure has been known for some time, although the theoretical framework is less developed than that of equilibrium statistical mechanics with its Boltzmann factors, etc. The question that has recently come to the fore is the origin of *complexity*. Why under the continuing rain of negentropy do the struc-

Low temperature

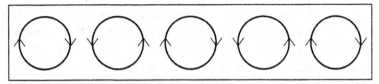

High temperature

Fig. 3.5.1. Rayleigh-Bénard flow. End-on view of fluid motion in a chamber whose top is kept cooler than its bottom. For a sufficiently large temperature difference heat is transferred convectively by fluid flow. The circles with arrows represent fluid motion as viewed from one end of the chamber.

tures become ever more complex? The most elaborate fluid flow pattern is dwarfed by the simplest living processes, which are in turn dwarfed by the delicate functional interdependencies of the mammalian brain. Even the question is not clearly defined: what is 'complexity'?

'Complexity' cannot be the same as negative entropy. Entropy is extensive—it doubles if the system size doubles—while complexity, whatever it is, should not be. A grass seed is a complex object, incorporating schemes for converting sunlight, water, etc. into ordered systems. Add such a seed to sterilized soil and it initiates far-from-equilibrium processes. Without the seed, the soil acted like Venus; the sunlight that struck it was simply reradiated at a lower temperature. With the seed, the sunlight produces cellulose, and much of its negentropy is captured before low temperature photons are returned to space. Doubling the amount of soil in which the seed is placed will double the negentropy capture—but this is not because the seed is any more complex. Similarly, *two* identical grass seeds produce a pattern no more complex than one. Hence complexity should not be extensive.

Definitions that have been offered for complexity focus on different aspects of the power of that seed. One class of definitions is related to *algorithmic information*; roughly, the algorithmic information associated with a number is the minimal length of a program for computing that number. The concept of algorithmic information is helpful for saying what one means by randomness, namely a random number should not be simple, should not have any easy way to be computed. For a living system, this kind of complexity translates into the length (or information content) of the instruction set in its genome. Other definitions consider a system to be complex if it is an unlikely outcome of a long series of choices. What are the odds that if you started with a soup of carbon, oxygen, hydrogen, etc., you would get a cat? (More precisely one looks at the probability of the sequence of steps that would lead to the cat.) Another definition is something I have called 'catalytic power.' The seed in a sense catalyzes the transition from soil, sunlight and air to cellulose. A catalyst is a path finder in phase space. For example, think of a metastable state of supercooled water (in contact with a reservoir of fixed temperature). What one needs to reach the stable state is an ice crystal. In terms of the phase space of the system, the metastable liquid states occupy a tiny fraction of the available phase space, most of which represents a solid. However, to reach that larger region it must find its way out, via a 'critical droplet,' in this case a sufficiently large ice crystal. The catalytic power of such a crystal can be measured according to schemes analogous to those mentioned: relative surface areas of bounding regions in phase space, barrier heights for effective potentials (as in Fokker-Planck descriptions) or instruction sets for producing crystals, all measure the effectiveness and subtlety of this catalytic process.

For any proposed definition, one would like to have theorems about how things tend to go from simple to complex, including measures of stability (as in the possibility that highly evolved creatures could induce a 'simplifying' nuclear holocaust). However, we don't even have a good general formalism for describing far-from-equilibrium steady states, much less criteria for transitions among them (thinking for example of biological species as steady states in an ecological dynamics). Furthermore, of the various definitions I've seen, none seems to me to give sufficient import to the distinction between self-reproducing objects and non-self-reproducing ones. Living (reproducing) systems have a cumulative and self-referential character that is not fully captured in the approaches mentioned.

In the notes below I give references to current literature and in Appendix A to this section I give an example in which structure develops far from equilibrium.

Appendix for this section (at the end of the chapter):
3.5.A   Heat conduction with breakdown

3.6 Notes and sources

Section 3.2. The prediction of long wavelength black body radiation left over from an early era of the universe is found in [ALPHER]. Steps to detect that radiation are reported in [DICKE], along with communications those authors had received from A. Penzias and R. Wilson who had encountered it in the course of studying noise in microwave antennas. Large scale structure of the universe and the treatment of its fractal nature can be found in [COLEMAN PH]. Contrary views, asserting large scale homogeneity, are found in [PEEBLES]. [JACOBS] deals with the metric perturbations due to mass irregularities that may be associated with that fractal structure. One theory for a fractal luminosity distribution is given in [SCHULMAN 86b]. The discovery of non-uniformity in the background radiation (other than that associated with the Doppler shift due to our own motion) is described in [LEVI].

Good references for general relativity are the texts [ADLER, DOMINGUEZ, SCHUTZ, STEPHANI, WALD] and [WEINBERG]. [ZEH]'s book treats many of the same questions that we do, although often from a different perspective. For example, I have little to say about quantum gravity, and those who would like a reference on that subject and its relation to questions about the directionality of time are advised to consult that book and [COLEMAN S]. Views on quantum gravity as a source of time asymmetry can be found in [PENROSE] and references within that article.

The "one foot" allusion is from Masechet Shabbat 31a.

Section 3.3. Time-symmetric, action-at-a-distance, electrodynamics is developed in [TETRODE, FOKKER, SCHWARZSCHILD, WHEELER 45] and [WHEELER 49]. On the mathematical formulation of two-time boundary value problems for equations with advance and retardation, see [BELLMAN, DRIVER 69, GRIMM, SCHULMAN 74a, 74b, 82, 95, TITCHMARSH 39, 48] and [WALTMAN]. The conjectured two-interval boundary value problem for action-at-a-distance electrodynamics is proposed and illustrated in [SCHULMAN 74a].

The example of time-like trajectories that never "see" each other was given by [HAVAS], and quoted in [DRIVER 63], where conditions are given for avoiding this situation. The retarded and advanced Green's functions can be found in [JACKSON]. A reference for Noether's theorem is [COURANT] and constants of the motion associated with the action of Eq. (3.3.1) are given in [ANDERSON].

The remarks on the path integral and the quantization of Lagrangian theories refer to work that appears in [FEYNMAN 42] and [FEYNMAN 65]. More on the early efforts of Feynman and Wheeler and their goals vis à vis the quantization of the electromagnetic field can be found in [SCHWEBER] and [GLEICK]. A quantization of the linear theory is in [SCHULMAN 95].

For a study of propagators (Green's functions) in general relativity, see [DE-WITT]. The possibility of constructing an action-at-a-distance theory of gravity was considered by Wheeler in 1949, although apparently the only record of this work is [WHEELER 79], in which he advocates field theory approaches, characterizing his enthusiasm by "no one gets religion like a reformed drunkard."

References on axiomatic bases for space and time are [ANDRE, AX, GOLDBLATT] and [MARCHISOTTO], the last a presentation of nineteenth-century work of the geometer M. Pieri.

The absorber theory of Wheeler and Feynman has generated interest over the years. The first problem is the derivation itself which in turn leads to the question, what if it's not quite true, what if the absorption is not complete? [HOGARTH] brought cosmology into the question, which is reasonable when one considers that photons can travel a long way before being absorbed, and indeed we detect photons from eras in which the radius of the universe was smaller. [HOYLE] claimed that the absorption could only work in a steady state universe. Other authors, among them [HOBART] and [SCHULMAN 80], speculated on the effect of a leftover, unabsorbed field. Finally, attempts were made to observe a failure of complete absorption [PARTRIDGE] by alternately exposing an antenna to the night sky and shielding it, figuring that the radiation reaction should be affected, in turn causing variation in what it took to maintain the alternating current in the antenna. No effect was seen, although there was disagreement over whether this proved that there was complete absorption, [PEGG 75a]. For a review see [PEGG 75b].

The calculation of radiation reaction from the difference of a particle's own advanced and retarded forces appears in [DIRAC]. The form he found had previously been used for radiation reaction by Lorentz. When the dynamical equations include this term (as in our equation Eq. (3.3.13)) a preacceleration is predicted. Even this degree of peculiarity is achieved only when outright blowups are neutralized by demands of future good behavior. (This is reminiscent of themes in this book.) For literature on this subject see [ANDERSON] and [ROHRLICH].

Appendix A to Section 3.3 on temporally non-local interactions uses material from [SCHULMAN 74a, 74b, 95]. The background material in the Appendix can be found in [BELLMAN]. See also [TITCHMARSH 39, 48], where he uses Fourier transforms on finite time intervals for functions that may grow exponentially in time. This is the key to the theorem that I quote from [BELLMAN] in the Appendix.

Section 3.4. In the text I assume familiarity with the terminology of particle physics. For background and for a fuller description, see [PERKINS]. The discovery of CP-violation in K-decay is reported in [CHRISTENSON]. The decay theory for the K system is given in [KABIR]. The properties of T as an *antiunitary* operator are discussed in [GOTTFRIED]. Complex eigenvalues for time translation operators as well as a definition of time reversal for unstable particles are presented in [SCHULMAN 70, 72]. A possible connection between the CP and the statistical arrows of time was considered by [NE'EMAN] and [AHARONY].

Section 3.5. The more troubling aspects of the biological arrow have to do with consciousness and our different perceptions of past and future. Digestion hardly seems more complicated than the usual thermodynamic arrow and even remembering the past rather than the future should be understandable by analogy with computers, devices for which one expects the past/future distinction to be purely physical. The spirit of our age is to consider consciousness to be explainable in physical terms. A few references with different approaches to achieving this sort of explanation are [CAIANIELLO, POSNER, SHAW] and [TRAUB]. A forum for diverse views on the subject (including 'non-reductionist' views) is [CHALMERS].

Rayleigh-Bénard flow is discussed in [BERGÉ] and [LICHTENBERG]. A classic text on non-equilibrium phenomena is [DE GROOT]. More recent work is that of [KEIZER]. A popular exposition on the planet Venus is in [LUHMANN].

'Complexity' has become a contemporary buzz word; nevertheless, it is worth threading through the hype. In many areas enough is known about putative complex systems to allow generalization and the development of organizing concepts. Of course there is a danger of idle speculation, particularly where the mechanisms of Nature's tricks are unknown. An early well grounded work is [GATLIN], which quotes Schrödinger's remarks on life being sustained by negentropy, but which recognizes the limitations of that concept for understanding the power of a seed. Her approach is to study DNA with the tools of communication theory. A recent work along these lines is [MANTEGNA]. In the past few years recognition of a certain unity in many phenomena of Nature and society have led to support of organizations like the Santa Fe Institute, whose avowed objective the study of complexity. There is a journal, 'Complexity,' devoted to this subject, whose initial issue contains introductory material, for example [GELL-MANN].

Algorithmic information is defined and developed in [CHAITIN 77, 87]. Complexity or depth, defined in terms of probabilities for the path leading to a state, is given in [LLOYD]. Other and related ideas can be found in [ZUREK] and [STEIN]. 'Catalytic power' as a measure of the effect of a seed arose from my work (joint with P. E. Seiden) on Conway's game of Life, [SCHULMAN 78, 79]. Part of our goal in studying this cellular automaton was to capture in a statistical mechanics sense its self-reproducing property. In the vast subsequent cellular automaton literature as well as in current work on complexity I don't find a resolution of this issue. Another matter that seems important, but which has drawn little interest in the complexity industry, is the self-referential nature of living systems. Intense exploration of these ideas flourished with the book of [HOFSTADTER]. The term "catalytic power" also suggests that one could learn from other concepts that have been found useful in chemistry.

Other approaches, rooted in mathematics and statistical mechanics, are being applied to natural systems, for example the concepts of fractals [MANDEL-BROT] and self-organized-criticality [BAK]. There is also other work, not quite so grandiose in scope, that takes one biological phenomenon at a time, and analyzes it with the tools we have been using. Here the literature is enormous. Two examples are the study of biological ratchets ([DOERING, ROUSSELET] and [LEIBLER]) and the laboratory/computer model studies of subsets of the brain in [TRAUB].

For detailed analytic development of a model for heat conduction see [GAVEAU 88]. The breakdown model in Appendix A to Section 3.5 is based on [GAVEAU 96].

Pattern formation in open systems has been studied in many contexts. Background for the galaxy figure can be found in [SCHULMAN 86a] with more detail in [SEIDEN]. For other examples, see [WALGRAEF] and (especially for cellular automata) [PERDANG].

References

ADLER 75	R. Adler, M. Bazin & M. Schiffer (1975) *Introduction to General Relativity*, McGraw-Hill, New York.
AHARONY 70	A. Aharony & Y. Ne'eman, (1970) Time-Reversal Symmetry Violation and the Oscillating Universe, *Int. J. Theor. Phys.* **3**, 437.
ALPHER 48	R. A. Alpher, H. A. Bethe & G. Gamow (1948) *Phys. Rev.* **73**, 803.
ANDERSON 67	J. L. Anderson (1967) *Principles of Relativity Physics*, Academic, New York.
ANDRE 74	J. Andre (1974) On Finite Non-Commutative Affine Spaces, *Math. Centre Tracts* **55**, 60.
AX 78	J. Ax (1978) The Elementary Foundations of Spacetime, *Found. Phys.* **8**, 507.
BAK 87	P. Bak, C. Tang & K. Wiesenfeld (1987) Self-Organized Criticality: An Explanation of $1/f$ Noise, *Phys. Rev. Lett.* **59**, 381.
BELLMAN 63	R. Bellman & K. L. Cooke (1963) *Differential-Difference Equations*, Academic, New York.
BERGÉ 86	P. Bergé, Y. Pomeau & C. Vidal (1986) *Order within chaos*, Wiley, New York.
CAIANIELLO 87	E. R. Caianiello (1987) *Physics of Cognitive Processes–Amalfi 1986*, World Sci., Singapore.
CHAITIN 77	G. J. Chaitin (1977) Algorithmic Information Theory, *IBM J. Res. Dev.* **21**, 350.
CHAITIN 87	G. J. Chaitin (1987) *Algorithmic Information Theory*, Cambridge Univ. Press, New York.

CHALMERS 96   D. J. Chalmers, S. Hameroff, A. Kaszniak, C. Koch, M. Schlitz, A. Scott, P. Stoerig & K. Sutherland (1996) Toward a Science of Consciousness 1996, Proc. to be pub. in *J. Consciousness Studies*. Conf. in Tucson, Arizona, April 1996.

CHRISTENSON 64   J. H. Christenson, J. W. Cronin, V. L. Fitch & R. Turlay (1964) Evidence for the 2π Decay of the K_2^0 Meson, *Phys. Rev. Lett.* **13**, 138.

COLEMAN PH 92   P. H. Coleman & L. Pietronero (1992) The Fractal Structure of the Universe, *Phys. Rep.* **213**, 311.

COLEMAN S 91   S. Coleman, J. B. Hartle, T. Piran & S. Weinberg (1991) *Quantum Cosmology and Baby Universes*, World Sci., Singapore.

COURANT 53   R. Courant & D. Hilbert (1953) *Methods of Mathematical Physics*, vol. I, Interscience, New York.

DE GROOT 62   S. R. de Groot & P. Mazur (1962) *Non-Equilibrium Thermodynamics*, North Holland, Amsterdam (Dover ed. 1984, New York).

DEWITT 60   B. S. DeWitt & R. W. Brehme (1960) Radiation Damping in a Gravitational Field, *Ann. Phys.* **9**, 220.

DICKE 65   R. H. Dicke, P. J. E. Peebles, P. G. Roll & D. T. Wilkinson (1965) Cosmic Black-Body Radiation, *Ap. J.* **142**, 414.

DIRAC 38   P. A. M. Dirac (1938) Classical theory of radiating electrons, *Proc. R. Soc. Lond.* A **167**, 148.

DOERING 94   C. R. Doering, W. Horsthemke & J. Riordan (1994) Nonequilibrium Fluctuation-Induced Transport, *Phys. Rev. Lett.* **72**, 2984.

DOMINGUEZ 88   R. Dominguez-Tenreiro & M. Quiros (1988) *An Introduction to Cosmology and Particle Physics*, World Sci., Singapore.

DRIVER 63   R. D. Driver (1963) A Two-Body Problem of Classical Electrodynamics: the One-Dimensional Case, *Ann. Phys.* **21**, 122.

DRIVER 69   R. D. Driver (1969) A "Backwards" Two-Body Problem of Classical Relativistic Electrodynamics, *Phys. Rev.* **178**, 2051.

FEYNMAN 42   R. P. Feynman (1942) The Principle of Least Action in Quantum Mechanics, Thesis, Princeton University

FEYNMAN 65   R. P. Feynman & A. R. Hibbs (1965) *Quantum Mechanics and Path Integrals*, McGraw-Hill, New York.

FOKKER 29   V. A. D. Fokker (1929) Ein invarianter Variationssatz fur die Bewegung mehrerer elektrischer Massenteilchen, *Z. Physik*, **58**, 386; (1929) *Physica* **9**, 33; (1932) *Physica* **12**, 145.

GATLIN 72   L. L. Gatlin (1972) Evolutionary Indices, in *Darwinian, Neo-Darwinian, and Non-Darwinian Evolution*, ed. L. M. Le Cam, J. Neyman & E. L. Scott, Univ. Calif. Press, Berkeley. (*Proc. 6th Berkeley Symp. Math. Stat. & Prob.*, 1970 & 1971, vol. V.)

GAVEAU 88   B. Gaveau & M.-A. Gaveau (1988) A Semi-Microscopic Model of the Heat Conduction Law, *Lett. Math. Phys.* **16**, 179.

GAVEAU 96   B. Gaveau & L. S. Schulman (1996) Master equation based formulation of non-equilibrium statistical mechanics, *J. Math. Phys.* **37**, 3897. See also, "A general framework for non-equilibrium phenomena: The master equation and its formal consequences," preprint

GELL-MANN 95   M. Gell-Mann (1995) What is Complexity?, *Complexity* **1**, 16.

GLEICK 92   J. Gleick (1992) *Genius: the life and science of Richard Feynman*, Pantheon, New York.

GOLDBLATT 87   R. Goldblatt (1987) *Orthogonality and Spacetime Geometry*, Springer, New York.

GOTTFRIED
66
K. Gottfried (1966) *Quantum Mechanics*, Benjamin, New York.

GOULD
96
H. Gould and J. Tobochnik (1996) *An Introduction to Computer Simulation Methods: Applications to Physical Systems*, 2nd ed., Addison-Wesley, Reading, MA.

GRIMM
70
L. J. Grimm & K. Schmitt (1970) Boundary Value Problems for Differential Equations with Deviating Arguments, *Aeq. Math.* **4**, 176.

HAVAS
49
P. Havas (1949) Bermerkungen zum Zweikörperproblem der Elektrodynamik, *Acta Phys. Austriaca*, **3**, 342.

HOBART
79
R. H. Hobart (1976) A Cosmological Derivation of Planck's Constant, *Found. Phys.* **6**, 473.

HOFSTADTER
79
D. R. Hofstadter (1979) *Gödel, Escher, Bach: an Eternal Golden Braid*, Basic Books, New York.

HOGARTH
62
J. E. Hogarth (1962) Cosmological considerations of the absorber theory of radiation, *Proc. R. Soc. Lond.* A **267**, 365.

HOYLE
95
F. Hoyle & J. V. Narlikar (1995) Cosmology and action-at-a-distance electrodynamics, *Rev. Mod. Phys.* **67**, 113.

JACKSON
75
J. D. Jackson (1975) *Classical Electrodynamics*, 2nd ed., Wiley, New York.

JACOBS
93
M. W. Jacobs, E. V. Linder & R. V. Wagoner (1992) Obtaining the metric of our Universe, *Phys. Rev.* D **45**, R3292; (1993) Green function for metric perturbations due to cosmological density fluctuations, *Phys. Rev.* D **48**, 4623.

KABIR
68
P. K. Kabir (1968) *The CP Puzzle: Strange Decays of the Neutral Kaon*, Academic Press, New York.

KEIZER
87
J. Keizer (1987) *Statistical Thermodynamics of Nonequilibrium Processes*, Springer, New York.

LEIBLER
94
S. Leibler (1994) Moving forward noisily, *Nature* **370**, 412.

LEVI
92
B. G. Levi (1992) COBE measures anisotropy in cosmic microwave background radiation, *Phys. Today* **45** (June), 17.

LICHTENBERG
83
A. J. Lichtenberg & M. A. Lieberman (1983) *Regular and Stochastic Motion*, Springer, New York.

LLOYD
88
S. Lloyd & H. Pagels (1988) Complexity as Thermodynamic Depth, *Ann. Phys.* **188**, 186.

LUHMANN
94
J. G. Luhmann, J. B. Pollack & L. Colin (1994) The Pioneer Mission to Venus, *Sci. Amer.* **270**, April, 90.

MANDELBROT
83
B. B. Mandelbrot (1983) *The Fractal Geometry of Nature*, Freeman, New York.

MANTEGNA
94
R. N. Mantegna, S. V. Buldyrev, A. L. Goldberger, S. Havlin, C.-K. Peng, M. Simons & H. E. Stanley (1994), Linguistic Features of Noncoding DNA Sequences, *Phys. Rev. Lett.* **73**, 3169.

MARCHISOTTO
92
E. A. Marchisotto (1992) Lines Without Order, *Math. Assoc. Amer. Mon.*, Oct., 738.

NE'EMAN
69
Y. Ne'eman (1969) The Arrows of Time, *Proc. Isr. Acad. Sci.*, No. 13.

PARTRIDGE
73
R. B. Partridge (1973) Absorber Theory of Radiation and the Future of the Universe, *Nature* **244**, 263.

PEEBLES
93
P. J. E. Peebles (1993) *Principles of physical cosmology*, Princeton Univ. Press, Princeton.

PEGG
75a
D. T. Pegg (1975) On a recent experiment to detect advanced radiation, *J. Phys.* A **8**, L60.

PEGG 75b   D. T. Pegg (1975) Absorber theory of radiation, *Rep. Prog. Phys.* **38**, 1339.

PENROSE 87   R. Penrose, Newton, quantum theory and reality, in S. Hawking & W. Israel (1987) *Three hundred years of gravitation*, Cambridge Univ. Press, Cambridge; especially § 3.11.

PERDANG 93   J. M. Perdang & A. Lejeune (1993) *Cellular Automata: Prospects in Astrophysical Applications*, World Sci., Singapore.

PERKINS 87   D. H. Perkins (1987) *Introduction to high energy physics*, 3rd ed., Addison-Wesley, Reading, Mass.

POSNER 89   M. I. Posner (1989) *Foundations of Cognitive Science*, MIT, Cambridge.

ROHRLICH 65   F. Rohrlich (1965) *Classical Charged Particles*, Addison-Wesley, Reading, Mass.

ROUSSELET 94   J. Rousselet, L. Salome, A. Ajdarl & J. Prost (1994) Directional motion of brownian particles induced by a periodic asymmetric potential, *Nature* **370**, 446.

SCHULMAN 70   L. S. Schulman (1970) Unstable Particles and the Poincaré Semigroup, *Ann. Phys.* **59**, 201.

SCHULMAN 72   L. S. Schulman (1972) Time Reversal for Unstable Particles, *Ann. Phys.* **72**, 489.

SCHULMAN 74a   L. S. Schulman (1974) Some Differential-Difference Equations Containing Both Advance and Retardation, *J. Math. Phys.* **15**, 295.

SCHULMAN 74b   L. S. Schulman (1974) On Deriving Irreversible Electrodynamics, in *Modern Developments in Thermodynamics*, ed. B. Gal-Or, Keter, Jerusalem.

SCHULMAN 78   L .S. Schulman & P. E. Seiden (1978) Statistical Mechanics of a Dynamical System Based on Conway's Game of Life, *J. Stat. Phys.* **19**, 293.

SCHULMAN 79   L. S. Schulman (1979) Catalytic Power, unpublished.

SCHULMAN 80   L. S. Schulman (1980) Formulation and Justification of the Wheeler-Feynman Absorber Theory, *Found. Phys.* **10**, 841.

SCHULMAN 82   L. S. Schulman (1982) Time-symmetric self-interaction, *Phys. Rev. D* **26**, 2934.

SCHULMAN 86a   L. S. Schulman & P. E. Seiden (1986) Percolation and Galaxies, *Science* **233**, 425.

SCHULMAN 86b   L. S. Schulman & P. E. Seiden (1986) Hierarchical Structure in the Distribution of Galaxies, *Ap. J.* **311**, 1.

SCHULMAN 95   L. S. Schulman (1995) Time displaced interactions: classical dynamics and path integral quantization, *J. Math. Phys.* **36**, 2546.

SCHUTZ 88   B. F. Schutz (1988) *A first course in general relativity*, Cambridge Univ. Press, Cambridge.

SCHWARZSCHILD 03   K. Schwarzschild (1903) *Nachr. Akad. Wiss. Göttingen Math. Physik. Kl.*, IIa; 128, 132, 245.

SCHWEBER 93   S. S. Schweber (1993) *QED and the men who made it: Dyson, Feynman, Schwinger and Tomonaga*, Princeton Univ. Press, Princeton.

SEIDEN 90   P. E. Seiden & L. S. Schulman (1990) Percolation Model of Galactic Structure, *Adv. Phys.* **39**, 1.

SHAW 88   G. L. Shaw & G. Palm (1988) *Brain Theory*, vol. I, *Advanced Series in Neuroscience*, World Sci., Singapore.

STEIN 89   D. L. Stein (1989) *Lectures in the Sciences of Complexity*, Addison-Wesley, New York.

STEPHANI   H. Stephani (1993) *General Relativity–An introduction to the theory*
93   *of the gravitational field*, 2nd ed., Cambridge Univ. Press, Cambridge
             (Trans. M. Pollock & J. Stewart).

TETRODE   H. Tetrode (1922) *Z. Physik* **10**, 317.
22

TITCHMARSH   E. C. Titchmarsh (1939) Solutions of some functional equations, *J. Lon-*
39   *don Math. Soc.* **14**, 118.

TITCHMARSH   E. C. Titchmarsh (1948) *Introduction to the theory of Fourier integrals*,
48   2nd ed., Oxford, London.

TRAUB   R. D. Traub & R. Miles (1991) *Neuronal Networks of the Hippocampus*,
91   Cambridge Univ. Press, Cambridge.

WALD   R. M. Wald (1984) *General Relativity*, Univ. of Chicago Press, Chicago.
84

WALGRAEF   D. Walgraef (1987) *Patterns, Defects and Microstructures in Nonequi-*
87   *librium Systems—Applications in Materials Science*, NATO ASI, Ser. B,
             Phys., vol. 121, Martinus Nijhoff, Dordrecht.

WALTMAN   P. Waltman & J. S. W. Wong (1972) Two Point Boundary Value Prob-
72   lems for Nonlinear Functional Differential Equations, *Trans. Am. Math.*
             Soc. **164**, 39.

WEINBERG   S. Weinberg (1972) *Gravitation and Cosmology: Principles and Applica-*
72   *tions of the General Theory of Relativity*, Wiley, New York.

WHEELER   J. A. Wheeler & R. P. Feynman (1945) Interaction with the Absorber as
45   the Mechanism of Radiation, *Rev. Mod. Phys.* **17**, 157.

WHEELER   J. A. Wheeler & R. P. Feynman (1949) Classical Electrodynamics in
49   Terms of Direct Interparticle Action, *Rev. Mod. Phys.* **21**, 425.

WHEELER   J. A. Wheeler (1979) Some Men and Moments in the History of Nuclear
79   Physics: The Interplay of Colleagues and Motivations, in *Nuclear Physics*
             in Retrospect, ed. R. H. Stuewer, Univ. Minnesota, Minneapolis.

ZEH   H. D. Zeh (1992) *The Physical Basis of the Direction of Time*, 2nd ed.,
92   Springer, New York.

ZUREK   W. H. Zurek (1991) *Complexity, Entropy and the Physics of Information*,
91   Addison-Wesley, Reading, Mass.

Appendix 3.2.A: General relativity on one foot

The space-time metric is

$$ds^2 = g_{\mu\nu}dx^\mu dx^\nu \tag{3.2.A.1}$$

with the usual summation convention. The Minkowski metric is $\text{diag}\,(1, -1, -1, -1)$ and in this Appendix I take $c = 1$. The Christoffel symbols are given by

$$\Gamma^\mu_{\alpha\beta} \equiv \frac{1}{2}g^{\mu\nu}\left[g_{\alpha\nu,\beta} + g_{\beta\nu,\alpha} - g_{\alpha\beta,\nu}\right] \tag{3.2.A.2}$$

The quantity $g^{\mu\nu}$ is the inverse of $g_{\mu\nu}$, i.e., $g^{\mu\alpha}g_{\mu\beta} = \delta^\alpha_\beta$. The comma (in Eq. (3.2.A.2)) is the ordinary derivative with respect to the coordinate: $f_{,\mu} \equiv \partial f/\partial x^\mu$. From the symmetry of $g_{\mu\nu}$ $(= g_{\nu\mu})$, follows the symmetry of $\Gamma^\mu_{\alpha\beta}$ in its lower indices. The Christoffel symbols are used to define covariant derivatives and are also known as the affine connection. The Riemann curvature tensor is built from them as follows

$$R^\alpha_{\mu\nu\beta} \equiv \Gamma^\alpha_{\mu\nu,\beta} - \Gamma^\alpha_{\mu\beta,\nu} + \Gamma^\rho_{\mu\nu}\Gamma^\alpha_{\beta\rho} - \Gamma^\rho_{\mu\beta}\Gamma^\alpha_{\nu\rho} \tag{3.2.A.3}$$

The Ricci tensor is a contraction of $R^{\alpha}_{\mu\nu\beta}$ and is given by

$$R_{\mu\nu} \equiv R^{\alpha}_{\mu\alpha\nu} \tag{3.2.A.4}$$

From the Ricci tensor one builds the curvature scalar

$$R \equiv g^{\mu\nu} R_{\mu\nu} \tag{3.2.A.5}$$

There is an enormous number of symmetries and identities satisfied by these quantities. The reader should also be aware that there are coordinate free formulations of differential geometry that bring out the geometrical significance of these objects and also help avoid problems associated with particular choices of coordinates, for example spurious singularities in $g_{\mu\nu}$. Another geometrical object to which we have occasion to refer is the Weyl tensor. It is given by

$$C^{\alpha\beta}_{\mu\nu} \equiv R^{\alpha\beta}_{\mu\nu} - \frac{1}{2} \left(g^{\alpha}_{\mu} R^{\beta}_{\nu} + g^{\beta}_{\nu} R^{\alpha}_{\mu} - g^{\beta}_{\mu} R^{\alpha}_{\nu} - g^{\alpha}_{\nu} R^{\beta}_{\mu} \right) + \frac{1}{6} R \left(g^{\alpha}_{\mu} g^{\beta}_{\nu} - g^{\beta}_{\mu} g^{\alpha}_{\nu} \right) \tag{3.2.A.6}$$

This is sometimes known as the conformal curvature tensor, reflecting the fact that it is unchanged when the metric undergoes a transformation $ds^2 \to d\bar{s}^2 \equiv f(x)ds^2$ with f a positive, coordinate dependent function.

The energy momentum tensor reflects the content of the space-time. For a collection of point particles at (3-) positions $\vec{x}_n(t)$, $n = 1, \ldots$, having 4-momentum p^{μ}_n, it is given by

$$T^{\mu\nu}(x) = \sum_n \frac{p^{\mu}_n p^{\nu}_n}{p^0_n} \delta^3(\vec{x} - \vec{x}_n) \tag{3.2.A.7}$$

The non-covariant looking energy in the denominator of Eq. (3.2.A.7) is matched by the non-covariance of the δ-function, so that $T^{\mu\nu}(x)$ does indeed transform like a tensor. Fields also contribute to $T^{\mu\nu}(x)$. For example, for the electromagnetic field $T^{\mu\nu}_{\mathrm{EM}} = F^{\alpha\mu} F^{\nu}_{\alpha} - \frac{1}{4} g^{\mu\nu} F^{\alpha\beta} F_{\alpha\beta}$.

Appendix 3.3.A: Temporally non-local differential equations

Action-at-a-distance electrodynamics and the linear, fixed time-displacement theory that was also considered in Section 3.3, involve mathematical techniques more familiar—at least for the retarded case—in control theory than in physics. Consider the equations

$$Dx(t) \equiv \left. \frac{dx(t')}{dt'} \right|_{t'=t} = ax(t) + bx(t - \tau) \tag{3.3.A.1}$$

and

$$Dx(t) = f\big(x(t), x(t - \tau)\big) \tag{3.3.A.2}$$

Here a, b and $\tau > 0$ are fixed constants and f is a well-behaved function. Eq. (3.3.A.1) is a special case of Eq. (3.3.A.2). Eq. (3.3.A.2) can be solved by giving a function $\phi(t)$ on an interval of length τ, say, $[0, \tau)$. Then for $\tau \leq t < t + \tau$, Eq. (3.3.A.2) becomes the first-order ODE (ordinary differential equation):

$$Dx(t) = f\big(x(t), \phi(t - \tau)\big) \tag{3.3.A.3}$$

This is integrated with the initial condition $x(\tau) = \phi(\tau)$. The process can be continued and yields a solution for arbitrary positive time. Since the solution on successive intervals is the integral of that on previous intervals, $x(\cdot)$ tends to become smoother. If ϕ is k

times differentiable, x on $[\tau, 2\tau)$ will be $k+1$ times differentiable. At $t = \tau$ we can define $x(t)$ so as to be continuous. However, its derivative need not be, since $D\phi(\tau)$ need not equal $f(\phi(\tau), \phi(0))$ (the latter being the derivative of x as dictated by the differential equation). However, at the next step, $t = 2\tau$, the derivative is continuous, but the second derivative need not be. And so it goes.

For the linear Eq. (3.3.A.1), $x(t) = \exp(zt)$ will provide a solution if the complex number z satisfies

$$z = a + be^{-z\tau} \qquad (3.3.A.4)$$

Calling $h(z) \equiv z - a - b\exp(-z\tau)$ the characteristic function, Eq. (3.3.A.4) is the demand that z be a root of $h(z)$. A polynomial in z and $\exp(\text{const}\cdot z)$ is known as an 'exponential polynomial.' Unlike the polynomials in z alone that arise from finite-order differential equations, an exponential polynomial generally has an infinity of roots. This makes perfect sense, since Eqs. (3.3.A.1–3.3.A.2) should be looked upon as infinite order, an aspect that is evident by writing $x(t - \tau)$ as $[\exp(-\tau D)x](t)$. Furthermore, this is consistent with the integration scheme (Eq. (3.3.A.3)) given in the general case. Giving a *function*, rather than a finite number of values, corresponds to the infinite order of the equation. Let $\{z_n\}$ be the set of roots of $h(z)$. Then anything of the form $\sum_n a_n \exp(z_n t)$ is a solution of Eq. (3.3.A.1). It is a theorem of Titchmarsh that *any* solution can be written in that form. This is proved by Fourier transformation.

The roots of characteristic exponential polynomials have been studied. Asymptotically, for retarded equations such as Eq. (3.3.A.2), their real parts go off to minus infinity, with imaginary parts (of both signs) growing as the exponent of the magnitude of the real part. To the right of any vertical line in the complex z-plane there can only be a finite number of them. The important question for control theory is whether any root lies to the right of $\operatorname{Re} z = 0$. This determines whether the motion is stable.

Our concern is with time-symmetric time-displaced equations, such as

$$D^2 x(t) + \omega^2 x(t) + \tfrac{1}{2}\alpha[x(t + \tau) + x(t - \tau)] = 0 \qquad (3.3.A.5)$$

or with the more complicated nonlinear equations of electrodynamics (cf. Eqs. (3.3.2–3)). To take the forward-integrating approach to Eq. (3.3.A.5), rewrite it as

$$x(t) = -\frac{2}{\alpha}\left[D^2 x(t - \tau) + \omega^2 x(t - \tau)\right] - x(t - 2\tau) \qquad (3.3.A.6)$$

It is clear that 2τ of data are needed. Now imagine that you have specified a function ϕ on $[0, 2\tau)$. Unless you impose special matching conditions, the value of the right-hand side of Eq. (3.3.A.6) at $t = 2\tau$ (evaluated with that ϕ) need not be the same as $\phi(2\tau)$. Thus the solution is not generally continuous. And it gets worse on successive time steps. Moreover, each iteration makes the solution two times less differentiable.

The substitution, $x(t) = \exp(zt)$, imposes on z that it be a root of the characteristic exponential polynomial, in this case

$$h(z) = z^2 + \omega^2 + \alpha \cosh(\tau z) \qquad (3.3.A.7)$$

The roots of this $h(z)$ go to infinity in all quadrants. (If z is a root, so are $-z$, z^*, and $-z^*$.) Writing $z = x + iy$, it is easy to see that for large $|z|$, $x \sim (2/\tau)\log y$ and $y \sim (2n + 1)\pi/\tau$. Since the hallmark of predictive instability is $x > 0$, one should not expect an initial value formulation (like Eq. (3.3.A.6)) to be useful.

Here is where we implement the advice given in the main text. We saw how the variational problem (based on Eq. (3.3.4)) guided us to a two-time boundary value problem with data given on separated intervals of length τ each, and with the solution realized between them. We will see that this will give us solutions that are well behaved in the intervening interval.

In the forthcoming method, we show explicitly the dependence of the solutions on the data (and *a fortiori* their sufficiency). We also demonstrate how two-time boundary conditions suppress troublesome modes—even to the point that perturbation theory (in α) is possible despite its singular nature (from the ODE standpoint).

We discretize the time to steps of length $\epsilon \equiv (1/N)$, for integer N. Consider the time interval $[0, T]$ and let $M \equiv NT - 1$ and $K \equiv \tau N$ (assumed to be integers). Let the time $t_n \equiv n\epsilon$ and define a column vector ξ by $\xi_n \equiv x(t_n)$ for $n = 1, \ldots, M$. We wish to find ξ, given data on $[-\tau, 0]$ and $[T, T + \tau]$. These data are two functions on time intervals of length τ. To use this in the discrete version of Eq. (3.3.A.5) we incorporate both functions in a column vector, ϕ, of length M. The first K components are the initial data (with time discretized), and the last K components the final data. (For simplicity we assume $T > 2\tau$.) When discretizing $D^2 x$, the equation for ξ_1 requires the last value of the initial data (call it x') and the equation for ξ_M requires the first value of the final data (call it x''). We combine these numbers in a single M-component column vector x_B ('Boundary') whose first component is x', last component x'', and all others zero. The discrete version of Eq. (3.3.A.5) takes the following form

$$\Lambda \xi = -\frac{1}{\epsilon^2} x_B - \frac{\alpha}{2} \phi \tag{3.3.A.8}$$

where

$$\Lambda = \Lambda_0 + \alpha V \ , \qquad \Lambda_0 = -\frac{1}{\epsilon^2} J + \omega^2 \mathbf{1}$$

$$J = 2 \cdot \mathbf{1} - R - L \ , \qquad V = \tfrac{1}{2}(R^K + L^K) \tag{3.3.A.9}$$

$$L = R^\dagger \qquad \& \qquad R = \text{Raising operator}$$

The Raising operator has ones just above the diagonal, zeros elsewhere. From Eq. (3.3.A.8), the solution is $\xi = -\Lambda^{-1} \left[x_B/\epsilon^2 + \alpha\phi/2 \right]$.

Perturbation in this framework relies on the identity

$$G = G_0 - \alpha G_0 V G \ , \qquad \text{where } G \equiv \Lambda^{-1} \quad \text{and } G_0 \equiv \Lambda_0^{-1} \tag{3.3.A.10}$$

The power series generated by this equation $[G = G_0 \sum_n (-\alpha V G_0)^n]$ involves powers of $V G_0$. Convergent perturbation theory requires that this have finite norm. We use the maximum-eigenvalue norm and recall the inequality $|AB| \leq |A||B|$ for norms. The norm of $R^K + L^K$ cannot be larger than two since all the individual operators do is displace pieces of a vector. For the norm of G_0 we need the smallest (in norm) eigenvalue of Λ_0. These eigenvalues are well known and are

$$\lambda_k = \omega^2 - \frac{4}{\epsilon^2} \sin^2 \frac{k\pi}{2(M+1)} \ , \qquad k = 1, \ldots, M \tag{3.3.A.11}$$

Unless the sine has small argument this gives large negative numbers. Therefore for the smallest values of λ_k we look to k small compared to M. In this case

$$\lambda_k = \omega^2 - \frac{k^2 \pi^2}{T^2} \tag{3.3.A.12}$$

The ϵ has disappeared, so that convergence is independent of N. The only trouble that can arise from Eq. (3.3.A.12) is when $T = k\pi/\omega$ for integer k. This is reasonable: G_0 integrates the ordinary oscillator ODE (with $\alpha = 0$) and these are the resonant values for its boundary value problem. At non-resonant T there is a non-zero minimum $|\lambda_k|$, so that for sufficiently small α perturbation theory converges.

Exercise 3.3.A.1. *Pure imaginary roots of the characteristic equation.* Because of time symmetry, the characteristic function $h(z)$ of Eq. (3.3.A.7) can generically have one or more pure imaginary roots, $z = iy$, with real y. Show that if τ is small there is but a single (positive) solution, given approximately by $y = \sqrt{\omega^2 + \alpha}$ (subject to $\omega^2 + \alpha > 0$). For large τ things can be more interesting. In that case, find the range and multiplicity of the band of roots.

Boundary conditions for action-at-a-distance electrodynamics

Our formulation of the boundary value problem for the linear case leads to a conjecture for the boundary value problem for action-at-a-distance electrodynamics. Suppose you want a solution for N-particles between two space-like surfaces, σ_1 (the earlier one) and σ_2. As for the linear case, you need particle trajectories prior to σ_1 and subsequent to σ_2. For σ_1 you need enough of each particle's trajectory that the light cone back from each particle's intersection with σ_1 finds a point on the trajectory of every other particle; an analogous condition holds for σ_2. The theorem one would like is that when all such trajectories are time-like, the resulting solution exists between the surfaces and consists of time-like trajectories. I do not expect this to be difficult to show for two particles and a short time interval between the surfaces. However, for more general results I expect complications. Even for retarded-only solutions, one should not underestimate the subtleties. For example, Havas has exhibited a pair of slower-than-light trajectories that are never on each other's light cones.[8]

Response to a perturbation: causality

In later chapters we consider whether, with boundary conditions in both past and future, a system's response to a perturbation is after the perturbation, before the perturbation, or both. Macroscopic causality corresponds to responses that only *follow* stimuli. Here we perturb the linear time-displaced equation.

Suppose the system is disturbed near time 0. The dynamical equation now has an inhomogeneous term:

$$D^2 x(t) + \omega^2 x(t) + \tfrac{1}{2}\alpha[x(t+\tau) + x(t-\tau)] = f(t) \qquad (3.3.\text{A}.13)$$

The disturbance is confined to small $|t|$; i.e., $f(t) = 0$ if $|t| > t_0$, with $t_0 \ll \tau$. We want a solution between times a_0 and b_0. We also need two other times a and b. They satisfy $a + 2\tau < a_0 \ll -\tau < 0 < \tau \ll b_0 < b - 2\tau$. As for the homogeneous problem, the data to be specified are the values of $x(t)$ on two intervals, each of length τ. In the book of Bellman & Cooke[9] a representation theorem is developed. For our purposes, this theorem ('6-10' in their numbering) simplifies to

> For times outside the interval $[-\tau, \tau]$, the solution $x(t)$ to Eq. (3.3.A.13) is given by $x(t) = \sum_n p_n(t) \exp(z_n t) + [\text{homogeneous solution}]$, with the numbers z_n the roots of $h(z) = z^2 + \omega^2 + \alpha \cosh \tau z$. The functions $p_n(t)$ are the residues of $\int_0^t f(t') \exp(-zt')dt'/h(z)$ at $z = z_n$.

Comments, amplification: 1. The homogeneous solution is a series in the same exponentials, but the coefficients are constant. (For degenerate roots the coefficients may be

---

[8] Imagine the figure $\times$ formed by light-like lines through the origin in the x-t plane. Then the time-like hyperbolas with the light lines as asymptotes are mutually invisible.

[9] See Section 3.6.

polynomials in t.) 2. The given form holds on $[a, b]$ and is a solution on $[a_0, b_0]$. 3. See Bellman & Cooke for conditions on f.

To avoid extraneous asymmetries take f real and symmetric about 0. Generally, for each $z_n = x_n + iy_n$ with $x_n > 0$ there are three other roots, and by combining contributions we have only two independent terms to deal with:

$$f_{1n} = \frac{\int_0^{t_0} f(t) \exp(-x_n t - iy_n t)dt}{k(+x_n + iy_n)}, \qquad f_{2n} = \frac{\int_0^{t_0} f(t) \exp(+x_n t - iy_n t)dt}{k(-x_n + iy_n)}$$

(3.3.A.14)

with $k(z) \equiv h'(z) = 2z + \alpha\tau \sinh(\tau z)$. For $t > t_0$, f_{1n} is the same as $p_n(t)$ and is the coefficient of $\exp(tx_n + ity_n)$; similarly f_{2n} is the coefficient of $\exp(-tx_n + ity_n)$. For $t < -t_0$ the signs are opposite; $-f_{1n}$ is the coefficient of $\exp(tx_n + ity_n)$, $-f_{2n}$ the coefficient of $\exp(-tx_n + ity_n)$.

We will not go into detail on appropriate boundary conditions except to demand that the solutions do not blow up for large $|t|$. We enforce this by adjusting the homogeneous solution. To avoid the blowup of $f_{1n}\exp(tx_n + ity_n)$ at positive times we add an identical term with opposite sign *for all times*. Such a term satisfies the *homogeneous* equation and may be added. As in the Wheeler-Feynman absorber theory this doubles the coefficient of $\exp(tx_n + ity_n)$ at negative times. Similarly for f_{2n}. The contribution of $\exp(\pm x_n t + iy_n t)$ to the solution is therefore

$$2[f_{2n}\theta(t) - f_{1n}\theta(-t)]\exp(-x_n|t| + iy_n t)$$

(3.3.A.15)

which is reasonable for a (piece of a) Green's function. For roots of $h(z)$ with $\operatorname{Re} z = 0$ there is no blowup and no condition to impose. In this case we can use the homogeneous solution as we wish.

Eq. (3.3.A.15) is a portion of the total solution, and includes acausal effects. Are they noticeable? Generally the real parts of the roots scale like $1/\tau$. If τ is small, as for radiation reaction, then the acausal effects will be confined to small intervals. For large τ there will be roots with moderate x_n whose effects can be observed well before and after the perturbation.

The result of this discussion is that for small τ we recover the usual form of causality as well as could be expected. For example, if one stipulates that a system is at rest initially and is later perturbed, then this will fix the homogeneous solution for the stable roots, in effect, inducing the causal Green's function. There will then be the usual causality, except for an anticipatory response from the unstable modes.

Appendix 3.5.A: Heat conduction with breakdown

In this Appendix I give an example of an open system that cannot settle down; it pulsates indefinitely, transmitting more heat than a static near-equilibrium counterpart.

Consider a row of K insulated boxes with the possibility of energy or matter exchange between adjacent boxes. Each box contains N particles. See Fig. 3.5.A.1. At the left and right ends of this row are two larger containers with far more particles. These are the reservoirs; label them 0 and $K + 1$. The average energy per particle in container k is E_k. From time to time each of the K interior containers interacts with its adjacent neighbors, for example by exchanging N_1 particles with each of them, where $1 \ll N_1 \ll N$. This interaction has the following effect:

$$\Delta E_k = \alpha\left[(E_{k+1} - E_k) + (E_{k-1} - E_k)\right] \qquad \text{for } k = 1, \ldots, K$$

(3.5.A.1)

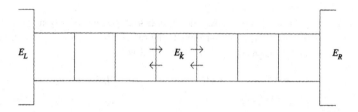

Fig. 3.5.A.1. Energy transfer along a row of containers. Energy is exchanged by each box with its neighbors. The boxes on the ends also exchange energy with their immediate neighbors, but are so large that they are not affected. Rapid transfer—temporary breakdown—can occur if the energy difference between adjacent small boxes exceeds a certain threshold.

with α a constant reflecting the particular energy exchange mechanism. The containers at the ends have energies per particle E_L and E_R (Left and Right), and when these differ there will be heat conduction along the chain. We assume the outside containers to be so large that their energy per particle is unaffected by the conduction. For an ideal gas, temperature (T) is proportional to energy and if it is appropriate to take short time intervals and a large number of closely spaced containers, Eq. (3.5.A.1) becomes Fourier's heat law:

$$\frac{\partial T}{\partial t} = D\frac{\partial^2 T}{\partial x^2} \tag{3.5.A.2}$$

with D a diffusion constant, reached by appropriate limits from α in Eq. (3.5.A.1). For the situation described above, the solution to Eq. (3.5.A.2) is

$$T(x,t) = T_L + \frac{x}{L}(T_R - T_L) + \sum_n a_n \sin\left(\frac{n\pi x}{L}\right) \exp\left[-Dt\left(\frac{n\pi}{L}\right)^2\right] \tag{3.5.A.3}$$

where L is the length of the chain, T_L and T_R are the boundary temperatures and the coefficients a_n depend on the initial conditions.

Exercise 3.5.A.1. Using the known form of the entropy as a function of temperature for a monatomic ideal gas, find the entropy production in the interior containers as a function of time for the discrete process (you'll have to give a time scale). In the steady state, by how much is the entropy lower than it would be if the interior containers were cut off from the ends? The second law demands that including the reservoirs yields overall entropy increase. Check this.

Now suppose that the walls between the small containers can maintain only a certain maximum temperature difference. If this difference is exceeded there is a temporary breakdown, and the contents of the pair of adjacent boxes mix freely. This introduces nonlinearity. After a breakdown the wall recovers rapidly. Obviously the discrete nature of the system is important and the precise outcome will depend on the various time scales—that of response to temperature difference, local equilibration, and wall recovery. Take the effect of breakdown (between boxes k and $k+1$) to be

$$\left.\begin{array}{c} E_k \\ E_{k+1} \end{array}\right\} \rightarrow \left(\frac{E_k + E_{k+1}}{2}\right) \tag{3.5.A.4}$$

where k can be any box number from 1 to $K-1$. The response will be roughly as follows: suppose that prior to the breakdown the energy profile was not too rough, but was steep enough for the breakdown threshold to be exceeded, perhaps as a fluctuation. (Thus $(E_R - E_L)/K$ would be about the same as the threshold.) A breakdown then occurs and the two adjacent boxes suffer the process described by Eq. (3.5.A.4). Subsequently the boxes on either side of them (numbers $k-1$ and $k+2$) will have even larger energy differences relative to the now-averaged amount in boxes k and $k+1$. Depending on the time scales, one or both of these boxes may suffer a breakdown. In any case the disturbance spreads rapidly to the ends of the system where it may bounce off or stop, since we do *not* permit the walls of the reservoirs (boxes '0' and '$K+1$') to break down. Typically there is a good deal of swishing back and forth until things settle down. Having settled down, there will be relatively large discontinuities at the edges with a smaller gradient in the middle. This reduces heat transfer and most of the activity in the system is the diffusive closing of the energy gap between the reservoirs and the boxes adjacent to them. As this occurs the overall energy profile again steepens until somewhere in the middle it becomes steep enough for another breakdown event.

In Fig. 3.5.A.2 I show the results of a computer simulation. The time scales and wall sensitivities are such that breakdown always occurs at the site of the largest discontinuity along the chain and full mixing (the process described by Eq. (3.5.A.4)) occurs immediately. What is shown is energy transfer as a function of time. The peaks correspond to breakdowns. The line near the bottom is the energy transfer rate when the energy profile is linear and is the steady state transfer rate. Following the first phase of a breakdown the transfer rate falls below this because through most of the chain the gradient is reduced and heat transfer is only occurring at the edges.

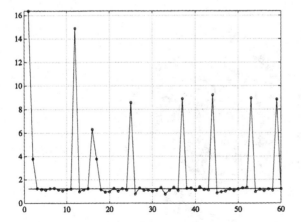
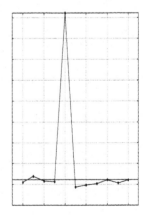

Fig. 3.5.A.2. Heat transfer rates, as a function of time. Peaks correspond to breakdowns. The right-hand figure shows a (rescaled) detail of that on the left for the time interval $40 \leq T \leq 50$. Note that following a burst the transfer rate drops below the 'minimum' associated with steady conduction (what would follow from the distribution Eq. (3.5.A.3), with the transients gone).

This example of nonlinear pulsating behavior depends on the temperature gradient, but it is not quite the structure of proteins nor the extraction of work from heat.[10] The open system does give rise to a structured behavior, but it only hints at the possibilities exploited by living systems.

One can construct richer systems using *cellular automata*, for example Conway's game of 'Life,' with its self-reproducing behavior. Cellular automata represent another level of abstraction in that they are automatically 'open;' there need be no conventional microscopic dynamics. Automata have remarkable pattern formation properties and have been used to model all sorts of systems, fluids, chemical reactions (including reaction-diffusion models where both spatial position and changing particle character are important), even stellar pulsations. An example of an application is shown in Fig. 3.5.A.3. This is a model of a disk galaxy. In such a galaxy the lovely spiral images that we see are the light from a small fraction of the galaxy, consisting of large, young stars. Typically, the forming of such stars is the result of a shock wave in the interstellar medium. And those shock waves arise from the explosion of just such stars going supernova. One can model the disk galaxy as a collection of regions about 500 light years on a side, each region 'alive' or not according to whether or not the matter in it is in the form of (newly) luminous matter. Then one can describe the transmission of luminosity with a probabilistic cellular automaton, in which living sites can induce life on the next time step to neighboring sites. The result—with the proper geometry—is Fig. 3.5.A.3. For a less telegraphic account of this work, see the references in Section 3.6. If you want to program this yourself, see [GOULD].

Fig. 3.5.A.3. Cellular automaton based model of a galaxy. See Section 3.6 for references.

[10] But it does resemble the sandpile model for self-organized criticality.

4

Correlating arrows of time

In the last chapter we enumerated 'arrows of time.' There was a subtheme concerned with which candidates made it to the list, which didn't, trying to eliminate arrows that were immediate consequences of others. Now the subtheme becomes the theme.

We are concerned with *correlating* arrows of time. Our most important conclusion will be that the thermodynamic arrow of time is a consequence of the expansion of the universe. Coffee cools because the quasar 3C273 grows more distant. We will discuss other arrows, in particular the radiative and the biological, but for them the discussion is a matter of proving (or perhaps formulating) what you already believe. For the thermo/cosmo connection there remains significant controversy.

As far as I know it was Thomas Gold who proposed that the thermodynamic arrow of time had its origins in cosmology, in particular in the expansion of the universe. Certainly there had been a lot of discussion of arrows of time before his proposal, but in much of this discussion you could easily get lost, not knowing whether someone was making a definition or solving a problem. Now I'm sure the following statement slights many deep thinkers, but I would say that prior to Gold's idea the best candidate for an explanation of the thermodynamic arrow was that there had been an enormous fluctuation. If you have a big enough volume and if you wait long enough, you would get a fluctuation big enough for life on earth. No one would lose patience during the long wait, since no one would be around. I don't believe this explanation, but mentioning it helps make Gold's suggestion look better.

In this Chapter I will discuss the correlation of various arrows, including criticisms both right and wrong (in my opinion) of Gold's proposal. I will give my justification of (cosmo $\Rightarrow$ thermo). This justification will be almost elementary. The 'almost' refers to subtleties in the argument that arise because the (thermodynamic) arrow of time is built into language, so that it is difficult to avoid circular arguments. Such avoidance will lead us to the use of boundary conditions that do not themselves impose an arrow. The 'elementary' refers to *not* needing to use speculative or even very recent physical ideas. I won't need quarks and I will certainly not invoke quantum cosmology. In my opinion this problem is basically one of statistical mechanics in an unconventional framework, namely a time varying geometry. For the purposes of understanding why coffee cools I don't think you need more than that. The mysterious or mystical nature of the psychological arrow of time tempts us to use high falutin' techniques where they aren't needed.

In Section 4.1 we give Gold's argument and some of the aforementioned criticisms. In the following section we give our justification for (cosmo ⇒ thermo). An essential point in our argument is the use of two-time boundary conditions.[1] But our explanation does not cover everything; in particular it does not account for the arena in which we do our statistical mechanics, namely the expanding (or contracting) universe. The response to other criticisms of Gold's ideas will be given in later chapters, when we explore more fully the consequences of demanding two-time boundary conditions. In Section 4.3 we go more deeply into two-time boundary value problems, the need for which was found in Section 4.2. This allows us in Section 4.4 to take up the argument in more subtle situations. Section 4.5 looks at the question of recovering causality in a time-symmetric framework. Finally in Section 4.6 we take up the important question of whether there can be experimental or observational tests of the ideas presented in this chapter.

Caveat and apologia: It is important to understand the limited nature of my goals. I am trying to explain one datum—the second law of thermodynamics as it is reflected in ordinary macroscopic physics—in terms of another—the expansion of the universe. I am not trying to explain the thermodynamics of gravity. We take the action of gravity as given. In the large there is expansion, in the small contraction. In my opinion our argument will solve the problem at the level that Boltzmann posed it, but it does not address certain ambitious contemporary questions. Indeed, *modulo* a single important subtlety, our demonstration is easy; but this does not disappoint me.

---

[1] In fact, this is more important than the actual expansion/contraction. As remarked in Section 4.2, double ended non-equilibrium boundary conditions are sufficient for most goals in this book.

4.1 The star-in-a-box

As observed in Chapter 2, the Boltzmann H-theorem and its analogs are only a prerequisite for the physical thermodynamic arrow of time. The issue is not whether systems move from uncommon to common states, but why, in our world, do the uncommon ones come first? *We* make the coffee and set in on the table to cool. We isolate the melting ice cube and check it 'later.' We ourselves are born and die in conformity with this arrow which in turn is in concert with the aging of the planet and the aging of the sun. So there's a bigger picture involved. At this point there are two questions you might ask: how does this big picture make itself felt in all the small interactions of our experience, and how did an arrow establish itself for this big picture?

Gold said, it's because the universe is expanding, and that expansion allows photons to get away. As he stated at a 1958 Solvay Conference:

> Let us take, for example, a star, and suppose we could put it inside an insulating box. ... when the star has been in the box for long enough (which in this case will perhaps be rather long), time's arrow will have vanished. ... now if we were to open for a moment a small window in our box, then what would happen? Time's arrow would again be defined inside the box for some time, until the statistical equilibrium had been reestablished. But what had happened when we opened the hole? Some radiation had, no doubt, escaped from the box and the amount of radiation that found its way into the box from the outside was incomparably smaller.
>
> The escape of radiation away from the system is, in fact, characteristic of the type of "influence" which is exerted from outside. ... The thermodynamic approach would be to explain that ... free energy can only be generated from the heat sources in the world by means of heat engines working between a source and a sink. There may be a variety of sources, but the sink is always eventually the depth of space, ...
>
> It is this facility of the universe to soak up any amount of radiation that ... enables it to define the arrow of time in any system that is in contact with this sink.

This illustrates the importance of outside influence and the way that influence establishes the particular direction chosen. Although I will complain below about the argument just given, I believe the fundamental thesis is correct. Others have denied the thesis outright, sometimes for reasons that I could understand (even if I disagreed) and sometimes, well, it's just that kind of subject. One objection, which is sufficiently direct to allow disproof is this. It follows from Gold's claim that in a contracting

universe the thermodynamic arrow of time should be reversed. If the biological arrow follows the thermodynamic one (which presumably it does)
this means that sentient creatures would always measure an expanding
universe. In any case, in an oscillating universe the arrow should first go
one way, then the other. Several authors were shocked at the astound-
ing events this would compel at the 'switchover,' the epoch at which the
expansion reversed. Here is Roger Penrose:

> Otherwise one would have to envisage, it seems to me, a mid-
> dle state in which phenomena of the normal sort (e.g. retarded
> radiation and shattering watches) would co-exist with phenom-
> ena of the 'time-reversed' sort (e.g. advanced radiation and self-
> assembling watches).

As will be shown later, Penrose's complaint is based on an error, an error
that arises from a basic weakness in ordinary language for discussing this
problem. Others, including Gold, have been plagued by this as well.

If you go step by step through the star-in-the-box argument, the entropy as a function of time should look like Fig. 4.1.1.

The star begins in a low entropy state at time zero. When confined,
it approaches equilibrium on a time scale of about 20 of the graph's
time units. At time 100 the window is opened and the entropy drops
precipitously. The window is shut (after 5 time units), and during the
recovery period the entropy climbs back to its equilibrium value. In the
figure the time scale for climbing back—the shape of the curve—is similar
to that of the initial equilibration. This need not be generally true. The
value to which the entropy returns also need not equal that with which
it started; presumably it would be less since the energy in the box is
reduced.

Does this show how we inherit the thermodynamic arrow from the
cosmological arrow? Unfortunately, not quite. Consider a variant of
the argument. Suppose that when you opened the box radiation entered
rather than left. Perhaps you're in a contracting universe or perhaps there
is a situation like that envisaged in Olber's paradox and the box contents
happen to be cooler than the prevailing outside temperature. Whatever
the reason, the system will again be thrown out of equilibrium. The
final entropy might be larger, but the essential feature, the fact that the
outside influence reestablished an arrow of time in the box, would still be
true.

This counterexample does not contradict the importance of outside
factors in imposing a local thermodynamic arrow, but it does show that
the choice of the direction of the arrow, and in particular the mechanism

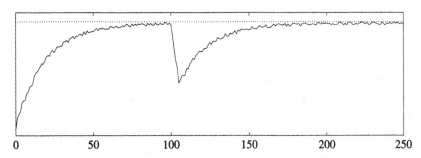

Fig. 4.1.1. Possible time dependence of entropy for the star-in-a-box. Initially entropy increases as the system approaches equilibrium. At time 100 (in the figure units) the window is opened and the entropy drops precipitously—not to its initial value, since opening the window does not undo the nuclear processes associated with the initial equilibration. After closing the window (at time 105), equilibrium is restored.

by which it has been imposed, is not a consequence of the *outgoing* feature of the radiation, or at least the foregoing demonstration is not adequate to establish that relation.

It is instructive to return to the above argument and ask, where did the arrow actually come from? Why do we naturally accept that the behavior should resemble that in Fig. 4.1.1? Why should the return to equilibrium *follow* the opening of the window? The answer is that the arrow of time in Gold's argument is the arrow of time of the *narrator*. In telling the story it is natural for us to assume that the response of the system follows, in our time sense, the opening of the window. This is the same assumption that is involved in our natural use of initial conditions in the posing of macroscopic, dissipative problems. Imagine an observer with an opposite running time sense and tell that observer that 50 years before and after the opening of the window the system is in equilibrium. The observer also knows that there is a period of one hour in the middle, during which the window is open. When do you think this observer will say the system is out of equilibrium? Before or after the opening? In whose time sense? And now even if you would claim that there could not be two opposite running observers (about which we speculate in Exercise 5.2.5), then your selection of which is the OK observer is the selection of the arrow, and again the star experiment is not what fixed the arrow.

This is an example of how in discussing the arrow of time ordinary language presents us with traps. In the foregoing argument the 'natural' use of *initial* conditions in formulating a problem is what gave the particular direction to the events described. As pointed out in Chapter 2, this use of initial conditions[2] is another way of formulating the thermodynamic arrow of time. So in the end the argument is circular.

What is the alternative? If initial conditions prejudice the answer, don't use them. If the claim is that the thermodynamic arrow is a consequence of the expansion of the universe, then one way to handle the problem would be to give partial boundary conditions at two different times. At one time the universe is in a contracted state, at the other, it is much larger. Then you would try to give the same kind of information at those two times, so that it would only be the expansion that could give rise to the arrow. Even better, since it might not be clear what 'same kind of information' means, you might consider an oscillating universe and then give the same kind of information for times a bit after the big bang and a bit (the same bit) before the big crunch. For example, you might propose local equilibrium 5 minutes into history and 5 minutes before the end of history. I am not concerned with whether local equilibrium is reasonable, only that the two conditions be qualitatively similar.

I would also like to clear the air with regard to what level of physical precision or expertise is needed for this quest. If it turns out that conclusions can be drawn when the boundary conditions are 5 minutes away from the endpoints (as suggested above), then it won't be necessary to know the equation of state of quark matter. Should we feel though that reasonable boundary conditions could only be guessed at times closer to the ends, then indeed more details of the physics would be necessary. As you will see from the arguments offered below, I don't think that happens. Finally, how much must we know of general relativity? Quantized general relativity? Something of the former, nothing of the latter. We need to take into account the change in the geometry; that's the expansion. I will also argue that for what we explain the formation of black holes (and maybe white holes) need not be taken into account. (The importance of entropy in the physics of black holes does suggest that for more ambitious questions it would be necessary to include these objects.)

---

[2] What is meant here by 'initial conditions' is *macroscopic* initial conditions, and the message of Section 2.6 is that the usual time sequence of macroscopic cause and effect is precisely the (thermodynamic) arrow of time. In Section 2.6, the thermodynamic arrow was characterized through differing rules for prediction and retrodiction. To see that this is the same as the time sequence of macroscopic cause and effect, again consider the ice-cube fable of Section 2.6. Imagine that at 2:30 p.m. a window shade is momentarily opened and a laser beam shined on the ice cube. By the rule for prediction, enhanced melting will only take place after 2:30. Now imagine this perturbation occurring at 1:30, but as before you are working with 2 p.m. information. If you use the *retrodiction* rule, no problem, the usual cause/effect sequence obtains. But if you use *all* 2 p.m. microstates (consistent with the 2 p.m. macrostate) and evolve them backward, you would find that the enhanced melting occurs only *before* 1:30.

In the next section I will formulate the problem of correlating arrows of time (thermo and cosmo) and discuss in general terms and with examples what I believe to be the solution.

4.2 The cosmological arrow aligns the thermodynamic arrow

The point of the star-in-a-box fable is that an outside influence—in particular the expansion of the universe—fixes our thermodynamic arrow of time. In this section I give an argument for that thesis that is free of the criticisms earlier noted. It also emphasizes two other aspects of the (cosmo ⇒ thermo) justification: (1,) the role of equilibration rates and metastability, and (2,) properties of matter. It will become clear that expansion alone is not the whole story; the expansion must be rapid compared to relaxation rates. Furthermore, it is a fact about matter that systems in condensed phases relax more slowly than those in diffuse phases. One should not be surprised that basic statistical mechanics notions, metastability and relaxation, enter. After all one is trying to explain the laws of statistical thermodynamics.[3]

First let's get the general relativity out of the way. A currently expanding FRW cosmology well describes the gross features of our universe. But on a finer scale it is *not* true that under gravitational forces alone everything moves away from everything else. For example, the earth-sun distance is relatively unchanged over billions of years, despite significant cosmological expansion. There are two useful coordinate system to describe this. One is our usual system in which the earth-sun distance is so many meters today and about the same tomorrow. In this coordinate system the quasar QC273 is more meters away tomorrow. The other coordinates are called *comoving*, and for these a time axis is the path of a freely moving particle at rest in the background metric of the universe. So if both QC273 and planet earth have zero velocity relative to the frame of the universe as a whole, then our separation (in this coordinate system) would be constant. For this system our distance to the sun is decreasing. The dynamical equations of motion in the comoving system are slightly different from the usual Newtonian laws. There is a small effective attraction between earth and sun—besides the usual—that makes the separation grow 'smaller' over the eons.

---

[3] As for the other theme of this book, quantum measurement theory, insufficient attention has been paid to the statistical mechanics side of what are essentially statistical mechanics problems. In the notes below are comments on the historical tribulations of the field.

This says that for purposes of studying thermodynamics the world is a three-dimensional box without walls, a box that either is treated as expanding, in the sense that non-interacting particles tend to move apart, or as not expanding, but then bound systems, such as meter sticks, grow smaller with time (in terms of their numerical coordinate values in this system). Furthermore, since the curvature of the universe turns out to be irrelevant for our thermodynamic arguments, when we actually discuss models I will take the 'box' to be a cube with periodic boundary conditions. Topologically this is a three-dimensional torus and the (spatial) period is the linear dimension of the universe.

Remark: For much of our discussion we will use roughly time-symmetric boundary conditions, as you might expect for an oscillating universe. But our arguments do not require an oscillatory universe or a big crunch. Rather that is a worst case scenario. *Even with cosmological time symmetry* you still get arrows in the appropriate direction. For indefinitely expanding cosmologies one does not have symmetry as a guide for selecting boundary conditions. However, whatever reasonable boundary conditions one took, the argument should *at worst* look like that which we give below for two time-symmetric conditions with incomplete equilibration of some modes.

The arena and the two-time boundary problem

Imagine then a gas of classical particles interacting by means of short range forces within a box. We take the box dimensions to be time dependent (expanding and then contracting). Since we wish to provide reasonable boundary conditions at two times—the beginning and end of our process—it will be easiest to imagine an oscillating universe. In that context, we will give similar macroscopic information, for example equilibrium, at both ends.

Making this last statement—macroscopic boundary conditions at both ends—precise brings us to the formulation of two-time boundary conditions.[4] In classical mechanics the way to do this is to use the coarse graining of phase space. This means you look for points in phase space that lie in the appropriate coarse grains at both times. To be clear let me introduce notation. Let Γ represent classical phase space for an N-particle system. As discussed earlier, a macroscopic characterization of a system means that its phase point lies in a subset Ω of Γ, where Ω is a union of *coarse grains*. Let the time evolution operator for the dynamics be a mapping $\phi: \Gamma \to \Gamma$. It is convenient to put a subscript t on ϕ to indicate the elapsed time for the map.[5] Thus if $\gamma \in \Gamma$ is the microstate at time 0, $\phi_t(\gamma)$ is the time t microstate. The same notation applies to sets. For

---

[4] In this chapter we only discuss classical and stochastic boundary value problems. Quantum mechanics will be treated later.

[5] For time-dependent dynamics one should indicate both starting and ending times.

example, a point made in Chapter 2 was that although Ω is made of coarse grains, $\phi_t(\Omega)$ may not be, and in fact this is the essence of entropy increase and loss of information.

Two-time macroscopic information means that a system has its microstate in a (coarse grain) Ω_0 at time zero, and in a (coarse grain) Ω_T at time T. To satisfy both conditions, it must be a point that is mapped from Ω_0 to Ω_T by the mapping ϕ_T. The time-0 characterization of the solution of our two-time boundary value problem is therefore $\Omega_0 \cap \phi_{-T}(\Omega_T)$, the intersection of the sets. See Fig. 4.2.1.

The basic correlation argument

We have N interacting particles in a 'box without walls.' There is a time interval, $[0, T]$ within which the box first expands and then contracts, more or less symmetrically, attaining maximum size at $T/2$. Later we will suppose that this interval begins a short while—perhaps 5 seconds or 5 minutes—after the big bang, and ends about the same time before the big crunch. To say that the system is in equilibrium at both ends

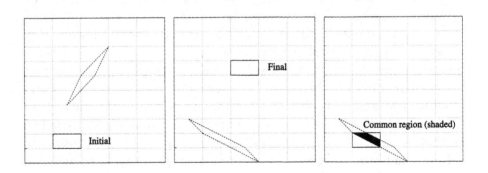

Fig. 4.2.1. Two-time boundary value problem for the cat map, with macroscopic data. The coarse grains are the 0.2×0.1 rectangles indicated by the dotted lines. On the left, the solid rectangle (a coarse grain) is shown with its image under ϕ, here taken to be a single step of the cat map. This rectangle is the initial macrostate. In the middle is shown another rectangle, the final state. The dashed parallelogram is its *preimage* under ϕ. The time-0 characterization of the solution to the two-time boundary value problem, $\Omega_0 \cap \phi_{-T}(\Omega_T)$ with $T = 1$, is shown in the right-hand figure, and is the shaded intersection of the rectangle and the parallelogram.

means that we have a coarse graining of the box and all the grains are occupied approximately in proportion to their (phase space) volume.[6]

For illustrative purposes I make assumptions about the equation of state of the particles. Suppose that for the original volume and total internal energy the system is mostly liquid. When the box expands much of the liquid evaporates, and the remaining liquid cools. Depending on details one might get solidification, but let's suppose the process does not go that far.

In general, when the box expands it will contain more coarse grains, more volume in its phase space. I first want to consider the case in which the box expands slowly and the system remains in equilibrium throughout. With slow expansion, although the system moves to new regions of phase space it will access the new grains in such a way that no occupied grain is emptier or fuller than another. This is a quasi-static expansion. If the contraction happens in the same (slow) way, then *this system never has an arrow of time.* If we didn't have a fancy 'box without walls,' but a conventional cylinder with a piston in it,[7] one would find that the evaporating and expanding gas performed work on the outside world ($\int pdV$); in the contraction phase we would say that the world performed work on the system to restore it to its initial state.

Next let us suppose instead that the expansion is more rapid than the system can follow, as illustrated in Fig. 4.2.2. The evaporation is slow. There will be symmetry breaking—the liquid cannot spread uniformly throughout the box. This can lead to a further failure to equilibrate in that the evaporated gas may not flow away from the liquid quickly enough to acquire uniform density in the box. Therefore, at any stage of the rapid expansion the entropy will be below the maximum consistent with the box volume at that stage. The system will be out of equilibrium. It will have an arrow of time. (I will elaborate on this in a moment.) To continue this scenario into the contracting phase we must consider two possibilities: although the initial expansion is rapid, as we approach the halfway point that expansion slows down enough to allow equilibrium to be established before the reversal; or it does not. The case where the expansion rate and the system-relaxation rate would not—under ordinary circumstances—give equilibrium at $t = T/2$ requires careful analysis. I don't want to slight its subtlety, so it will be treated separately, below.

---

[6] This is not an important assumption. In later examples (e.g., Fig. 4.3.2) we drop it.
[7] In this discussion I blithely and conventionally ignore the future conditioning about which so much fuss was made above. Why the usual initial conditions arguments can be made here will be briefly justified below, and in the next section a fuller explanation given.

For now we assume equilibrium at the midpoint and get on with showing how this system has an arrow of time.

For the case in which there is equilibrium in the middle, the two-time boundary condition (equilibrium at $t = 0$ and $t = T$) is easy to satisfy (eliminates relatively few[8] microstates), and the above scenario, which was based on our intuitive initial conditions predisposition, goes through unchanged.

We next elaborate on the statement that there is an 'arrow of time' during the rapid expansion. Consider a time during the interval $[0, T/2]$ at which the entropy is well below its maximum and let's assume that not only is the evaporation slow, but that the expansion is fast enough so that the gas cannot fill space uniformly. Then there are regions of high gas density and regions of low density, with gas flowing from one to the other. If we put a non-expanding container around a volume in

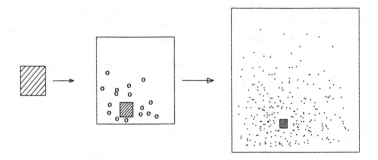

Fig. 4.2.2. Schematic diagram of an expanding box initially containing substance entirely in the liquid phase. The box on the left represents the initial condition and the stripes indicate the liquid phase. (In the figure we do not use periodic boundary conditions for the box.) The next two pictures represent later stages. The expansion is rapid enough so that the evaporation cannot keep up, and the system is far from equilibrium. In particular there is a breaking of translational symmetry and the liquid occupies a particular subregion of the box. In the illustration the expansion is so rapid that there is a density gradient in the evaporated gas. Although steadily relaxing, the system remains out of equilibrium.

---

[8] The fraction eliminated is the product of the fractions separately associated with the macrostates at the two times. Without midpoint equilibration there is greater reduction. As we show below, in the case of midpoint equilibration those states eliminated by the future boundary condition have no particular early-time features; correlations are lost. As such, their selection is pseudo-random—not random, but not distinguishable from random.

which there were both high and low density regions, the densities would have the opportunity to equalize and the entropy of this isolated system would increase until it reached local equilibrium.[9] Note too, that for our purposes non-expanding boxes are reasonable objects, just as the earth-sun distance or meter sticks are little affected by the overall expansion of the universe. This shows that the inability of the local dynamics to follow the expansion has imposed an arrow of time in distant portions of the box. The continuing failure of the liquid to reach equilibrium maintains a flow of gas away from its (symmetry broken) location. This gas is itself unable to keep up with the expansion and as it spreads into new regions carries the message that back where it came from there was a big system well below *its* maximum entropy.

Remark: In the previous paragraph, a thermodynamic arrow was shown to arise from the density gradient and its relaxation. This is conceptually simple, but not realistic, because of the relatively small range of distances within which this would work. With a richer system many other ways arise for the arrow to be manifested. For example, the escaping particles can be molecules that are stable in the liquid, but unstable in the low density gas. After the molecules evaporate, they can be captured and held at low density in a container. Their subsequent breakup into atoms provides another manifestation of the arrow. Yet more accuracy is possible when there are complex structures (e.g., leaves with chlorophyll) for exploiting the energy of the escaping particles.

Before getting to important open points (what happens after $T/2$ and what if there isn't equilibrium at $T/2$), I want to identify the physical systems that play the role of the liquid and gas in the foregoing example. In practice, stars do the job. Other systems are involved, but it is enough to see that stars have the qualities needed. In the early universe one had neither stars nor galaxies. However, expansion was so fast that matter did not stay in equilibrium (or never was). Much matter became gravitationally bound in states far from equilibrium. Thus a ball of hydrogen and a lot of empty space has far lower entropy than a ball of nuclear-stable iron and the decay products (neutrinos, etc., from the processes leading from H to Fe) spread throughout space. A burning star is a long lived metastable state,[10] like our slowly evaporating liquid. It sends out messengers, carrying its disequilibrium to distant places. For the earth, the most important such message is the stream of high temperature photons that we get from our sun (and ultimately discharge at lower temperature). Elsewhere I've discussed how a temperature gradient can produce structure. How to make precise the steps from photon flux to

---

[9] There is a two-time boundary condition way of saying this. See the next section.

[10] Stars don't live forever either. For a continuing supply of stars, I turn to galaxies, which, because *they* constitute a local density variation, give rise to star formation. If galaxies do not last long enough and a stage reached for which no low entropy system is available, then indeed this system will no longer have an arrow.

the complexity of life is an exciting contemporary goal. But there is little doubt that having accounted for a continuing source of negentropy, one that reaches us locally, the physicist's job, *qua* physicist, of accounting for the thermodynamic arrow, is done.

A question that might be raised is the entropy of the gravitational fields associated with the expansion/contraction. I won't worry about defining this, because it is irrelevant to our argument. In particular we take it as *given* that the box is expanding or contracting. In that sense, we can think of the system of interest (the contents of the box) as open, acted upon by outside forces. However, for us the important issue is *not* whether total entropy (including those outside forces) increased, or even whether the entropy of our system increased (as in the quasi-static expansion). The issue is whether the contents of the box can stay in equilibrium during the expansion, whether its entropy equals or is less than the maximum allowed for the given state of expansion of the box.

With this in mind I also relegate black and white holes to a minor role insofar as our own thermodynamic arrow of time is concerned. Black holes, notwithstanding the beautiful general ideas that compel us to associate an entropy with them, function as another kind of sink. As to white holes, there is no evidence for their existence. Bear in mind though that I am making no statement about possible exotic effects in regions where gravitational forces are dominant, such as collapsed objects. Also the present discussion concerns thermodynamics—the effect of singularities on possible cosmological evolution, such as the speculations mentioned in Chapter 3, are not considered here.

Remark: Under gravity distant masses become more distant and nearby masses closer (subject to other forces, such as electromagnetism). The state toward which our metastable stars and galaxies tend (with an initial-conditions-only specification) is thus not a uniform distribution of iron, but isolated masses that are deeply bound. These can be black holes or dense alternatives in which interparticle forces are not overwhelmed by gravity. This means that an early state of the universe in which matter is uniformly spread may not have been the highest entropy state to which gravitational forces could take it. With this possibility one can again pose the two-time boundary value problem, and the arrow will emerge by virtue of the endpoints being far from equilibrium, both because of material homogeneity and because they will be further driven from equilibrium by the process of expansion. From the physical standpoint, the expansion of the cosmos and the attraction of nearby objects are governed by the same law and combine to provide both the disequilibrium and the ever increasing arena in which the consequences of that disequilibrium are felt. These considerations have two implications. First, for the thermodynamic arrow in our universe it is true not only that we are expanding but that the boundary states may also be far from equilibrium with respect to their homogeneity. You still have (cosmo ⇒ thermo), but 'cosmo' now includes not only the fact of expansion but the fact of early and late time disequilibrium. Second, if one could construct cosmological models with coalescence of matter but without expansion (presumably much different from our universe), one might get an arrow without expansion.

Remark: A thermodynamic arrow that arises from *both* endpoint disequilibrium *and* expansion/contraction does not perturb the main thrust of this book (although it would make Gold's identification of the physical mechanism incomplete). Our point is the introduction of our extended 'Cosmological Principal:' not only is our spatial position not special, our temporal direction isn't either. You would still have the observational predictions of Section 4.6; you would still have the possibilities envisaged in Chapter 8. In fact, our illustration, the symmetric cat movie of Fig. 4.3.2, *only* involves endpoint disequilibrium—there is no contraction or expansion.

Yet another enormous entropy repository is the cosmic background microwave radiation. If we make the usual assumptions about numbers of particles in the universe (massive and massless, bosons and fermions), and if entropy is roughly proportional to particle number, these photons represent a substantial fraction of the total entropy. Is this important for our arguments? It is not. Recall that these photons are left over from the era when the electrons combined with positively charged particles and the universe became transparent. In other words there was a significant decoupling of matter and radiation. It is exactly the fact that for most purposes this radiation *is* decoupled that makes it unimportant for our argument. It is as if in our liquid-in-a-box argument there were an additional substance present, a light and barely interactive atom that was initially dissolved in the liquid. By equipartition, it will have high velocity. For our arguments we needed the box to expand much faster than the molecules constituting the liquid could keep up with. This rapid expansion can still be slow compared to the light-atom velocity, so that as the box expands the collection of light atoms is in equilibrium.[11] Since this atom interacts weakly, our previous arguments are unaffected.[12] Returning to low energy photons in the universe, the passage of a star from being mostly hydrogen to being other things takes place irrespective of the background radiation, and our use of the star's emitted photons is also indifferent to the presence of all those low energy photons.[13] The

---

[11] The gas atoms as a separate entity are in equilibrium. They are out of equilibrium with the slower molecules that constituted the liquid.

[12] With two-time boundary conditions and chaotic dynamics, even a weak interaction will affect the exact paths taken. However, macroscopic dynamics will not be affected. The justification is the same as that which lies behind our ability to use ordinary thermodynamic arguments in the presence of a future boundary condition. That is, there *is* a selection of states, but because the future condition is far away, it is quasi-random with respect to macroscopic variables at the present. Averaging over the set of acceptable microstates is the same as averaging over a random set, despite their precise determination. Of course one also may have here phenomena like the 'butterfly effect' that dramatizes the sensitivity of nonlinear systems. By our arguments this perturbation is below the level of a single coarse grain and as such a macroscopic dynamical theory for it would call this noise or a fluctuation and could describe it by stochastic differential equations.

[13] Since these photons keep up with the expansion of the universe, they are expanding quasi-statically. As remarked in the light atom in the liquid-in-a-box example, and

situation is similar to the negative temperature experiments mentioned in Chapter 2, in which the lattice and spin degrees of freedom are substantially decoupled.

Let us now deal with the other side of the midpoint, the interval $[T/2, T]$. I claim the same thing happens as in the interval $[0, T/2]$, but reversed. By the symmetry of our boundary conditions this is obviously true. I make this assertion tongue in cheek, since this is the kind of statement that is 'obvious' only after you are convinced of it. For this reason, I want to go into greater detail, with the help of examples, on the consequences of time-symmetric or two-time boundary conditions. In the next sections I present examples of dynamical systems with boundary conditions defined at more than one time, and following that return to complete the arguments. In particular I will discuss the case where the system does *not* reach equilibrium by time '$T/2$,' along with the possibility of experiments or observations allowed by this way of looking at things.

Remark: (Further to a note earlier in this section.) There is another asymmetry that has been used here and it highlights why photons played a role in the star-in-a-box argument. It is an asymmetry in the properties of matter, namely relaxation is slower from more condensed phases. In the evaporating liquid example we assumed that even with a moderate expansion rate the liquid would fail to stay in equilibrium once the expansion had begun. On the other hand, we could easily contemplate that in the latter stages of expansion the free gas particles would reach equilibrium. Even in the next sections, when the expansion is so fast as to preclude midpoint equilibration, it will be assumed that in the condensed phase relaxation is slower. It is a property of matter that it can get stuck in condensed phases for a long time. Escape from the condensed phase is generally accompanied by the emission of less interactive objects whose relaxation mainly depends on their flying apart. For the molecules, the low mutual interaction sets in once they are sufficiently distant from one another, so that we are using the fact that material interactions tend to be local in space.

Remark: The foregoing demonstration did not require an oscillating universe, it was just that in such a context it was easier to guess unprejudiced boundary conditions. Also, since the boundary conditions can be symmetric there is a clear sense of having deduced something. In an expanding universe the same exercise can be contemplated. For example, one could assume equilibrium at both ends, and then show a preference for a certain arrow. The arrow of time would be present during the intermediate rapid expansion. It would be asymmetric (going in the direction of the expansion) because it is easier to find microstates that approach equilibrium exceptionally rapidly in their diffuse state than those that do so from their condensed state. With the periodic cosmology, when the midpoint time is insufficient for equilibration, the rare states are not those that

---

as we will see below, having one component equilibrated is no bar to normal looking relaxation in other components of a system. Of course the *precise* dynamical motions that arise from satisfying a two-time boundary condition depend on *all* components, but almost all such evolutions look essentially the same. Note too that the photons of the background radiation do not seem to be fully relaxed insofar as their positions are concerned, i.e., they appear to preserve elements of primordial spatial structure.

prematurely find equilibrium, but those that don't reach equilibrium, but turn around and go back. This is illustrated in the next section.

Remark: Our proof takes symmetric boundary conditions and uses expansion to derive a thermodynamic arrow. With asymmetric boundary conditions the arrow need not follow expansion and could even go in the opposite direction. For example, early computations of increasing entropy on successive oscillations of the universe arose from using initial conditions only. Even with two-time boundary conditions, boundary asymmetry could overwhelm the effect of expansion. Such asymmetry would be the source of this system's arrow. The justification for that asymmetry would then represent the origin of the thermodynamic arrow for this system. Our proof of (cosmo ⇒ thermo) is therefore valid in the absence of a separate reason for a boundary condition sufficiently asymmetric to overwhelm the effects of expansion. (However, the 'natural' use of initial conditions would *not* constitute such a justification, since that's what you're trying to prove.)

Remark: Consider the liquid/gas in a box (Fig. 4.2.2). With future boundary conditions in which the box is so small that only the condensed phase can appear it may seem that putting time symmetric boundary conditions would change nothing at earlier times. That is not (in general) so. If expansion is rapid enough, there is a symmetry breaking involved in selecting *which* microstates actually do appear, as the rapid expansion forces the system away from equilibrium. Symmetrically there is a state selection near the end. Effectively this provides a nontrivial pair of boundary conditions. (Of course if expansion is slow there may be no departure from equilibrium and indeed no arrow.) Another way to see this is to consider what would happen in a more familiar process—a box *with* walls, a piston for example. With *initial* conditions, the system will be compressed at the end, but in general compressed at a higher temperature. One might also vary the original problem by having the condensed phase be one of several long-lived metastable phases.

Remark: Don't forget the 'missing mass' problem. As discussed in Chapter 3, it is believed that the matter we can now account for only constitutes 10% of that in the universe. Clearly there is significant potential for the trashing of the foregoing arguments. However, the fact that this matter (whatever its form—Jupiters, WIMPs, ...) has remained stubbornly unnoticed should allow our arguments to survive for the same reason that the weak coupling to the photon background did not affect our conclusion.

4.3 Classical two-time boundary value problems

In this section we use the cat map to study variations on the theme of two-time boundary conditions. Our goal is to explain why many arguments of the previous section could be phrased in initial value terms and also to prepare for the more subtle case in which there are slow modes for the two-time boundary condition. First we reproduce the standard, initial conditions only, relaxation properties.

In Section 2.2 we defined the cat map to be the following transformation of the unit square into itself:

$$
\begin{aligned}
x' &= x + y && \mod 1 \\
y' &= x + 2y && \mod 1
\end{aligned}
\tag{4.3.1}
$$

Alternatively

$$\xi' = \varphi(\xi) = M\xi \quad \text{mod } 1, \qquad \text{with} \qquad \xi = \begin{pmatrix} x \\ y \end{pmatrix} \quad \text{and} \quad M = \begin{pmatrix} 1 & 1 \\ 1 & 2 \end{pmatrix}$$

$$(4.3.2)$$

A collection of points started in a small region of the unit square rapidly spreads, and, if thought of as a gas, soon reaches equilibrium.[14] The rate for this process is the Lyapunov exponent, which for this dynamical system is the logarithm of the larger eigenvalue of the matrix M, namely $\log\left[(3 + \sqrt{5})/2\right] \approx 0.962$. The equilibration can also be followed by calculating an entropy for the system, thinking of the unit square as classical phase space. The square is divided into coarse grains, i.e., disjoint subsets $\{\Delta\}$ whose union is the entire square. (We assume the areas of the Δs to be equal.) If the total number of points is N and if n_ℓ of them are in coarse grain ℓ, then the entropy S is

$$\frac{1}{k}S = -N \sum \rho_\ell \log \rho_\ell \qquad \text{with} \qquad \rho_\ell = n_\ell/N \qquad (4.3.3)$$

and k a constant. A convenient coarse graining is a set of rectangles obtained by partitioning the x-axis into N_x regions and the y-axis into N_y regions. Since the coarse grains are interpreted as our macroscopic characterization of a state, it is reasonable to take as an initial condition a collection of points entirely in one of these rectangles. The evolution of such a collection of points was shown in Fig. 2.2.3. As the points evolve they spread throughout the unit square. In Fig. 4.3.1 we show what happens to the entropy when all points are started randomly within a single grain. We present three situations: 100 points with a 5×5 coarse graining, 500 points with the same grains and 500 points with a 10×10 coarse graining.

Exercise 4.3.1. Two finite size effects are evident in the figure. Exercise (2.2.2) explored the difference between log(number of grains) and the fluctuating maximum of S/k as a function of the number of points. Now find the relation between grain size and total equilibration time.

Remark: 'Phase space' is used above to mean the individual point phase space, the unit square. But if one thinks of these points as atoms, the full *system* phase space is the tensor product of many unit squares.

---

[14] For the analogy with the previous section these points can be thought of as the evaporating molecules with all the evaporation taking place in a single time step. The fact that the liquid itself does not spread with the expansion (and breaks translational symmetry) motivates starting all the points in a single cell. Future, symmetric boundary conditions demand that the points end in a single cell as well.

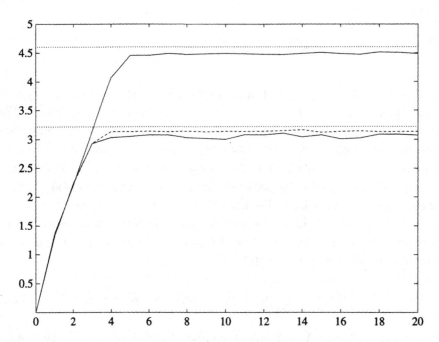

Fig. 4.3.1. Time dependence of the entropy, varying the number of points and the grain size. In all cases, the initial state has its points randomly distributed in a single grain. The lower solid curve is for 100 points and 5×5 grains. The dashed lower curve uses the same coarse grains (5×5) but 500 points. Note the closer approach to the dotted line immediately above them. This is $\log 25$, the theoretical maximum for 25 grains. For a 10×10 coarse graining (upper, solid curve), equilibration takes longer and the maximum is larger.

A two-time cat movie

We next produce solutions to the two-time boundary value problem for the cat map. This will be displayed in the form of a movie, a display that may at the same time be obvious and disconcerting.

The scheme for handling two-time boundary conditions was presented in the previous section. If you want the macroscopic state Δ_{17} at time 0 and the macroscopic state Δ_3 at time T, than the set of points satisfying this is precisely $\Delta_{17} \cap \varphi_{-T}(\Delta_3)$, where φ_1 is the cat map and φ_t is its tth power. For the purpose of producing Fig. 4.3.2 (the 'movie'), I used the following procedure: a large number, say $\mathcal{N}$, points within Δ_{17} were randomly generated. These were evolved forward T time steps and tested for whether or not they were in Δ_3. About $\mathcal{N}/(N_x N_y)$ passed the test. This provides a random subset of $\Delta_{17} \cap \varphi_{-T}(\Delta_3)$.

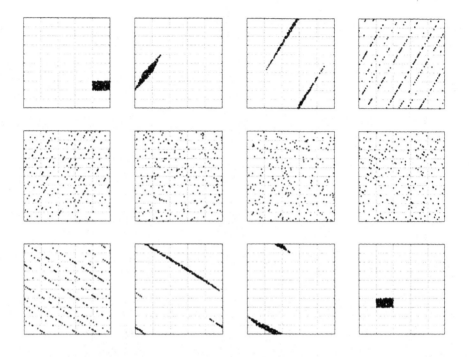

Fig. 4.3.2. History of 250 points that begin in a single coarse grain at time 0 and end in a single coarse grain at time 19. Selected times are shown according to the following scheme: top row (left to right), [0 1 2 4]; second row, [5 8 11 14]; third row, [15 17 18 19].

Aside: This 'random' subset will be homogeneous and symmetric for a random number generator that is homogeneous and without correlations. The large Lyapunov exponent (≈ 0.962) means that accuracy is rapidly lost. For the computational software used (MATLAB), $M^{23}M^{-23}$ gives unity to all digits, while the 24th power gives errors.

Fig. 4.3.2 exhibits the evolution of a set of points defined in this way.

There are two messages from this display, both of them 'obvious.' First, when Fig. 4.3.2 is compared with Fig. 2.2.3, it is seen that in both figures the initial stretching and dispersal of the box is essentially the same. The future boundary condition has no perceptible influence. Second, the coalescence of the points into a coarse grain at the end of the time period is essentially the same process. The only difference is that for the inverse transformation the eigenvectors are interchanged and the unstretching direction is orthogonal to the stretching direction. The symmetry can be seen in another way. Fig. 4.3.3 shows the entropy as a function of time for the process. Again the features to notice are (1,) the similarity of the initial equilibration to unconditioned equilibration and (2,) the similar shapes of the reversed end of the curve and the beginning of the curve.

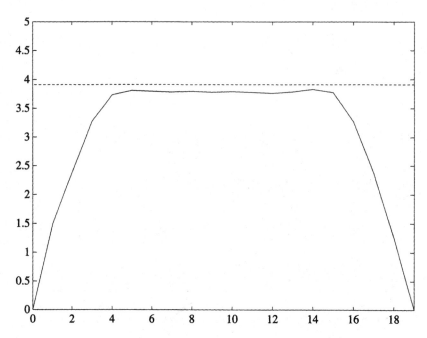

Fig. 4.3.3. Time dependence of the entropy with two-time boundary conditions.

The fact that the initial relaxation looks normal despite the future conditioning is what justifies the ordinary (initial value) arguments that we used in the last section when we demonstrated that there is no arrow with slow expansion and there is an arrow with rapid expansion. Furthermore, the symmetry of the termination process, despite, by the way, a slightly different dynamics, is what justifies our arguments for the interval '[T/2, T]' in that same section. It is easy to understand qualitatively what is going on. The initial behavior looks normal because the future is 'forgotten.' It is many relaxation times away from the present state and the conditioning it imposes at early times, albeit precise, is still effectively random with regard to the early behavior. I call this *crypto-conditioning* because its effects are hidden—until the time evolution makes them manifest. 'Hidden' also implies unmeasurable, which is to say the system behaves in the ordinary way during the early times. In the Appendix to this section we give an analytic approach to the two time cat map problem. That formalism provides the sharp version of these remarks. The slight difference between the forward and backward cat map makes another point. Exact time symmetry plays no role in the (cosmo ⇒ thermo) argument, so that the issue of lack of time-reversal invariance in the weak interactions (Section 3.4) is not relevant to our main theme. All that's needed is roughly similar equilibration in both directions.

Exercise 4.3.2. By extension of the cat map arguments to general classical systems, show that specific heats (for example) are the same, with and without future conditioning. Note that in the usual specific heat calculation one uses the partition function, a sum over *all* states. Now only a limited subset is used.

Exercise 4.3.3. In Section 4.2 we imagined a non-expanding container inserted in the system, big enough to include a density gradient before the container was closed. We then argued that during the period that the box is closed the density gradient would level off and the entropy of this isolated system increase. Show how the argument can go through with two-time boundary conditions by taking the total interval T long compared to the relaxation time and making a model of an evaporating liquid and a container, within the cat map world. Using two-time boundary conditions show that if the isolating of the container takes place sufficiently near to one or the other end of the time interval the usual argument goes through, with the appropriate arrow of time.

A local arrow in the absence of full equilibration

Our next order of business is to see what happens when the time period T is *not* sufficient for the system to come to equilibrium by time $T/2$. In Fig. 4.3.4 we show the entropy dependence with 10×10 coarse graining for 250 points with various (or no) future coalescence times (as can be read from the graph). As usual the result is obvious once you've seen it. The effect of the future conditioning is not felt when that future is distant relative to the relaxation time. When that conditioning is too soon, the system cannot equilibrate normally and never attains the equilibrium entropy.

Notice that for the minimum return time shown (7) there *are* solutions to the two-time boundary value problem, it's just that the future conditioning means that they appear not to equilibrate normally. Their microscopic dynamics (the cat map) is obeyed perfectly, but the collective behavior appears to be strange (*if* you suffer from initial conditions prejudice).

For small T, the future conditioning prevents the system from accessing equally all regions in phase space and ordinary equilibration cannot be expected. Therefore it would be incorrect blithely to use ordinary, initial-conditioning based, arguments to predict the system's behavior. This observation takes us in two directions. First we will later (in Section 4.6) take up the question of whether there are physically observable effects (in the laboratory or in the cosmos) related to this. Second, we next see what it takes to recover our (cosmo $\Rightarrow$ thermo) reasoning.

Although the systems in Fig. 4.3.4 do not all reach equilibrium, they do show intuitively understandable behavior. Recall the rationale for the

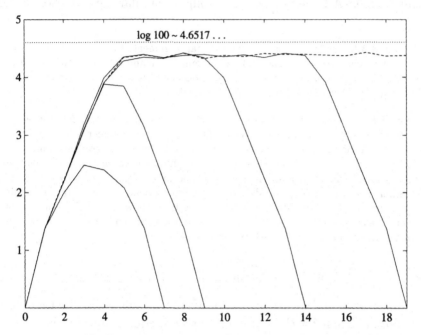

Fig. 4.3.4. Time dependence of entropy with one- and two-time boundary conditions. In all cases, 250 points start in a particular 0.1×0.1 region of the unit square. In one case (the dashed line), no further conditions were imposed. The four solid curves show the entropy for points that were required to return to the original region at times that can be read off the curves (7, 9, 14, & 19).

second law of thermodynamics: if you start a system in a rare macroscopic state[15] and if it has sufficiently potent dynamics, it is most likely to move to regions with less rare macroscopic states.[16] The same here. The initial state (all in a single grain) is more rare than the constraints alone require, so the points spread, reaching, at the midpoint, the condition that is least unlikely, consistent with returning to a single grain at the later time.

The other observation needed to recover our (cosmo $\Rightarrow$ thermo) argument is that the world is more complex than the cat map. In particular there are dynamical processes with many time scales. Although the overall conditioning time, T, may be short for one process, it may be long for others. Physically, we are usually more interested in the others, and

---

[15] A 'rare' macrostate is a small coarse grain. Alternatively, if (as in Section 2.2, '*Defining entropy*') grains have been subdivided so that all have the same size, a rare macrostate is a grain for which few others have unmeasureably small differences.

[16] As remarked in Chapter 2, this gives the second law the status of a tautology.

for them we can use our previous (cosmo $\Rightarrow$ thermo) reasoning with little change. Let me give an example (which may even be true). Let the big-bang-to-big-crunch time be called $\mathbb{T}$. Suppose $\mathbb{T}$ is short on the scale of galactic dynamics; that is, when the galaxies are thought of as a gas of interacting particles (each atom a galaxy), it cannot both come to equilibrium in time $\mathbb{T}/2$ and also meet a future boundary condition in which it must coalesce in a certain way. So galactic behavior is dynamically correct, but apparently unlikely. This would not prevent subsystems with shorter lifetimes from coming to equilibrium, and for them our reasoning goes through without a hitch. But a stronger statement can be made. Even as the slow system moves towards its less than maximally disordered midpoint, it increases in entropy and this increase can continue to feed other processes with shorter lifetimes. We will give a physical example of this in the next section.

In Section 5.2 we will give a formalism within which the above assertion can be verified in a precise and general way. In the present section we use the cat-map world to study examples with multiple time scales. We do this for a system with two uncoupled degrees of freedom. The more realistic situation, where they are coupled, and in particular where the existence of one depends on the decay of the other (as in the production of unstable isotopes in a star), will be discussed in an exercise.

For the case of unrelated processes a cat map implementation is easy. We define a speedy cat. Instead of the matrix M of Eq. (4.3.2), use

$$M_{\text{speedy}} = \begin{pmatrix} 9 & 7 \\ 14 & 11 \end{pmatrix} \tag{4.3.4}$$

As for the usual cat map, M_{speedy} is of determinant one, measure preserving and a candidate for dynamics. Its larger eigenvalue is $10 + \sqrt{99} \approx 19.95$, which is larger than that of M. Now let each point have *four* coordinates, the usual (x, y) and another pair (u, v) (in the unit square) mapped by M_{speedy} (and the mod 1 operation). For both we use the same coarse grains. If T is a time for which the usual cat coordinate does not equilibrate (with two-time conditioning) but the speedy cat does, then if one looks only at the (u, v) part of the dynamical system things will again look normal. Standard relaxation arguments can be used, notwithstanding the fact that the (x, y) density has not equilibrated.

Exercise 4.3.4. Check this.

Exercise 4.3.5. One can also develop a multiple-time-scale, *coupled*-coordinate cat-map model. Let the points have the usual coordinate (x, y). Simulate the creation of additional degrees of freedom by occasionally allowing two points to break into several pieces of a new character (and allow the inverse process as well), for example, if the two points are within a given distance from one another. Such a coupled system could be used to demonstrate that slow processes, even if not fully equilibrated, can drive faster processes in (nearly) the normal fashion.

Finally, we give a cat-map model of an expanding universe and see how in this context the expansion can outpace the equilibration. We use the analog of comoving coordinates so as to leave the cat-map dynamics intact. With these coordinates meter sticks shrink, since the size of the universe remains constant. This implies that the expansion of the system (universe) leads to a shrinking of the coarse grains (with these coordinates). In Fig. 4.3.5 is shown entropy as a function of time, with initial conditions only, for the usual cat map with grain size decreasing. At each time there is a coarse graining and the dashed line shows the maximum

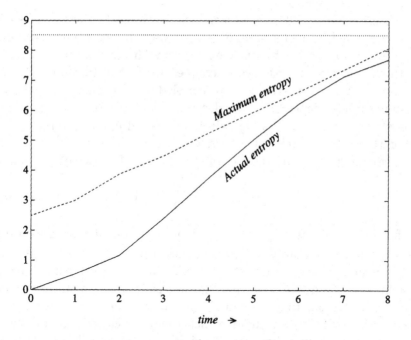

Fig. 4.3.5. Entropic history of 5000 points starting from a region of size 0.05×0.05, with exponentially shrinking grain size. Initially the grains are $(1/4) \times (1/3)$. On successive time steps, they are cut finer, with grain area $\sim \exp(-t/1.482\Lambda)$ where t is the step number and Λ the Lyapunov exponent (≈ 0.962). The dashed curve is the maximum entropy at the indicated time ($= \log(\# \text{ grains})$). The solid curve is the actual entropy of the expanding system. Compare the slope of this curve to that in Fig. 4.3.4. Entropy grows slowly, since the grain shrinkage is comparable to the cat map expansion. At yet later times entropy levels off to $\log 5000$ (≈ 8.5), a finite size effect not relevant to our considerations (at $t = 8$ the number of grains is 3234).

entropy for that coarse graining, namely $\log(N_x N_y)$. The solid line plots the entropy at each time, calculated with the coarse graining *for that time.* By using an exponential expansion rate with exponent greater than the cat map Lyapunov exponent the (inverse) grain size stays ahead of the cat map expansion and the entropy remains lower than the maximum allowable for the given stage of expansion.

Aside: How NOT to get an arrow. Many steps in our argument as well as in the usual justifications of the second law of thermodynamics require what I referred to above as a 'sufficiently potent dynamics.' Sometimes though the dynamics are not effective at moving the system into all the corners of phase space. A case in point is the harmonic oscillator, which would be otherwise tempting to use in two-time boundary value problems because of its easy solvability. The oscillator serves not merely as an example of a poorly equilibrating system, but because integrable systems can be recast as oscillators using action angle variables, it is the archetype of such systems.

Appendix for this section (at the end of the chapter):
4.3.A   Doubly constrained cat map: analytic treatment

4.4 Arrows of time with incomplete equilibration

The purpose of the previous sections was to educate the intuition for two-time boundary value problems and to treat the case in which the time interval between the boundaries is long compared to all relaxation times of the system. We also began to consider the effect of two-time boundary conditions when there is insufficient elapsed time between the boundaries for equilibration of all intermediate processes. As noted in Section 4.2, in this case the imposition of two-time boundary conditions does *not* reduce to the imposition of initial conditions only. However, we did see in Section 4.3 that although the slowest processes were affected by the boundary conditions, for faster ones one could once again use initial-value or final-value formulations only. In this section we use a richer physical example than the cat map to justify the thermodynamic arrow in the case where some relaxation times are long compared to the boundary value time.

In fact, there *are* many time scales in the universe. For the longest it may be true that the future condition prevents them from relaxing normally. Such an effect is evident in Fig. 4.3.4. If a system must return to an unlikely state by time (say) 7, while its relaxation time is on the order of 5, then it does not relax normally. However, for subsystems with shorter time scales, our early correlation of arrows arguments go through. To see this in a physical example, we embellish the system used in Section 4.2. There we allowed a liquid to evaporate in an expanding box and required that at a much later time, when the box had shrunk, the evaporated gas condense to its original state.

We now suppose that the gas comes off as diatomic molecules, molecules that remain bound when they are part of the liquid, but dissociate in free space. Let the spontaneous dissociation rate be low, but still fast on the scale of the overall evaporation.[17] The two-time boundary value problem now demands that all the molecules be in liquid form (with the same energy per molecule) at times 0 and T. Moreover, as before, the box is small at those times but grows rapidly and subsequently shrinks rapidly. Again from Fig. 4.3.4 it is evident that during expansion molecules will initially escape with great enthusiasm, but after a while will slow down. To an initial conditions dogmatist this slowdown would look like unlikely concatenations of molecular scatterings at the surface of the liquid, but with a grander perspective it is simply what it takes to satisfy the future boundary condition. But now consider the relative number of dissociated and undissociated molecules. Since the molecules come apart quickly (relative to the overall boundary condition time), by detailed balance they recombine quickly. Two comments are in order about this last assertion. First detailed balance is a consequence of time reversal invariance, but as remarked previously, exact time reversal invariance is not the important thing; rather what matters is an approximate equality of equilibration times for the dynamics and its inverse.[18] Second, we know that detailed balance applies to microscopic processes, but it doesn't put Humpty Dumpty together again. In more sophisticated terms: the recombination rate depends not only on the matrix element of the Hamiltonian, but also on phase space factors. You need to have the two atoms coming at each other. That part of the story is taken care of by the shrinking box. The point is that with the box expanding, they dissociate as if there is no future condition, because their recombination (at the microscopic level, once you know the atoms are headed for each other) is also rapid. The dissociation process is thus described in the same terms and with the same time constants as if there were no future boundary condition to meet. Note too that as one approaches the midpoint you may have dissociations taking place whose recombinations only occur well on the other side of the midpoint. If one goes to yet more realistic systems one could have localized regions in which an arrow persists into the reversed region. This will be a relatively rare occurrence since the overall driving disequilibrium (the evaporating liquid) is itself winding down and turning around. Therefore its production of new molecules is much reduced in this era. Neverthe-

---

[17] This does not imply that the molecules do not evaporate easily. The overall rate for equilibration of the liquid also depends on the total amount of liquid.

[18] Even that's not important if all you want is arrows in the two directions, but not necessarily symmetric ones.

less, if one goes back to talking about the universe, this kind of thing seems possible when time scale of the shorter processes—provided they occasionally occur—is on the order of the time remaining to the midpoint.

These arguments can be stated more abstractly and with less appeal to intuition. We do this for stochastic two-time boundary value problems in Section 5.2. See especially Eq. (5.2.12) and the discussion around it. Our argument in the present section depends on having several time scales. The slowest variables—hence the collection of all variables—do not access the entire phase space. Nevertheless, if you treat the rapidly relaxing variables as evolving on their own, with the slower ones treated as constraints, then the rapid ones act like normally equilibrating systems in a time varying field, notwithstanding the future condition they must satisfy.

4.5 Causality

'Causality' is a philosopher's plaything. You can spend 16 pages on definitions before getting to what it was you wanted to ask. Even in physics it takes many guises, from dispersion relations to conditions on commutators of fields.

This book makes statements on the relation between the microscopic and the macroscopic. Here we study how the ideas we introduce, for example cryptic conditioning on the future, can affect the form taken by *macroscopic* causality. We examine the requirement—or perhaps the breakdown—of the notion that macroscopic effect follow macroscopic cause. With conventional statistical mechanics this is a consequence of the version of the second law of thermodynamics given in Section 2.6. The use of *initial* conditions implies this form of causality, *by definition*.

We do *not* take initial conditions as the automatic way to formulate problems. Rather, we advocate unprejudiced boundary conditions on a time varying cosmology and derive the approximate validity of the initial conditions formulation from the time variation of the cosmology. Macroscopic causality now requires proof.

To gain perspective, recall the occurrence of the time derivative of the acceleration in the equations of motion for a radiating, charged particle (cf. Eq. (3.3.13)). To suppress a runaway mode a future boundary condition is imposed. This leads to 'preacceleration' effects on a time scale of 10^{-23} s, which is a breakdown of causality.

At the level of classical and stochastic mechanics, our demonstrations of the previous sections and of Chapter 5 show that causality is indeed recovered. This is implicit in the normal equilibration pattern for the two-

time boundary value problem for the cat map, as illustrated in the figures
of Section 4.3. One can also formulate this by applying a perturbation to
the system at some time and asking whether the response is only before,
only after or both. 'Only after' would represent conventional causality.
Such a perturbation can be modeled by applying a different map at a
particular time step. For example, suppose we compare

*50 cat-map steps, and a condition that all points start and end in the
rectangle* $[0.2, 0.2 + 10^{-10}] \otimes [0.3, 0.3 + 10^{-10}]$

with

*5 cat-map steps, followed by something else (e.g., M_{speedy} of Eq. (4.3.4)),
followed by 44 cat-map steps, with the same boundary condition and total
number of points as before.*

If, prior to step number 6, the two motions are the same at the macroscopic level (for some coarse graining), then causality is recovered.

In fact, prior to the perturbation, the motions will be the same (unless
the coarse graining is too fine) for the same reason that the equilibration
of the cat map looks normal if the future condition is far away. Namely,
satisfying the future condition can be accomplished by very small changes
in initial location and as far as the initial collection of points is concerned
one pseudo-random subset equilibrates in the same way as another.

Aside: You could check the foregoing assertion with a 'movie' of the sort shown in Fig.
4.3.2. I haven't bothered, since I know that $10^{10} \ll 2.6^{50}$.

Exercise 5.2.2 formulates a precise version of this statement for stochastic dynamics. Appendix A to Section 3.3 takes up related questions.

For quantum mechanics we later make the point that distant future
conditioning *can* have effects in the present. For this reason there can be
breakdowns in causality. In Chapter 10 we consider experimental tests
of our quantum theory and one such test involves possible indications of
advanced effects.

4.6 Weakly directional
thermodynamics: observational tests

With two-time boundary conditions, systems can behave differently from
what would be expected with initial conditions only. If our world has future constraints, would there be experimental consequences today? From
what we have seen in the past few sections this could occur for processes
with relaxation rates on the same scale as that of the interval separating

the boundary conditions,[19] presumably the big-bang to big-crunch time that we designate $\mathbb{T}$. What I find attractive about this is that it represents cosmological probe that is conceptually different from those usually examined and relies intimately on thermodynamics or deviations from normal equilibration. The kinds of slow processes that I will discuss are nuclear decays and large scale dynamics.

Slow nuclear processes

There are unstable nuclides with lifetimes on the order of 10^{10} years, 10^{11} years and longer. This is comparable to what one might guess is the lifetime of the universe (or at least you wouldn't guess that it's less). If you take a sample of one of these (e.g., Re^{187} or Sm^{147}) in your laboratory and watch it decay, will it obey the initial-condition-oriented exponential, or will it do something else? Wheeler thought it would do something else, namely the number undecayed at any time would be the sum of a decaying exponential and a growing exponential.[20] If for convenience the beginning and end of the bounding time interval[21] are taken to be $\pm\mathbb{T}/2$, and if those conditions both require N_0 of this species at those times, then the number around at time t would be

$$N_0 \frac{\cosh(\Gamma t)}{\cosh(\Gamma\mathbb{T}/2)} \qquad (4.6.1)$$

In Fig. 4.6.1 this behavior is contrasted with the usual exponential, starting from time $-\mathbb{T}/2$ for various values of the product $Q \equiv \Gamma\mathbb{T}$. For the solid line the decay time is half the 'lifetime of the universe.' If an observer's value of the nuclide's lifetime is based on decay measurements, it would be necessary to have decay information from widely different eras before a discrepancy could be noted. (This should be clear from the dotted line in the figure.) Given the availability of ancient rocks and the possibility of deducing nuclear properties from them, this approach seems promising.

Unfortunately, the justification for this cosmology-modified decay is problematic. Both Wheeler and I criticized the original proposal. If the actual transition time (the time interval between having the Re^{187}

---

[19] Later in the book we note that quantum mechanics is different: satisfying two-time boundary conditions can become *more* difficult as the separation time grows.

[20] I can't follow his argument for the functional form. In Section 5.2 we show that the 'cosh' follows from the lowest approximation to certain doubly conditioned statistics. But the conclusion depends on additional properties of the process and is not generally true.

[21] For this example the early bounding time should be the epoch of creation of the nuclides.

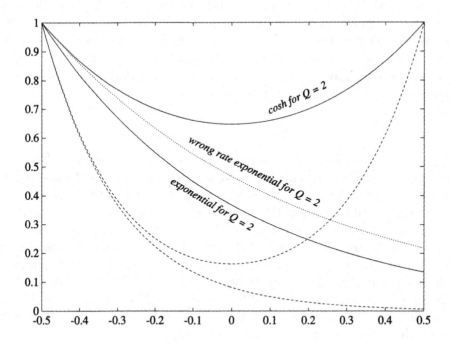

Fig. 4.6.1. Exponential versus hyperbolic cosine decay laws. For the solid lines (one is cosh, one is exp), $Q \equiv \Gamma T = 2$. For the dashed lines, $Q = 5$. The dotted line is an exponential that matches the slope of the (solid line) hyperbolic cosine. It would be the curve expected by an observer whose value of the nuclear decay parameter was based on an early measurement of that same hyperbolic cosine.

and having its decay products) is short, the boundary condition imposes constraints on the overall average, but not on small selected samples. Wheeler phrased his doubts in terms of questions about what was the suitable time constant to use in the rate equations, the nuclear lifetime or perhaps what I just called the actual transition time. This highlights the difference between a process being slow and a process being unlikely. The distinction is not always sharp (as in diffusion) but as we will see in our next proposal, there are processes that are intrinsically slow, with no possible short cuts to confuse the issue.

Remark: See Exercise 5.1.4 for an example of this distinction.

Galactic dynamics

Think of the galaxies as atoms and their dynamics as the dynamics of a gas. When near each other the force law is Newtonian; at larger distances one needs modifications, including allowance for the expansion of the universe. (When very near there are inelastic collisions.) People who study 'large scale structure' make models of this sort. A feature of this gas is that it evolves toward inhomogeneity. There are clumps, voids and a hierarchical structure typical of fractal sets. For this grand system the long range nature of the forces compels the maximum time scale to be the size of the system itself (divided by c). Therefore if the future big crunch time is not significantly greater than the time already elapsed since the big bang, we could expect future conditioning constraints to operate on galactic dynamics.

In principle one could look for such constraints in the following way. Examine a (large) region of space and measure a statistical property of the galaxies it contains, perhaps something like the entropy we used for the cat-map points. Again, in principle, you can use your measured values of galactic positions and velocities to calculate the later positions of the galaxies and the later value of the entropy-like quantity. Do this for the actually measured values and do it for slightly smeared values, i.e., introduce slight deviations. Now compare to what actually occurs in Nature. You may find that for a variety of smeared values your entropy-like quantity increases more than for the actual values. This would mean there are precise correlations in the actual values, and that by virtue of those precise correlations the system heads for a future condition that occupies less than the maximal volume of phase space. This is what is happening in Fig. 4.3.4. The system that is required to reconverge at (the early) time 7 ceases to enjoy a maximal rate of entropy increase rather early in its development in order to meet the stringent boundary condition.

As just described the measurement is not possible, since it requires comparisons across cosmologically separated times. Instead one can examine samples at various distances from us, reflecting different stages in the history of the universe. Use these to get enough information about the earliest samples to predict the later values of some entropy-like quantity. Probably your data will already have enough noise to obviate the need to smear. If the actually measured values (of the entropy-like quantity for the closer samples) are lower than the computed values there would be evidence for correlations.

Astrophysicist friends tell me that the foregoing is not presently feasible. With the wealth of new astrophysical data pouring in these days, one can hope that a clever and sensitive measure can be found. A successful observation along these lines would not only tell us about the spatial structure of the universe, but would in a sense be a reading from the future.

Other celestial phenomena

Galactic and stellar observations offer other ways that future constraints could reveal themselves to statistical analysis. There is information available about velocity distributions for a variety of celestial objects, from comets in the solar system to stars in a globular cluster to clusters of galaxies within a supercluster. In each case one would need *a priori* estimates of what they *ought* to be doing in the absence of future constraints and have sufficiently good observational data to confirm or deny the prediction. It would also help if one had some idea of the changes in the distribution function that a plausible future constraint would impose (for example less evaporation than initial randomness alone would predict). Of the systems mentioned, globular clusters probably offer the best opportunity. Their associated time scales may be long; moreover, quite a bit is known about them, dynamically as well in terms of the structures of their constituent stars. Gravitational collapse is another phenomenon that could be affected by future constraints. However, to reach a definite conclusion about this one would need a theoretical analysis in which conventional causality did not play a role.

4.7 Notes and sources

Section 4.0. The idea that the thermodynamic arrow of time arose from a gigantic fluctuation leads to an amusing form of solipsism. From the standpoint of entropy, I, sitting at my keyboard typing these lines, am a pretty big fluctuation. A tree that I remember seeing is also a big fluctuation. It would be a smaller fluctuation, entropy-wise (recalling that entropy is *extensive*), not to have the tree, but to change my brain slightly and create the memory of that tree. Therefore, in terms of likely or unlikely fluctuations (and that's what entropy measures) it would be far more likely that the tree doesn't exist. You, reading this, should similarly doubt the existence of the writer.[22] I've been making this argument, as a form of joke almost, for many years, although I'm surely not the first to propound it. In fact the creationist suggestion that dinosaur 'bones' are simply complex mineral structures placed in the ground to confound or test us, is a variation on this joke.[23]

---

[22] Doubts about the existence of readers may be better justified.

[23] About a month after writing this, I had a discussion with a colleague who felt righteously altruistic (toward the scientific community) in foregoing the considerable income he felt he could have by testifying for the 'creationists.' It was his judgment that the most probable history of the world was that prior to some moment—perhaps 5756 years ago—entropy ceased to decrease as you move backward in time, and began to increase. This would mean that, as in the argument I characterize as a 'joke,' dinosaur bones and geological strata are more likely to be spontaneous creations than reliable guides to history. Why this moment of entropy minimum was

Based on Section 2.6, it is possible to see that this argument is circular. The discussion in the last paragraph is based on the premise: what is most likely is what is. In less oracular terms: given a macrostate, the things that happen to it are the things that happen to most associated microstates. What we saw in Section 2.6 is that this is the way you *predict*. It is *not* the way you *retrodict*. And in fact this distinction is a restatement of the second law of thermodynamics. So saying that the dinosaur bones were 'more likely' to be artifacts than to be guides to history, is to misapply the relation between likelihood and phase space volume. It is precisely the second law of thermodynamics that connects these concepts for future events. It does not connect them for past events, quite the contrary.

However, *if* your explanation of the second law is that we are experiencing a large, spontaneous fluctuation, then the reasoning goes through. But you cannot use the phase space–likelihood relation to argue in favor of this explanation, since we would accept that argument for past events *only* if that is indeed the explanation. With this caution, one can now reconsider the fluctuation argument. And here is where common sense enters. The events of the past show self-consistency. Similar dinosaur bones are found in many places. They are found in geological strata whose formation we understand by the same physical laws that hold today. Among all possible fluctuations those showing this internal consistency would be rare. Therefore I reject the idea that we are the result of an unconstrained fluctuation.[24] (What I *do* believe to be the source of the thermodynamic arrow is of course the subject of the present chapter.)

I expect that the aforementioned 'common sense' argument can be formalized. Just as the second law formalizes much of what we take as common sense about predictions (e.g., the phase space–likelihood relation for unfair dice), so one could imagine laws governing the emergence of complexity. Then the complex, interrelated world around us bespeaks minimum development times. However, I don't want to leave the impression that until we know such laws the issue remains open. It does not take a major advance in science for me to be convinced that I wasn't born yesterday.

---

years in the past rather than milliseconds, was justified by not having the present epoch be special, just as we assume our position is not special. As discussed below, entropy-based reasoning is valid only if the thermodynamic arrow is a consequence of a fluctuation. Moreover, my colleague may be overestimating the acceptance he would find in the fundamentalist community. I remember, as a student, proposing a similar explanation of dinosaur bones (*sans* entropy analysis) to a clergyman of that stripe. He did not feel that this idea advanced the case for his reading of Genesis.

[24] My use of probabilistic reasoning in this context goes beyond the re-statement of the second law of thermodynamics in Section 2.6 (the retrodiction/prediction distinction). It is related to time-of-death statements in forensic medicine. You want to know how long has a system been undisturbed or how long a process has been going on. Because of structure and correlations that you know exist in the present macrostate your hypothetical retrodicted earlier macrostates may not be able to provide those correlations if you assume equal *a priori* probabilities in a hypothetical too recent macroscopic state.

Section 4.1. Gold presented his idea at a Solvay conference [GOLD 58] and in [GOLD 62]. My guess is that this led to discussions with Bondi and Hoyle on the Wheeler-Feynman absorber theory, a framework that could provide a link between cosmology and a radiative arrow of time. [HOYLE] argues that the absorber theory only works in a steady state cosmology. (Recall that the steady state theory was put forth by Bondi, Gold and Hoyle.) Gold and Bondi also organized one of the first physics oriented conferences on the subject of time, [GOLD 67]. My work on the circularity of Gold's argument and the correct way to establish an arrow is [SCHULMAN 73]. I should mention the role of a mathematician colleague, Andrew Lenard, in this. It was in explaining Gold's idea to him, and in trying to maintain his standard of rigor and clear thinking that I realized that there was a problem. The 'switchover' complaint against Gold is in [PENROSE], and others in the same vein are [DAVIES] and [WEINBERG]. My response is [SCHULMAN 91].

It is remarkable that in the years following the discovery of the expansion of the universe no one related the two great physical arrows of time. Certainly people like Wheeler, Feynman and Einstein were aware of the problem—the first two quote a related work of the last in their absorber theory paper, [WHEELER 45]. With the wisdom of hindsight it seems an obvious connection. Even now there is opposition, but whether Gold is right or wrong in detail, most students of this field agree that the cosmos plays a role in setting the thermodynamic arrow.

Section 4.2. From my pleas for more direct involvement of statistical mechanics notions in arrow of time arguments and (later in the book) in the quantum measurement problem, one might feel that the field is in need of a public relations campaign. Problems falling on its turf have been coopted by everyone from quantum cosmologists to particle physicists to philosophers. Similarly, thermodynamics has had a turbulent history. [TRUESDELL] characterizes (nineteenth century) thermodynamics as, 'accursed by misunderstanding, irrelevance, retreat and failure.' [GOODSTEIN] opens his book with a caveat: 'Ludwig Boltzmann, who spent much of his life studying statistical mechanics, died in 1906, by his own hand. Paul Ehrenfest, carrying on the work, died similarly in 1933. Now it is our turn to study statistical mechanics. Perhaps it will be wise to approach the subject cautiously.' A good source for the history of the field is [BRUSH].

Two-time boundary conditions were studied by [COCKE, SCHRÖDINGER, SCHULMAN 73, 74a, 74b, 76, 77a, 77b, 80, 91] and [WHEELER 79].

Particle dynamics in comoving coordinates is presented in [PEEBLES]. To illustrate that the earth-sun distance does not change over cosmological times, [MISNER] embellishes the expanding balloon analogy. Pennies are glued to the balloon. Their centers separate, but they do not change size. Although the size of a penny is a consequence of electromagnetic forces, under pure gravity, masses near each other do not fully participate in the cosmological expansion. [MISNER] cites [NOERDLINGER]. For recent work see [ANDERSON].

The association of entropy with black holes is due to [BEKENSTEIN 73, 74].

Section 4.3. The need to go beyond harmonic oscillator dynamics to recover the foundations of statistical mechanics has long been appreciated. However, it's not enough to go a little beyond the linear regime, as [FERMI], Pasta & Ulam realized when they discovered solitons. On the other hand, with appropriate collections of oscillators relaxation processes *can* be studied, as in [HUERTA]. Specifically for two time boundary value problems, the limitations of oscillators are found in [SCHULMAN 73, 77b]. The analytic approach to two-time boundary value problems of Appendix A to Section 4.3 is based on [SCHULMAN 77a].

Section 4.6. Use of contemporary data for evidence of future conditioning was discussed by [SCHULMAN 77a] and [WHEELER 79]. It was Wheeler who first suggested using radioactivity, but the only document I can now find predating my 1977 paper (in which I mention his work) is an indirect remark in [WHEELER 75]. Reservations about the radioactivity measurement idea are in [WHEELER 94] and in [SCHULMAN 77a].

As remarked in Section 4.2, once you use remote two-time conditioning, expansion or contraction is no longer essential for the physical consequences in Section 4.6 and in Chapter 8. All you need is disequilibrium at (say) the early time, and statistical time symmetry. The latter is an extension of the Cosmological Principal: our time direction is no more special than our location. The disequilibrium could be due to the very early evolution of the universe or it could be the result of later expansion, say 5 minutes after the big bang. This remark is mainly of conceptual (rather than physical) interest, since the universe is in fact expanding.

References

ANDERSON 95 J. L. Anderson (1995) Multiparticle Dynamics in an Expanding Universe, *Phys. Rev. Lett.* **75**, 3602.

BEKENSTEIN 73 J. D. Bekenstein (1973) Black Holes and Entropy, *Phys. Rev.* D **7**, 2333.

BEKENSTEIN 74 J. D. Bekenstein (1974) Generalized second law of thermodynamics in black-hole physics, *Phys. Rev.* D **9**, 3292.

BRUSH 83 S. G. Brush (1983) *Statistical Physics and the Atomic Theory of Matter from Boyle and Newton to Landau and Onsager*, Princeton Univ. Press, Princeton.

COCKE 67 W. J. Cocke (1967) Statistical Time Symmetry and Two-Time Boundary Conditions in Physics and Cosmology, *Phys. Rev.* **160**, 1165.

DAVIES 77 P. C. W. Davies (1977) *The Physics of Time Asymmetry*, Univ. California Press, Berkeley.

FERMI 55 E. Fermi, S. Pasta & S. Ulam (1955) Studies in nonlinear problems, I., Los Alamos Report, LA-1940. Reprinted in *Collected Works of E. Fermi* (1965) **2**, 978, Univ. Chicago Press, Chicago.

GOLD 58 T. Gold (1958) The Arrow of Time, in *Proc. 11th Solvay Conf., Structure and Evolution of the Universe*, Stoops, Brussels.

GOLD 62 T. Gold (1962) The Arrow of Time, *Am. J. Phys.* **30**, 403.

GOLD 67 T. Gold (1967) *The Nature of Time*, Cornell Univ. Press, Ithaca, New York.

GOODSTEIN 75 D. L. Goodstein (1975) *States of Matter*, Prentice-Hall, Englewood Cliffs (Dover ed., corrected, 1985).

HOYLE
95
F. Hoyle & J. V. Narlikar (1995) Cosmology and action-at-a-distance electrodynamics, *Rev. Mod. Phys.* **67**, 113.

HUERTA
69
M. A. Huerta & H. S. Robertson (1969) Entropy, Information Theory, and the Approach to Equilibrium of Coupled Harmonic Oscillator Systems, *J. Stat. Phys.* **1**, 393.

HUERTA
71
M. A. Huerta & H. S. Robertson (1971) Approach to Equilibrium of Coupled Harmonic Oscillator Systems. II, *J. Stat. Phys.* **3**, 171.

MISNER
73
C. W. Misner, K. S. Thorne & J. A. Wheeler (1973) *Gravitation*, Freeman, San Francisco.

NOERDLINGER
71
P. D. Noerdlinger & V. Petrosian (1971) The effect of cosmological expansion on self-gravitating ensembles of particles, *Ap. J.* **168**, 1.

PEEBLES
80
P. J. E. Peebles (1980) *The Large-Scale Structure of the Universe*, Princeton Univ. Press, Princeton.

PENROSE
79
R. Penrose (1979) Singularities and time-asymmetry, in *General Relativity: An Einstein Centenary Survey*, ed. S. W. Hawking & W. Israel, Cambridge Univ. Press, New York, page 597.

SCHRÖDINGER
32
E. Schrödinger (1932) Sur la theorie relativiste de l'electron et l'interpretation de la mecanique quantique, *Ann. Inst. H. Poincaré* II, 269.

SCHULMAN
73
L. S. Schulman (1973) Correlating Arrows of Time, *Phys. Rev.* D **7**, 2868.

SCHULMAN
74a
L. S. Schulman (1974) Some Differential-Difference Equations Containing Both Advance and Retardation, *J. Math. Phys.* **15**, 295.

SCHULMAN
74b
L. S. Schulman (1974) On Deriving Irreversible Electrodynamics, in *Modern Developments in Thermodynamics*, ed. B. Gal-Or, Keter, Jerusalem.

SCHULMAN
76
L. S. Schulman (1976) Normal and Reversed Causality in a Model System, *Phys. Lett.* A **57**, 305.

SCHULMAN
77a
L. S. Schulman (1977) Illustration of Reversed Causality with Remarks on Experiment, *J. Stat. Phys.* **16**, 217.

SCHULMAN
77b
L. S. Schulman & R. Shtokhamer (1977) Thermodynamic Arrow for a Mixing System, *Int. J. Theor. Phys.* **16**, 287.

SCHULMAN
80
L. S. Schulman (1980) Formulation and Justification of the Wheeler-Feynman Absorber Theory, *Found. Phys.* **10**, 841.

SCHULMAN
91
L. S. Schulman (1991) Models for Intermediate Time Dynamics with Two-Time Boundary Conditions, *Physica* A **177**, 373.

TRUESDELL
80
C. A. Truesdell (1980) *The Tragicomical History of Thermodynamics 1822-1854*, Springer, New York.

WEINBERG
72
S. Weinberg (1972) *Gravitation and Cosmology: Principles and Applications of the General Theory of Relativity*, Wiley, New York.

WHEELER
45
J. A. Wheeler & R. P. Feynman (1945) Interaction with the Absorber as the Mechanism of Radiation, *Rev. Mod. Phys.* **17**, 157.

WHEELER
75
J. A. Wheeler (1975) Conference Summary: More Results than Ever in Gravitation Physics and Relativity, in *General Relativity and Gravitation*, ed. G. Shaviv & J. Rosen, Wiley, New York.

WHEELER
79
J. A. Wheeler (1979) Frontiers of time, in *Problems in the Foundations of Physics*, ed. G. Toraldo di Francia, North Holland, Amsterdam.

WHEELER
94
J. A. Wheeler (1994) Time Today, in *Physical Origins of Time Asymmetry*, ed. J. J. Halliwell, J. Pérez-Mercader and W. H. Zurek, Cambridge Univ. Press, Cambridge.

Notes on the exercises

Exercise 4.3.2 Appeal to the ergodic hypothesis. The usual justification for using the partition function is that the actual system (the one in your laboratory) rapidly visits many regions of phase space, so that the properties of particular but typical systems are given by phase space averages. The future conditioning says that only a certain few of the microstates are available. But if the future conditioning is many relaxation times away the points selected will still be random with respect to measurements at the early times. This is another aspect of the normal equilibration in the initial phases of, say, Fig. 4.3.3.

Exercise 4.3.3 Here is a suggestion for a cat-map-world model of evaporating liquid and a temporarily closed container. Think of the liquid as being in the corner of the box, say in $0 \leq x < \ell$, $0 \leq y < \ell$, with $\ell \ll 1$ Call this region $\mathcal{L}$. You want a steady stream of newly 'created' gas particles to leave $\mathcal{L}$ during early times and to flow back (with annihilation) at late times (approaching T). This can be modeled by combining several two-time boundary value problems starting and ending at times within the interval $[0, T]$. Next you need a model of a container within which to temporarily isolate some points and allow them to come to equilibrium. This could be a small box within the original one with its own internal dynamics, perhaps a variant of the cat map. With appropriate rules and parameters, this system should show the usual causality, that is, there will be a smoothing of the density gradient only *after* the isolation, and continuing past the recoupling. (The large Lyapunov exponent though will make this difficult to do numerically.) The reason is that for sufficiently large T the time at which a given point is absorbed by the liquid is independent of whether or not it was in the container or outside of it (during the container's isolation). This independence justifies the usual initial-conditions-only reasoning for the early times. If the isolation time is close to T, the density smoothing will happen in the opposite time direction. A different but related effect was seen in [SCHULMAN 77b].

Appendix 4.3.A: Doubly constrained cat map: analytic treatment

Two-time boundary value problems can be difficult, but for the cat map there is an analytic approach using the Fourier series methods developed in Appendix B to Section 2.2. We here use the notation of that Appendix.

Let A and B be (measurable) subsets of the unit square and, as in the computer simulations in the text, we want the system to start in A and end in B. The set satisfying this boundary conditions is $A \cap \varphi_{-T}(B)$. The unnormalized density is therefore

$$\rho(\xi) = \chi_A(\xi)\chi_{\varphi_{-T}(B)}(\xi) \tag{4.3.A.1}$$

One time step later this density is

$$\rho(\varphi(\xi)) = \chi_{\varphi(A)}(\xi)\chi_{\varphi_{-T+1}(B)}(\xi) \tag{4.3.A.2}$$

As in Appendix B to Section 2.2, the Fourier coefficients for the characteristic functions of the sets A and B are designated a_ν and b_ν. (There is one difference between here and there: the set B is now a final state, whereas there both A and B were propagated forward and their overlap studied.) The Fourier transform of ρ is essentially a convolution of the characteristic functions of A and B. Thus

$$\rho(\xi) = \sum_\sigma e_\sigma(\xi) \sum_\nu a_\nu b_{\varphi_T(\sigma-\nu)} \tag{4.3.A.3}$$

The first exercise is to see what ρ looks like at time 0. If T is large, $\varphi_T(\sigma - \nu)$ will be large, except for $\nu = \sigma$. Recall that $b_0 = \mu(B)$, the measure of B. If B is reasonably smooth, the Fourier coefficients of χ_B drop off. Therefore the application of high powers of φ to the label of b in Eq. (4.3.A.3) means that small numbers will appear for all values of ν except $\nu = \sigma$ (i.e., in Eq. (4.3.A.3) either the b factor is small or the a factor or both). Therefore, to a good approximation, $\rho \approx \chi_A$. Similarly, for small t, $\rho(\varphi_t(\xi)) \approx \chi_{\varphi_t(A)}(\xi)$. This is the 'crypto' aspect of crypto-conditioning. How can there be deviations from the approximate equality? Clearly some of the large-label Fourier coefficients must be large. By standard Fourier series arguments, this means that B must have features on the scale of λ^{-T}. This is how the future is 'forgotten' with the time constant of the cat-map relaxation. For collections of points, as in the computer simulations, the true density is a sum of delta functions—whose Fourier coefficients do not drop off, so one must rely on phase cancellation.

What would it take to notice the crypto-conditioning? If you had an observable $\mathcal{O}$ with the ability to discriminate on the scale λ^{-T}, then when you evaluated it (with $\int \rho \mathcal{O}$), its Fourier coefficients times those of B could give non-zero results.

It is easy to see what time evolution does to ρ and how one gets dynamical symmetry as one moves backward *from* the final time. Under time evolution the ν on a_ν evolves under φ, while the coefficient label for b loses powers of φ. Finally it is the coefficients b that give the density its dominant functional dependence; of the as, only a_0 survives.

5

Two-time boundary value problems

Although the variational principles of classical mechanics lead to two-time boundary value problems, when dealing with the real world *everyone* knows you should use initial conditions. Not surprisingly, the eighteenth-century statements of classical variational principles were invested with religious notions of teleology: guidance from future paths not taken could only be divine. The Feynman path integral formulation of quantum mechanics makes it less preposterous that a particle can explore non-extremal paths; moreover, it is most naturally formulated using information at two times. In the previous chapter, use of the two-time boundary value problem was proposed as a logical prerequisite for considering arrow-of-time questions. Perhaps this is no less teleological than Maupertuis's principle, except that we remain neutral on how and why those conditions are imposed.

In this chapter we deal with the technical aspects of solving two-time boundary value problems. In classical mechanics you get into rich existence questions—sometimes there is no solution, sometimes many. For stochastic dynamics the formulation is perhaps easiest, which is odd considering that this is the language most suited to irreversible behavior. Our ultimate interest is the quantum two-time problem and this is nearly intractable.

Later in the book I will propose that the universe is most simply described as the solution of a two-time boundary value problem. A natural reaction is to wonder whether this is too constraining. Given that our (lower entropy) past already cuts down the number of possible microstates, are there sufficient microstates to meet a future condition as well? To this end, in Appendix A to this section I exhibit some of the numbers. It is hardly an existence proof, but it justifies the remark that losing a few million powers of ten does not make a dent in phase space.

The results of that appendix are summarized as follows. In a cubic centimeter of air there are about $10^{(10^{20})}$ microstates, where $2\pi\hbar$ is the discretization scale for phase space. If the gas is initially in $(1/4)^3$ of the one cubic centimeter volume (i.e., each linear dimension is reduced by four), the possible microscopic states are reduced by $4^{(10^{20})}$. On the other hand, if we demand that the gas was in $(1/2)^3$ of the volume at one time and again at a later time (two linear reductions by factors of two), then the reduction in phase space is the same $4^{(10^{20})}$. Thus two-time boundary conditions may seem to demand amazing things, but they are on the same scale of improbability as conventional initial conditions. One can also make a system satisfy constraints at many times. For example, with reference to our cat movie, Fig. 4.3.2, one could place the points in a particular box at times 20, 40, 60, etc. Because of the continuum phase space, any finite number of sufficiently time-separated conditions can be met. For quantum dynamics, such issues are more difficult to study.

Appendix for this section (at the end of the chapter):
5.0.A   Counting states in a cubic centimeter of air

5.1 Classical dynamics

Even at the simplest level the two-time boundary condition for classical mechanics has fascinating features. Take one particle moving in a potential. Require that at time 0 it be at a and at time T it be at b. There may be no solution to this problem, there may be one, there may be many. There may non-generically be an infinity of solutions. There may be exponential growth, with T, of the number of solutions. In Appendix A to this section I give a few examples. Here I show a useful way to think about this which links with our earlier work on the cat map. This is a demonstration of how with chaotic dynamics the number of solutions to the two-time boundary value problem grows exponentially—in fact the growth is related to the Lyapunov exponent.

The demonstration effectively uses the concept of Lagrangian manifold (although it will not be necessary to make formal definitions). At time 0 the particle is at $x = a$, but it is allowed all initial momenta. Go into phase space, which in this case is the plane. The set of possible initial position/momentum states is the straight line passing through $x = a$ and parallel to the p axis. See Fig. 5.1.1. Now allow this set of points to evolve in time. For a potential whose effect is not too drastic, the high p points move rapidly to the right, the large negative p points move rapidly to the left. For smaller $|p|$ the potential will have greater influence on the motion, curving some points upward, some down (i.e., accelerating or decelerating). After a time T the original line will be shifted and distorted. The boundary value problem then looks for the intersection of this distorted curve with the line of possible final conditions, namely $x = b$, $p =$ anything. If, at the intersection, the distorted curve is vertical, you have a focal point or caustic. In Fig. 5.1.1 is a case where there will be three solutions to the boundary value problem, but no focussing.

When the motion is chaotic and the system confined (in space) to a finite region, the curve winds around on itself and there will be more intersections with the set of possible final states. This is schematically illustrated in Fig. 5.1.2. For chaotic motion a pair of points whose momenta differ slightly (Δp) will generally be separated by a distance $\Delta p \exp(\Lambda T)$, with Λ the Lyapunov exponent. Since the spatial volume is finite, this lengthening curve cannot run off to spatial infinity and must wind around. An interval of the set of initial states will have lengthened by the factor $\exp(\Lambda T)$, generally giving rise to a number of intersections of that order.

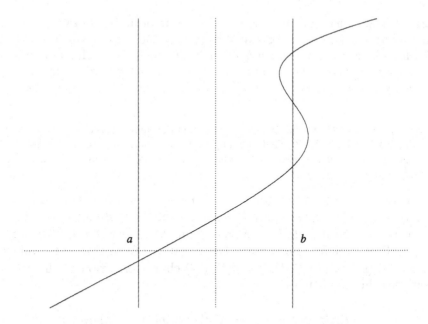

Fig. 5.1.1. Phase plane. The initial state must lie on the vertical line based on a; the final state must lie on the vertical line based on b. The curved line represents the evolute of the line based on a at a later time T. That is, each phase space point (a, p_0) evolves under the dynamics to another point (x, p), the set of which is depicted as the curve. To satisfy the boundary condition find intersections of the curve with the line based at b. In the figure this occurs at three p values.

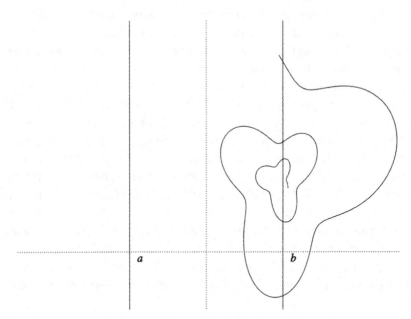

Fig. 5.1.2. Phase plane. The time T evolute of initial states is depicted as a winding curve, intersecting several times with the line of possible final states. Each intersection is a solution of the two time boundary value problem. The winding curve shows the evolute of an initial interval with $x = a$ and $p_0 \le p \le p_0 + \Delta p$.

Exercise 5.1.1. For the harmonic oscillator, find the curves analogous to those depicted in Fig. 5.1.1 and Fig. 5.1.2. That is, find $p(x;t)$, the functional form for the set $\{(x, p)\}$ defined in the caption to Fig. 5.1.1. What peculiarity is shown by the caustics in this case?

A variation on this argument can be formulated in the language of Section 4.2. Call phase space Γ. If you want the system to be in $\Omega_0 \subset \Gamma$ at time 0 (whether Ω_0 is a coarse grain, as in Section 4.2, or not), in $\Omega_T \subset \Gamma$ at time T, and if your dynamical law from time 0 to time T is a mapping $\phi : \Gamma \to \Gamma$, then the time 0 characterization of the set that satisfies this condition is $\Omega_0 \cap \phi^{-1}(\Omega_T)$.

Exponential growth of the number of solutions

If the original and final phase space volumes are v, then it is not difficult to see that the number of paths connecting these grows exponentially in time at a rate governed by the Lyapunov exponent (or several exponents if there are several positive growth rates).[1] If the total accessible phase

---

[1] 'Path' now means a bundle of paths, since the solution set is thickened by v.

space volume is V, the number of connecting paths is $(v/V)\exp(\Lambda T)$, with T the elapsed time and Λ the Lyapunov exponent. To see this, take the original phase space volume, v, and let it evolve for time T. Although its measure is preserved, it now has a linear dimension on the order $\ell\exp(\Lambda T)$, where $\ell \equiv v^{-1/2d}$ is a characteristic dimension of the original volume, d is the coordinate space dimension, and I am assuming that there is only one positive Lyapunov exponent. We want the number of intersections of this long 'noodle' with another phase space volume (also of size v). By assumption paths are confined to a finite volume V, so the noodle winds around quite a bit and its transverse dimensions are very small. To know how often the noodle hits v, thicken the noodle by ℓ in all transverse directions and replace the target volume by a point. The number of intersections is now the number of times the point finds itself in the thickened noodle (which may often overlap with itself). For chaotic motion this number should be about the same everywhere in V and is the ratio of the volume of the thickened noodle to V. The thickened noodle has volume $\ell\exp(\Lambda T)\cdot\ell^{2d-1}$, which, when divided by V is $(v/V)\exp(\Lambda T)$, as claimed. For situations with more than one significantly positive Lyapunov exponent, the same formula emerges (with Λ replaced by the sum of those exponents), but the argument uses hypersurfaces instead of noodles. (See Section 5.4.)

What happens if the dynamics are not ergodic? Then you could give boundary conditions in which the two sets were mutually inaccessible, and there would be no solution. On the other hand, if the sets *were* mutually accessible, you could expect relatively large numbers of connecting paths.

In summary, the more chaotic the dynamics the more easily solvable the two-time boundary value problem. Yet another way to see this is not to count two-time solution paths, as we just did, but to estimate total phase space volume of solutions. If the solution set is characterized by $\Omega_0 \cap \phi^{-1}(\Omega_T)$, then with mixing dynamics one gets $|\Omega_0||\Omega_T|/V^2$ as the relative volume of the two-time solutions (with $|\Omega|$ the volume of Ω). This shows that the phase space estimates at the beginning of this chapter extend to interacting systems.

Classical field dynamics

What about continuum equations? Electromagnetism? Electromagnetism with particles coupled? The Navier-Stokes equation? General relativity? This list reflects ascending mathematical difficulty. Electromagnetic two-time boundary value problems can be formally solved, although with a bad choice of boundary conditions you will get into trouble. (See Exercise 5.12 below.) The Navier-Stokes equation of fluid dynamics is different—it can be dissipative and a demand that it yield recurrent solutions may be impossible to satisfy. However, the Navier-Stokes equation is an average over an underlying microscopic particle dynamics. If those dynamics are ergodic and the time interval long enough, then there are microstates that evolve from any initial macroscopic fluid state to any final one of the same energy. However, since Navier-Stokes dynamics and exact particle dynamics are different, this physical evolution will not reflect a solution to the Navier-Stokes equation. Thus a confined N-particle system will have a Poincaré recurrence (even without ergodicity), but a dissipative Navier-Stokes equation that may be thought to model the (large-)N-particle system will never return to its initial condition.

Exercise 5.1.2. *Free-field two-time boundary value problems.* The free field is equivalent to a set of uncoupled oscillators and a state of the system can be written as a sum of the oscillator fields. Therefore a two-time specification of free fields should have a solution. Consider two functions of bounded support whose supports are separated by a distance R. Let T be the time separation. (a.) What happens when you try to impose this argument on the scalar wave equation with $cT < R$? (b.) What if $cT > R$? (c.) For one space dimension you can write solutions to the wave equation as $\phi(x, t) = a(x-ct)+b(x-ct)$ for arbitrary (sufficiently smooth) a and b. Examine the two-time boundary value problem in this context.

General relativity is not dissipative (if dissipation is not introduced for the matter content), but the equations themselves are more complicated. Approaches to handling the continuous dynamics come to mind (for example discretizing) but I know of no work by physicists or mathematicians in which multiple time conditions are placed on these nonlinear partial differential equations.

As mentioned in Chapter 3, it has been claimed that the singular behavior of the gravitational field equations can be used to derive an arrow of time. I want to point out that nonlinearity and singularity by themselves do not fix an arrow. As an example, I use a an exercise in classical mechanics appearing in Appendix A to this section. Consider a particle of mass unity in one dimension with the potential $V(x) = -\exp(-2x)/2$. Here is a solution (for $t < t_0$) to the equation of motion:

$$x(t) = \log\left[\frac{1}{k}\sinh k(t_0 - t)\right] \qquad (5.1.1)$$

k is a real parameter, and[2] is related to the energy by $E = k^2/2$. As t approaches t_0 (from below) the particle runs off to $-\infty$ in finite time. Due to the singularity, this system has an arrow of time.[3]

---

[2] ... as can be seen by letting $t \to -\infty$.

[3] With a different sign for a square root, there is another solution with $t \geq t_0$, having the reversed behavior. Each solution has an intrinsic asymmetry.

The arrow, however, is an arrow of the solution, not the dynamics. For this example it is easy to produce *symmetric* solutions, despite the singularity. One such solution[4] is

$$x(t) = \log\left[\frac{1}{\kappa}\cos\kappa t\right] \qquad (5.1.2)$$

with κ real and related to the energy by $E = -\kappa^2/2$. Eq. (5.1.2) gives a doubly singular, but perfectly symmetric, solution.

Exercise 5.1.3. In Section 4.3 we studied the two-time boundary value problem for the cat map, including analytic methods presented in an Appendix. A similar explicit calculation is possible for the Kac ring model that was used for our exposition of irreversibility in Section 2.2. Suppose you demanded that 70% of the balls are white at time 0 and 70% white at a later time T—in contrast to the 50% that the system would have in its equilibrium state. This could be imposed as a constraint on the set of active sites as well as on the ball sequences. Show that if the parameter 'δ' is required to take the value α at both initial and final times and that if the fraction of 'active' sites is μ, then the expected value of δ at time t is $\alpha\cosh[\gamma(t - T/2)]/\cosh(\gamma T/2)$, with $\exp(-\gamma) = 1 - 2\mu$ and $\mu < 1/2$ for convenience.

Exercise 5.1.4. Referring to the system and boundary conditions of the previous Exercise, suppose that at a time near the beginning a small stretch of balls is found to be entirely white. Does the 'δ' associated with this subset decay exponentially or with the hyperbolic cosine law? What happens for the analogous boundary condition problem for the cat map? That is, consider a subset of points with boundary condition times that are *not* distant on the scale of the relaxation time. This exercise has bearing on the experiments and observations proposed in Section 4.6.

Classical systems with temporal non-locality

Time symmetric electrodynamics (Section 3.3) is a classical theory, as is its linear simplification. For these theories not only do we have the physical arguments of Chapter 4 to motivate the use of two time boundary conditions, but as shown in Chapter 3, such use is mathematically natural. In Appendix A to Section 3.3 we solved the linear two-time boundary value problem, including study of the response to a perturbation, an issue related to causality in theories with future boundary conditions.

Appendix for this section (at the end of the chapter):
5.1.A Multiplicity of solutions for classical mechanics

---

[4] This solution is related to the previous by analytic continuation and also to the function given in Exercise 5.1A.1. The parameter t_0 has disappeared because $t = 0$ is taken to be the time at which x is maximum, $-\log\kappa$.

5.2 Stochastic dynamics

It turns out to be surprisingly easy to impose final conditions on a stochastic process. That is, it's easy to compute the distribution function—I don't know how to win the Lottery. It is ironic that stochastic dynamics, which naturally lend themselves to time-asymmetry, nevertheless allow easier imposition of future boundary conditions than classical or quantum mechanics.

Let X_t be the random position (or state) of a system at time t. Its transition matrix or propagator is the conditional probability

$$G(x, t; y) = \Pr\left(X_t = x \mid X_0 = y\right) \qquad (5.2.1)$$

If we focus on a particular interval for a time step, Δt, and assume that the transition probabilities are time-translation invariant, then it is useful to go back to the notation of Section 2.4, namely

$$W(x \leftarrow y) \equiv W_{xy} \equiv \Pr\left(X_{t+\Delta t} = x \mid X_t = y\right) \qquad (5.2.2)$$

For our two-time boundary value problem, we want the system to be in state α at $t = 0$ and β at $t = T$. Call this event $\mathcal{E}$. Two questions arise: what is the probability distribution of X_t for $0 \leq t \leq T$, and what are the *effective* transition probabilities? To formulate this we define an effective distribution function, to be called $\Gamma(x, t; \mathcal{E})$. In a short series of equalities we establish the main properties of future conditioned stochastic processes:

$$\begin{aligned}
\Gamma(x, t; \mathcal{E}) &\equiv \Pr\left(X_t = x \mid \mathcal{E}\right) \\
&\stackrel{a}{=} \frac{\Pr\left(X_t = x \text{ and } X_T = \beta \mid X_0 = \alpha\right)}{\Pr\left(X_T = \beta \mid X_0 = \alpha\right)} \\
&\stackrel{b}{=} \frac{\Pr\left(X_T = \beta \mid X_t = x\right)\Pr\left(X_t = x \mid X_0 = \alpha\right)}{\Pr\left(X_T = \beta \mid X_0 = \alpha\right)} \\
&\stackrel{c}{=} \frac{G(\beta, T - t; x)G(x, t; \alpha)}{G(\beta, T; \alpha)}
\end{aligned} \qquad (5.2.3)$$

Equality a is the conditional probability identity. Equality b uses the Markov property. That is, once the system is at x at time t, the likelihood that it reaches β at T does not depend on its history prior to t. Step c invokes the definition of G and uses time-translation invariance (so that $\Pr(X_T = \beta | X_t = x)$ depends only on $T - t$).

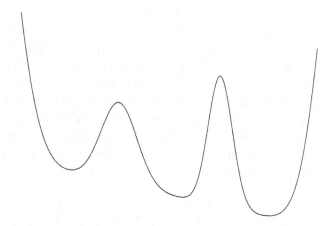

Fig. 5.2.1. Potential with three minima. A random walk can be defined in the potential with the rule that the walker (or particle) is more likely to step downhill than uphill. The particle will then spend most of its time in the minima. An overall effective transition rate between the minima can be used to replace this walk by a three-state stochastic process. Transitions between the extreme left and right minima cannot take place directly and the effective rate will mostly depend on the barrier height.

If one does not consider the continuous time limit, $\Delta t \to 0$, then there is no loss of generality in taking $\Delta t = 1$ and t an integer. In that case

$$G(x, t; y) = (W^t)_{xy} \qquad (5.2.4)$$

and Eq. (5.2.3) becomes

$$\Gamma(x, t; \mathcal{E}) = \frac{\left(W^{T-t}\right)_{\beta x} (W^t)_{x\alpha}}{(W^T)_{\beta \alpha}} \qquad (5.2.5)$$

Another quantity of interest is the *effective* transition rate. With future conditioning the relative likelihood of various transitions is affected; the closer to the end the more the effect. In the following equation we both define and evaluate the effective transition rate, based on a calculation similar to that of Eq. (5.2.3):

$$\widetilde{W}_{xy} \equiv \Pr\left(X_{t+1} = x \mid X_t = y \ \& \ \mathcal{E}\right) = W_{xy} \frac{G(\beta, T - (t+1); x)}{G(\beta, T - t; y)} \qquad (5.2.6)$$

Exercise 5.2.1. Prove this.

Exercise 5.2.2. Adapt Eq. (5.2.6) to Brownian motion. Assume that the unit time interval is short compared to t and T. With diffusion coefficient D, the usual propagator is $G(x, t; y) = (4\pi Dt)^{-1/2} \exp\left(-(x - y)^2/4Dt\right)$. Find the effective transition probabilities for a random walk starting from $x = 0$ and required to end (at $t = T$) at $x = \xi$.

A great deal can be done with this formalism. As a first application I'll examine a situation along the lines of those considered in Sections 4.2–4.4. Suppose there is more than one basic rate, for example what you would get for motion in the potential shown in Fig. 5.2.1. You can think of the dynamics in the potential as a random walk with hills. A hill in this context means that there are smaller probabilities for going uphill than downhill. We'll take an even easier case where we don't watch the detailed steps, but just have an overall transition rate for going from minimum 1 to minimum 2 or back; similarly for minima 2 and 3. The matrix W takes the form:

$$W = \begin{pmatrix} 1-a & b & 0 \\ a & 1-b-c & d \\ 0 & c & 1-d \end{pmatrix} \qquad (5.2.7)$$

Thus the probability of going from state 2 to state 1 in unit time is b. If you start a system in state 1 and ask for the probability that it has gotten to states 2 or 3 at time t, the distribution function has the form shown in Fig. 5.2.2 for the parameter values indicated. What is plotted is simply Eq. (5.2.4), $W^t(x,1)$, for $x = 1, 2, 3$. State 1 rapidly empties, 2 fills and 3 feeds off 2 and begins its filling at the slower rate governed by its transition probabilities.

The appropriate measure of dispersal for this system is not the entropy, '$-\sum p \log p$,' but a related quantity that is an alternative thermodynamic potential. It is the entropy relative to the stationary distribution for this stochastic process, which is the normalized right eigenfunction of the matrix W. (See the discussion in Section 2.4, including conditions for uniqueness of the largest eigenvalue.) If $\bar{p}_x$ is the stationary distribution, this relative entropy is defined by $-\sum p_x \log(p_x/\bar{p}_x)$. In Fig. 5.2.3 the relative entropy is plotted.[5] There is steady progress toward equilibrium. Note that because this system admits a stationary distribution for which $W(x \leftarrow y)\bar{p}_y = W(y \leftarrow x)\bar{p}_x$, it satisfies detailed balance.

Imposing the future constraint

We next take up evolution with future conditioning. Fig. 5.2.4 shows the evolution of the probability distribution for a system with the same fundamental transition probabilities as Figs. 5.2.2 and 5.2.3, but conditioned on having all samples return to state 1 at the final time, $T = 100$. It is

---

[5] The maximum relative entropy is 0 and is achieved for $p = \bar{p}$. One can also evaluate $-\sum \bar{p} \log \bar{p}$, in effect a property of the stochastic process. For the dynamics of Fig. 5.2.3, this number is 0.8516.

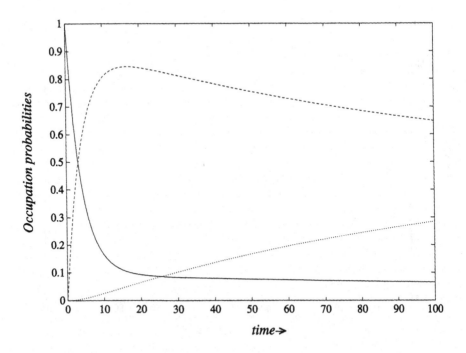

Fig. 5.2.2. Occupation probabilities as a function of time for three states with transition probabilities $W(2 \leftarrow 1) = 0.2$, $W(1 \leftarrow 2) = 0.02$, $W(3 \leftarrow 2) = 0.005$, and $W(2 \leftarrow 3) = 0.005$. The system starts in state 1 and proceeds without future conditioning. The solid line is the probability of being in state 1, the dashed line, 2, and the dotted line, 3.

fully symmetric—even more so than the corresponding plot for the cat map (Fig. 4.3.3).[6] Notice that the 1–2 equilibration sets in quickly, but that movement into the state 3 is suppressed. From the standpoint of an initial value mind-set this would be strange indeed. It would not be so strange as the scenario painted in Section 4.1—clocks fragmenting and reassembling. But there would not be the normal equilibration process. Fig. 5.2.4 thus represents a response to those who felt that Gold's thesis was invalidated by the wild behavior it predicted at the 'switchover.' There is no wild behavior at the switchover. There is equilibrium for fast variables and there is a failure to equilibrate normally for degrees of freedom whose time scales are comparable to that of the future conditioning. In fact, as we shall see below, the decay out of a metastable state is approximately given by the hyperbolic cosine mentioned in Section 4.5.

---

[6] In that example there are finite-size effects and not fully symmetric boundary conditions.

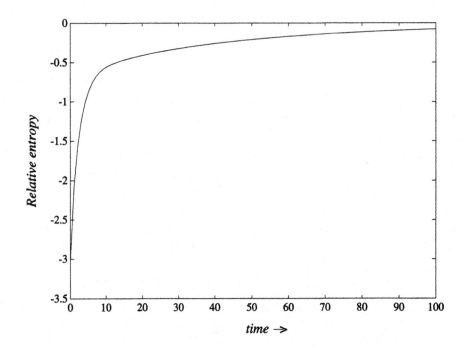

Fig. 5.2.3. Relative entropy (see text for definition) as a function of time for the time evolution shown in the previous figure.

In Fig. 5.2.5 we show the relative entropy for the history depicted in Fig. 5.2.4. Naturally it is symmetric in time. It is also well below maximum (which is 0, by definition).

Of relevance too is a boundary value problem with slightly different transition probabilities. We now take the passage out of state 1 to be the slowest process, rather like the irregularities in the galactic matter distribution that drive the formation of stars. In Chapter 4 we argued that even if the slowest processes could not equilibrate (in time for the future constraint), nevertheless, faster *dependent* processes would look normal, at least during the expansion. We make this point here using a process with the transition probabilities given in the caption to Fig. 5.2.6. For this process, 1→2 is slow, while 2→3 is fast. With no future conditioning (Fig. 5.2.6) probability goes slowly from 1 to 2, but quickly from 2 to 3.[7] In Fig. 5.2.7 we show the same process with future conditioning. Aside from the fact that with future conditioning less total probability enters

---

[7] Since the 2→3 and 3→2 rates are equal, their asymptotic occupancies are too.

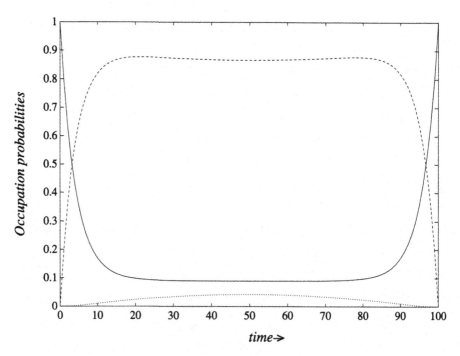

Fig. 5.2.4. Same fundamental transition probabilities as for Fig. 5.2.2, but conditioned on returning to the state 1 at time $T = 100$.

the states 2 and 3, during the early stages (up to time 20, say) the *relative* occupation of states 2 and 3 looks the same with or without conditioning.

Yet another way to see this is to look at the 'effective transition rate' defined in Eq. (5.2.6). The comparison of the previous paragraph showed that the relative 2-3 occupation was unaffected (at early times) by the fact that the slow process (1→2) could not come to equilibrium in the conditioning time ($T = 100$ in the example). Looking at the *effective transition rate* is a more demanding test in that it asks about the moment-by-moment dynamics. By showing that the effective rate is substantially the same as the usual initial-condition rate, the fact that the slowest process is not in equilibrium—nor headed there in the 'normal' way—becomes irrelevant. This is behind our assertions in the last chapter that stellar evolution and the associated life-giving rain of photons would not be affected by a future constraint that might limit in some way large scale motion and structure in the universe. In the next figure, Fig. 5.2.8, we show the *ratio* of the effective transition rate to the usual rate for the process of Figs. 5.2.6 and 5.2.7. Note that during what corresponds to

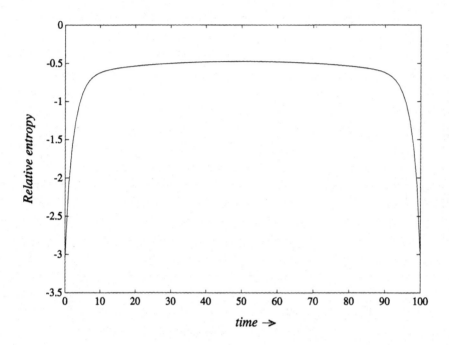

Fig. 5.2.5. Relative entropy (see text for definition) as a function of time for the time evolution shown in the previous figure.

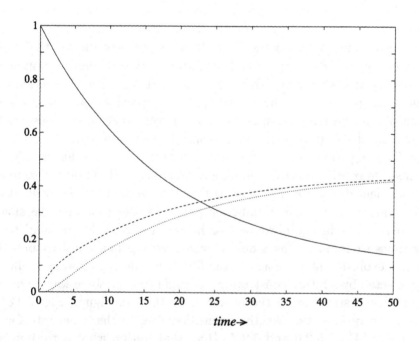

Fig. 5.2.6. Occupation probabilities as a function of time with initial conditions only and transition probabilities $W(2 \leftarrow 1) = 0.05$, $W(1 \leftarrow 2) = 0.01$, $W(3 \leftarrow 2) = 0.25$, and $W(2 \leftarrow 3) = 0.25$.

the expansion phase the ratio is close to unity, validating the arguments of the previous chapter.[8]

Fig. 5.2.8 in fact vastly underestimates the proximity to unity in richer systems and reflects the limitations of looking at a 3-state 'universe.' Later in this section, using the spectral decomposition of the transition matrix, an analytic expression will be found (Eq. (5.2.12)). For states even moderately connected by the dynamics, the ratio $\widetilde{W}/W$ will be extremely close to unity.

Note the value of stochastic dynamics for explicit display of the intuitive notions to which I appealed in Chapter 4. The virtue of the cat map is its resemblance to an ideal gas, coupled with its large Lyapunov exponent, making it rapidly accomplish the equilibration expected of chaotic systems. The disadvantage of the cat map lies also in its large single, Lyapunov exponent: it is difficult to simulate multiple rate processes.

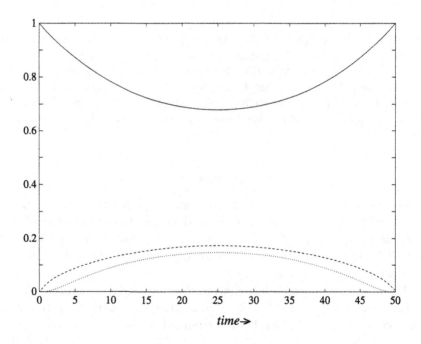

Fig. 5.2.7. Same process as in Fig. 5.2.6, but conditioned to return to state 1.

[8] The lack of symmetry about the midpoint is due to the asymmetry in the definition of $\widetilde{W}$. An opposite time-sense observer would examine the oppositely defined $\widetilde{W}$.

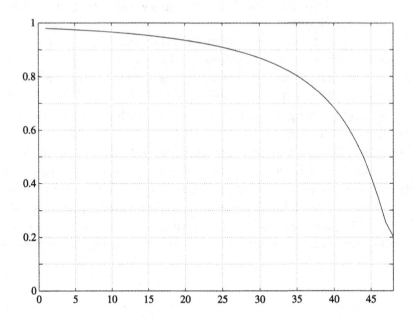

Fig. 5.2.8. Ratio of the effective transition rate to the usual transition rate, for the 2→3 transition. The parameters for this process are given in the caption to Fig. 5.2.6. The fact that the illustrated ratio is near unity at early times indicates that although galactic dynamics may be affected by a future constraint, derivative processes, such as stellar formation and evolution, are not so affected. For more complex systems this ratio is generally much closer to unity. See Eq. (5.2.12).

Non-autonomous systems

Since our system of interest is the cosmos, and since we are treating the matter within it as an open, time dependent system (expansion and possible contraction externally imposed), we should consider stochastic processes that are *not* time-translationally invariant. How does this affect our formulation and conclusions?

We now have a transition matrix that depends on time: $W_{xy}(t) = \Pr(X_{t+1} = x \mid X_t = y)$, but we still assume the Markov property. The basic equation, Eq. (5.2.3), is only changed in the last line, so that

$$\Gamma(x, t; \mathcal{E}) = \frac{G(\beta, T; x, t) G(x, t; \alpha, 0)}{G(\beta, T; \alpha, 0)} \tag{5.2.8}$$

Now the propagator requires two time arguments and can no longer be written as a power of 'W,' but is a time-ordered product.

A non-autonomous stochastic process suitable for studying an expanding/contracting model is one for which the sense of the largest transitions change their preferred direction. In the language of our three-state example, one might take 1→2 larger just after $t = 0$ and 2→1 larger just before $t = T$ (with T also playing the role of conditioning time). If the relaxation time for this process is still on the order of T, we are back in the situation studied in Figs. 5.2.6–5.2.8.

In practice, the effect of time dependence is to make it easier to satisfy two-time boundary conditions requiring occupancy of 1 at both 0 and T. 'Easier' means the conditions are satisfied by a larger fraction of the sample space. As a result, our qualitative conclusions are unchanged and effective transition rates (as illustrated in Fig. 5.2.8) are even closer to their unconditioned values.

General results using the spectral
decomposition of the transition matrix

Thermodynamics in the richly conditioned world we have been discussing will be different from what we are used to. The development of that thermodynamics in ways more complex than can be immediately intuited (as we did in Chapter 4) may be accomplished by formalizing the calculations of the present section. I now give two examples of this, deriving in a general way the 'cosh' that arose in Section 4.5, and obtaining an expression for the 'effective transition rate,' Eq. (5.2.6).

To this end, we return to a more systematic study of Eq. (5.2.3) or Eq. (5.2.5). The stochastic matrix W has well-known properties, some of which were alluded to in Chapter 2. In particular we can use its spectral decomposition[9] to write

$$W = \sum_\alpha \lambda_\alpha P_\alpha \qquad (5.2.9)$$

For any stochastic matrix there is always the eigenvalue $\lambda_0 = 1$, with its constant left eigenvector. If this constant is set to 1, the associated right eigenvector is a normalized stationary distribution, $\bar{p}$, and for systems satisfying detailed balance this is the vector that enters the detailed balance relation. It follows that the projection for the eigenvalue 1 is $P_0 = \bar{p} \cdot (1, \ldots, 1)$, or $(P_0)_{xy} = \bar{p}_x$. All eigenvalues $\{\lambda_\alpha\}$ are on or inside the unit circle. In general the possibilities for the spectrum are extensive, but if there is a slowest relaxation mode, with the next rate being somewhat faster, the spectrum satisfies $1 > \lambda_1 \gg |\lambda_2| \geq \ldots$. Under this

[9] Without detailed balance a Jordan form may be required.

circumstance

$$W^t = P_0 + \lambda_1^t[P_1 + \mathcal{R}] \tag{5.2.10}$$

with the remainder $\mathcal{R}$ involving $(\lambda_2/\lambda_1)^t P_2$ and other projections. To evaluate $\Gamma(x, t; \mathcal{E})$ from Eq. (5.2.5) we require $(W^t)_{x\alpha}$, $(W^{T-t})_{\beta x}$ and $(W^T)_{\beta\alpha}$. Since our goal is to recover the 'cosh' (and to see that it is an approximation), we concentrate on the x and t dependencies. As such, $(W^T)_{\beta\alpha}$ is an overall multiplicative constant. For symmetric behavior, we take $\alpha = \beta$. Using Eq. (5.2.10), after a few steps, this gives

$$\Gamma(x, t; \mathcal{E}) = \text{const} \cdot \left(P_0 + \lambda_1^{T-t}[P_1 + \mathcal{R}]\right)_{\alpha x} \left(P_0 + \lambda_1^t[P_1 + \mathcal{R}]\right)_{x\alpha}$$

$$= a(x) + b(x)\left[\lambda_1^t + \lambda_1^{T-t}\right] + c(x)\lambda_1^{-T} + \mathcal{R}'$$

$$\tag{5.2.11}$$

The substantive step in deriving Eq. (5.2.11) is the use of detailed balance for P_1. To justify this, introduce the matrix $S_{\alpha\beta} \equiv (1/\sqrt{\bar{p}_\alpha})W_{\alpha\beta}\sqrt{\bar{p}_\beta}$. By the detailed balance of W, S is symmetric and Hermitian. Its spectral expansion is trivially related to that of W and all the projections (for S) are Hermitian. Since we assume λ_1 is real and nondegenerate, its projection is real and therefore symmetric. It follows that P_1 satisfies detailed balance. In Eq. (5.2.11), a, b and c are functions of x only. The remainder term, $\mathcal{R}'$, is built of several pieces, but has the property of declining rapidly away from $t = 0$ and $t = T$. (This was an explicit assumption.) *From Eq. (5.2.11) it follows that $\Gamma(x, t; \mathcal{E})$ is the sum of a constant plus* $\cosh[\log \lambda_1(t - T/2)]$. For a decay scenario one might take $\lambda_1^{T/2}$ small (but not too small), $x = \alpha$, and $\bar{p}_\alpha$ small. In that case the constant would also be small. Note too that $\mathcal{R}'$ is *not* in general zero, and would yield corrections to the 'cosh' approximation.

The normalcy of contemporary transition rates in the face of future conditioning relies on $\widetilde{W}/W \approx 1$. This was illustrated in Fig. 5.2.8. For richer dynamics and with the above spectral assumptions, it is easy to show that

$$\frac{\widetilde{W}_{xy}}{W_{xy}} = \frac{\bar{p}_\beta + P_1(\beta, x)\lambda_1^{T-t-1}}{\bar{p}_\beta + P_1(\beta, y)\lambda_1^{T-t}} \tag{5.2.12}$$

(where β is the final state and detailed balance is not used). For the case of interest, $\lambda_1^T = O(1)$ for large T, so that the extra power of λ_1 in the denominator is irrelevant. The only difference then between numerator and denominator is in the second argument of P_1. This translates into a possible difference in the left eigenfunction associated with λ_1. For $1 - \lambda_1 \ll 1 - |\lambda_2|$ this left eigenvector generally only takes two values, their respective supports representing different phases (in a sense) of the system. Thus the ratio will be very close to unity when x and y are in the same 'phase.' However, when they are not, it is also the case that W_{xy} is itself small. This confirms the remarks made earlier. It will also be interesting to check these ideas without the spectral assumptions of the current discussion.

Exercise 5.2.3. Show that if λ_1 is real and $T \log |\lambda_1| \gg 1$, then for small t, $\Gamma(x,t;\mathcal{E}) \approx G(x,t;\alpha)$. Show too that for small $T - t$, $\Gamma(x,t;\mathcal{E}) \approx G(x,T-t;\beta)$. This is a precise form of the intuitive remarks in Chapter 4 about time symmetry and the normalcy of equilibration despite the future conditioning.

Exercise 5.2.4. These methods can be used to study *causality*, as defined in Section 4.5. Suppose we give future boundary conditions and want to compare two dynamical evolutions. In one of them the stochastic matrix W is used on all time steps. In the other, a different matrix, say W', is used on step t_0. The system can be viewed as having been perturbed at t_0. Causality, in the macroscopic sense, means that until t_0 the two evolutions show the same behavior, with the effects of the perturbation $W' - W$ only being felt subsequent to t_0. Develop the appropriate modification of Eq. (5.2.3) to cover the non-autonomous (time-translation non-invariant) evolution and show that under the conditions of the previous exercise causality is recovered if t_0 is close to 0. You can also derive reverse causality if t_0 is near the end.

Remark: Some results in this section require use of detailed balance for the stochastic dynamics. For a general open system (which is how we view the universe when we treat gravity as an external force) appropriate modification would be needed. The qualitative features arising from future conditions and from the variety of time scales should be unchanged.

Exercise 5.2.5. *Observers with opposite time sense.* With the accumulated experience of Sections 4.3, 5.1 and 5.2, we can pose one of the more science-fiction-like questions that has been raised when discussing arrows of time. Can the universe simultaneously contain more than one macroscopic time sense? This might be answered affirmatively with escalating degrees of the bizarre. Least dramatic: there are regions with opposite time sense, but they are separated by vast expanses in equilibrium. Next: the systems are in mild contact and in some sense can observe one another. Most extreme: both arrows occur for the same collection of particles, perhaps through different choices of coarse grains. By posing two-time boundary value problems for model systems (e.g., variations of the cat map) one can turn these speculations into well-defined questions. I won't admit to having spent much time on this, but I'd say that the least radical idea is easy to implement. To the extent that I've tried to set two-time boundary conditions to accomplish the next level, I found that solutions tended to be those intermediate motions for which the systems did not much interact.

5.3 Quantum dynamics

The quantum two-time boundary value problem is difficult. It has no *a priori* natural definition, proposed formulations resist solution, and, one of the main simplifying features of other dynamical contexts—the fact that the further separated in time the boundaries are, the easier to find a solution—works in the opposite direction. As the separation increases, it becomes *more* difficult to find solutions. And, as if that were not enough, it is not clear (to me anyway) what influence indistinguishability of particles has.

In this section I will pose and examine two such problems, two-time localization of wave packets and the defining of boundary values in terms of subspaces of Hilbert space.

Two-time wave packet localization

As an imitation of the classical two-time problem, one can try to have a particle's wave packet well localized around a particular position at one time and around another position at a later time. One also has in mind that if you would like to define a two-time problem in terms of coarse grains, it is likely that spatial localization will be a property of naturally defined coarse grains.[10] It remains to formulate 'well localized.' We define this in terms of the spatial standard deviation.

Here then is the problem. A particle has coordinates in $\mathcal{M} \subset \mathbb{R}^n$. It evolves under a *fixed* Hamiltonian,[11] H. At a given time t_1 we wish the wave function to be well localized about a point $x_1 \in \mathcal{M}$. At a later time t_2 we make the same demand about a point $x_2 \in \mathcal{M}$. Between those times the wave function evolves under H. Define the spread, $S(x_0; \phi)$, of a function ϕ about a point $x_0 \in \mathcal{M}$ as follows:

$$S(x_0; \phi) \equiv \int_{\mathcal{M}} \phi^*(x)(x - x_0)^2 \phi(x) dx \qquad (5.3.1)$$

Suppose the time t_k wave function is ψ_{t_k}, so that the wave functions at different times are related by $\psi_{t_2} = \exp[-iH(t_2 - t_1)/\hbar]\psi_{t_1}$. (This assumes a time independent H.) Then our two-time localization problem is defined to be the minimization of the quantity

$$W \equiv S(x_1; \psi_{t_1}) + S(x_2; \psi_{t_2}) \qquad (5.3.2)$$

Since everything but the wave function has been given, it is the choice of wave function that allows us to reduce W. To focus on this we pick a fiducial time t_0 and relate the wave function at all other times to this one, using $\exp(-iHt/\hbar)$. Without loss of generality, let $t_0 = 0$. Thus $\psi_t = \exp(-iHt/\hbar)\psi_0$. It follows that W depends on ψ_0 as well as on x_k and t_k $(k = 1, 2)$, and of course on H. Since for now only ψ_0 is to be varied we write

$$W(\psi_0) = S(x_1; e^{-iHt_1/\hbar}\psi_0) + S(x_2; e^{-iHt_2/\hbar}\psi_0) \qquad (5.3.3)$$

---

[10] Compare comments in the opening of Chapter 6 on locality and coarse graining.
[11] People sometimes look for a Hamiltonian with best possible multi-time localizations. That is not our question, although we later consider several Hamiltonians and for each seek the best localization consistent with it.

The Heisenberg position operator, with 0 as its fiducial time, is

$$\hat{x}(t) \equiv e^{iHt/\hbar} x e^{-iHt/\hbar} \tag{5.3.4}$$

Comparing Eq. (5.3.1) and Eq. (5.3.3), it is obvious that the operator $\exp(-iHt/\hbar)$ can be applied to the 'x' rather than to the ψ or ϕ, so that W becomes

$$\begin{aligned}
W(\psi_0) &= \int_{\mathcal{M}} \psi_0^*(x) \left\{ (\hat{x}(t_1) - x_1)^2 + (\hat{x}(t_2) - x_2)^2 \right\} \psi_0(x) \\
&= \left\langle \psi_0 \middle| (\hat{x}(t_1) - x_1)^2 + (\hat{x}(t_2) - x_2)^2 \middle| \psi_0 \right\rangle
\end{aligned} \tag{5.3.5}$$

The problem posed is to find the ψ_0 that minimizes W. From Eq. (5.3.5) it follows that this means looking for the minimum or ground state of the operator

$$\widehat{W} \equiv \left(\hat{x}(t_1) - x_1 \right)^2 + \left(\hat{x}(t_2) - x_2 \right)^2 \tag{5.3.6}$$

Remark: In this form, an obvious extension of our localization problem is to look for other moments of $\hat{x}(t)$, or more generally other functions of $\hat{x}(t)$. What is not included in the present formulation is the use of higher powers of the wave function, for example $\int_{\mathcal{M}} |\phi(x)|^4 dx / [\int_{\mathcal{M}} |\phi(x)|^2 dx]^2$. Such definitions are not natural to Hilbert space, but they do arise in condensed matter applications.

The operator $\widehat{W}$ is the sum of the squares of two Hermitian operators and positive. Call its smallest eigenvalue W_0. If it happens that those operators have a commutator that commutes with both of them, W_0 is given by the absolute value of that commutator. We relegate the proof to the following exercise.

Exercise 5.3.1. Let $K \equiv A^2 + B^2$ with A and B self adjoint and $[B, A] = i\gamma$ with γ a positive real scalar. Show that $A - iB$ is a lowering operator for K and that the lowest eigenvalue of K is γ. What is the extension of this result when γ is an operator that commutes with both A and B?

The simplest application is to the one-dimensional harmonic oscillator $H = p^2/2m + m\omega^2 x^2/2$, with $x \in \mathbb{R}$. For the Heisenberg operators we have the explicit solution

$$\hat{x}(t) = \hat{x} \cos \omega t + \hat{p} \frac{\sin \omega t}{m\omega} \tag{5.3.7}$$

where $\hat{x}$ and $\hat{p}$ are the time-0 operators. Using $[\hat{x}, \hat{p}] = i\hbar$, one obtains

$$[(\hat{x}(t_1) - x_1), (\hat{x}(t_2) - x_2)] = \frac{i\hbar}{m\omega} \sin \omega (t_2 - t_1) \tag{5.3.8}$$

It follows from Exercise 5.3.1 that

$$W_0 = \frac{\hbar |\sin \omega (t_2 - t_1)|}{m\omega} \tag{5.3.9}$$

Exercise 5.3.2. How can it be that this is independent of x_1 and x_2?

Exercise 5.3.3. Find the eigenfunction associated with W_0.

Exercise 5.3.4. For any two self adjoint operators, A and B, the generalized uncertainty relation is the following. Let ψ be a fixed state; for any operator Q, write $\langle Q \rangle \equiv \langle \psi | Q | \psi \rangle$. Let $\Delta A \equiv \left\{ \langle [A - \langle A \rangle]^2 \rangle \right\}^{1/2}$, and similarly for B. Let $[A, B] = iC$. Then $\Delta A \Delta B \geq \frac{1}{2} |\langle C \rangle|$. This is proved using the Schwarz inequality. Prove Eq. (5.3.9), using the generalized uncertainty relation and fact that the arithmetic mean of two numbers is equal to or greater than the geometric mean.

Eq. (5.3.9) covers three important cases: $\omega^2 > 0$, $\omega = 0$ and $\omega^2 < 0$. For real positive ω, W can be kept small no matter how large $|t_2 - t_1|$. In fact, for certain non-zero time differences it can be zero. The case $\omega = 0$ is the free particle result and we have the limiting form $W_0 = \hbar |t_2 - t_1|/m$. This is a basic localization limitation due to wave packet spreading. As $|t_2 - t_1|$ grows, so does W_0. The growth is diffusion-like in that the least spread ('Δx' $\sim \sqrt{W_0}$) grows like the square root of the time difference.[12] The case of pure imaginary ω is not as exotic as it sounds in terms of potential models. It also represents local path dynamics for a classically chaotic system or motion on a space of negative curvature. In these situations wave packet spreading is exponentially rapid: the 'sine' becomes the hyperbolic sine. (This is a place where classical chaos has the effect on a quantum system that you would have expected.)

The results I have just described extend to situations more general than the harmonic oscillator. Consider a pair of particles, each moving in $\mathbb{R}^3$ with a potential V between them, depending only on their relative coordinate. *For the relative coordinate alone*, large $|t_2 - t_1|$ behavior of W_0 will depend only on whether V has bound states. If there is a bound state, put the particles in it. No matter how bad this choice is for the given x_1 and x_2, the growth of W_0 in time ultimately stops. On the other hand, if there is no bound state, then for large enough time (interval) the diffusive behavior takes over. Of course for the center of mass coordinate of the particles, if they are in no confining external potential, there will be the previously calculated free particle spreading, except that the effective mass (the denominator in Eq. (5.3.9)) will be larger. Similar results hold for smoothly varying potentials. See Appendix A to this section.

Indistinguishable particles

A question that arises later is the effect of the indistinguishability of identical particles. I do not have an answer to this question. Partly it is not clear what is the appropriate definition of localization. When there is a particle near spatial position A and an identical particle near

---

[12] From the solution to Exercise 5.3.3 it is seen that the optimal packet *shrinks* and then expands. Not all wave packets expand all the time.

spatial position B, you don't want to include the distance between A and B in your spread. Eliminating this effect for many particles introduces consistency problems (at least for the ways I've proposed). It may be that intrinsic clumpiness definitions are better (for example the form $\int_{\mathcal{M}} |\phi(x)|^4 dx / [\int_{\mathcal{M}} |\phi(x)|^2 dx]^2$). Ultimately—as discussed later in this book—the two-time boundary conditions should have a physical origin. Once that is understood the correct formulation would be fixed by physical considerations.

Despite the absence of a clear and simple result like Eq. (5.3.9), I expect that for fermions two-time localization will be *more* difficult than it is for distinguishable particles, while for bosons it would be less difficult. This is based on the same features that lead to what one calls 'exchange forces' in magnetism, namely the tendency for fermions to effectively repel—from statistics alone, even if there are no dynamical forces between them. This is illustrated in Fig. 5.3.1, which shows the absolute value of the *two-particle* wave function for two one-dimensional particles. This wave function is built from two single-particle wave functions (with coordinates x_1 and x_2), centered on points that are within the (one-particle) wave function overlap. The same one-particle wave functions are used for both fermion and boson cases. The left figure is the boson wave function and the density is highest along the line $x_1 = x_2$. The right figure uses the same one-particle wave functions, but shows the (normalized) fermion wave function. Now the maximum density occurs in two places, symmetrically disposed about the line $x_1 = x_2$. By any measure of localization, the fermion wave function is more spread than that of the bosons.

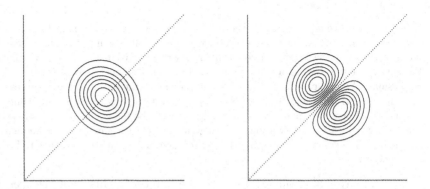

Fig. 5.3.1. Contour plot of the absolute value of the wave function for bosons (left side) and fermions. Both cases use the same one-particle wave functions. The one-particle coordinate space is one dimensional and the dotted lines represent coincidence of the particles.

Subspace boundary value problems

We next consider a more abstract form of the two-time boundary value problem. It is mathematically cleaner, and makes no reference to spatial localization.[13] The boundary condition is a demand that the wave function falls into particular subspaces of the Hilbert space at the boundary times.

Again, we specify a Hamiltonian, H. It acts on an N-dimensional Hilbert space $\mathcal{H}$. Consider the demand that at time $t = 0$ the system lie in a subspace $\mathcal{H}_0$ and that at a later time $t = 1$ it lie in $\mathcal{H}_1$. (One can scale H so that $t = $ '1' is long or short.) Let P_k, $k = 0, 1$, be the projections from $\mathcal{H}$ to $\mathcal{H}_k$ and let their dimensions be N_k. Let $U = \exp(-iH/\hbar)$. Our two-time boundary condition is the stipulation of a vector $\psi \in \mathcal{H}$ such that

$$P_0\psi = \psi \quad \text{and} \quad P_1 U\psi = U\psi \qquad (5.3.10)$$

We first show that if $n \equiv N_0 + N_1 - N > 0$, then there is at least an n-dimensional space of solutions to Eq. (5.3.10). To see this write ψ in a basis for which P_0 is diagonal and projects on the first N_0 components, $\psi = \sum_{\ell=1}^{N_0} a_\ell |e_\ell\rangle$. This evolves, in time, to $U\psi = \sum_{\ell=1}^{N_0} U a_\ell |e_\ell\rangle$. Let the basis in which P_1 is diagonal be $|f_k\rangle$, with P_1 projecting on the *last* N_1 vectors. Then the requirement Eq. (5.3.10) becomes

$$\sum_{\ell=1}^{N_0} a_\ell \langle f_k | U | e_\ell \rangle = 0, \qquad k = 1, \ldots, (N - N_1) \qquad (5.3.11)$$

In the worst case, this gives $N - N_1$ constraints on N_0 parameters (a_ℓ) with an $n = N_0 - (N - N_1)$ dimensional space of solutions. Depending on the rectangular matrix, $\langle f_k | U | e_\ell \rangle$, $k = 1, \ldots, (N - N_1)$, $\ell = 1, \ldots, N_0$, this space could be larger.

Remark: Later we will emphasize the role of a complex environment in the solution of two-time quantum boundary value problems. Note that a complex environment does not directly affect the parameter counting argument just given. In lowest approximation the environment enters as a cross product; if the environment is described by an M-dimensional Hilbert space, then with the notation above, the initial and final spaces would have dimensions $N_0 M$ and $N_1 M$. If $N_0 + N_1 < N$, multiplying by M will not change the inequality. Where the environment *can* help is in the boundary value problem we next consider, in which small deviations are allowed from perfect subspace containment. As we shall see in Section 7.1, enlarging the environment allows those deviations to be made extremely small.

For the physical applications to be made later in this book, the requirement Eq. (5.3.10) is too stringent. We soften it in the following way: ask

---

[13] This is a mixed blessing in that the physical two-time boundary value problem may arise from spatial localization constraints.

only that there be a vector that *almost* succeeds. That is, we seek a normalized vector $\psi \in \mathcal{H}$ such that

$$\| P_1 U P_0 \psi \|^2 > 1 - \varepsilon \tag{5.3.12}$$

In effect, we wish to maximize $\| P_1 U P_0 \psi \|^2$. Define

$$A \equiv P_1 U P_0 \tag{5.3.13}$$

Our goal is therefore to find the largest eigenvalue and associated eigenvector of (the $N_0 \times N_0$ matrix)

$$B \equiv A^\dagger A = P_0 U^\dagger P_1 U P_0 \tag{5.3.14}$$

It is easy to show that the spectrum of B lies on the interval $[0, 1]$. As we will see in Chapter 7, it often happens that although the average, $\operatorname{Tr} B / N_0$, is significantly different from either 0 or 1, B has eigenvalues close to the extremes. In some examples it even turns out that *most* eigenvalues are at the edges. Furthermore, the proximity of approach can be remarkable. In one example in Chapter 7, although $N_0 \approx 40$, there are eigenvalues of B that differ from 1 by as little as 10^{-11}. (The Hamiltonian in that case is suitable for a decay problem. P_0 is the projection on undecayed states while P_1 is either P_0 or its complement.)

In this section I will look at the problem in a general way, and extend the analysis performed for the decay model to a more general situation. I will argue that the sudden transitions recounted in the decay model can occur in general contexts.[14] We consider the case where the parameter counting is borderline, i.e., where $P_0 + P_1 = 1$ and $P_0 P_1 = 0$. If h is defined as the restriction to $\mathcal{H}_0$ of $P_0 H P_0$, $\tilde{h}$ the restriction to $\mathcal{H}_1$ of $P_1 H P_1$, with similar definitions for the rectangular matrices connecting $\mathcal{H}_0$ and $\mathcal{H}_1$, the total Hamiltonian takes the form

$$H = \begin{pmatrix} h & C^\dagger \\ C & \tilde{h} \end{pmatrix}$$

A vector $\psi \in \mathcal{H}$ is written $\psi = \begin{pmatrix} x \\ Y \end{pmatrix}$ with x the restriction of $P_0 \psi$ to $\mathcal{H}_0$. Y is the restriction of $P_1 \psi$. Our initial condition is $Y(0) = 0$. Integrating the Schrödinger equation, $i\dot{\psi} = H\psi$, allows us to express Y in terms of x:

$$Y(t) = -i \int_0^t ds\, \tilde{u}(t - s) C x(s) \qquad \text{with} \qquad \tilde{u}(t) \equiv \exp(-it\tilde{h})$$

This is substituted into the Schrödinger equation for x:

$$\dot{x}(t) = -ihx(t) - \int_0^t ds\, C^\dagger \tilde{u}(t - s) C x(s) \tag{5.3.15}$$

---

[14] This is relevant for a description of 'quantum jumps.' References in Section 5.4.

In the decay model of Section 7.1, the form of C allows Eq. (5.3.15) to become $\dot{x} = -ihx - \alpha\eta\eta^{\dagger}x$, with η an N_0-vector, making $\eta\eta^{\dagger}$ a one-dimensional projection. Here we only use a general property, one that refers to the irreversibility that is present when systems are coupled to an environment. In Section 2.5 we worked with kernels similar to $C^{\dagger}\tilde{u}(\tau)C$ and found that for τ larger than a characteristic microscopic time scale the kernel vanished. If the time scale of interest is greater than that microscopic time scale, then $\tilde{u}$ acts essentially like a δ-function (in τ). We therefore assume that on a long enough time scale Eq. (5.3.15) takes the form[15]

$$\dot{x}(t) = -ihx(t) - dd^{\dagger}x(t) \qquad (5.3.16)$$

with d proportional to $C^{\dagger}$, by a proportionality factor depending on the relative time constants. Now $dd^{\dagger}$ is a Hermitian non-negative matrix and therefore has a spectral decomposition $dd^{\dagger} = \sum_{\alpha} r_{\alpha}\tilde{P}_{\alpha}$, with $r_{\alpha} \geq 0$, and $\tilde{P}_{\alpha}$ a projection. Thus Eq. (5.3.16) can be rewritten as

$$\dot{x} = -ihx - \sum_{\alpha} r_{\alpha}\tilde{P}_{\alpha}x \qquad (5.3.17)$$

In the absence of special structure, the overall decay rate would be $\Gamma = 2\,\mathrm{Tr}\,dd^{\dagger}/N_0$. This can be seen by writing

$$\frac{d}{dt}|x|^2 = -2x^{\dagger}dd^{\dagger}x = -2\sum r_{\alpha}x^{\dagger}\tilde{P}_{\alpha}x$$

'Absence of special structure' will be taken to mean $x^{\dagger}\tilde{P}_{\alpha}x = \dim(\tilde{P}_{\alpha})/N_0$, from which our decay rate assertion follows.

A priori there is no reason for all the r_{α} to equal one another. On the contrary, I have done numerical exploration with not-unreasonable Hamiltonians and found that the decay of particular initial conditions may be much faster or much slower than the average. For example, take the Hamiltonian h of the undecayed system plus environment to be a random matrix in the spirit of Wigner's treatment of nuclei. (Specifically, I used a matrix from the Gaussian orthogonal ensemble (GOE).) For the interaction Hamiltonian connecting the excited levels with the ground state *cum* decay products, I used rather different statistics. Taking a cue from results that will be derived in Chapter 9 of this book, I took the elements of the matrix d to be random, but distributed according to the Cauchy distribution.[16] One then observes that $x(t) = \exp(-Qt)x(0)$,

---

[15] This is not the most general form. The loss of information involves collusion with the coupling, so that the product form need not be preserved even with a short memory.

[16] The Cauchy distribution is of the form $C_a(x) = a/[\pi(a^2 + x^2)]$, with a a parameter.

with $Q = ih + dd^\dagger$, so that, recalling the definition of x, one has that
the A of Eq. (5.3.13) (with $P_1 = P_0$) is given by $A = \exp(-Qt)$. This
makes it easy to find the spectrum of B (of Eq. (5.3.14)) and to address
the existence of vectors satisfying Eq. (5.3.12). For the average decay one
looks at $\operatorname{Tr} B$ and indeed one finds that its dropoff (for the Hamiltonian
described above) is exponential. However, there are also modes that decay
almost immediately as well as others that lag significantly behind the
average. These provide solutions to Eq. (5.3.12). For this Hamiltonian,
the diagonalized $dd^\dagger$ showed a wide spread in eigenvalues, as well.

A similar theme has been noted in studies of the 'quantum dynamical
suppression of classical chaos.' One has a system whose classical dynamics
are ergodic. If the quantum dynamics spread a wave function through
Hilbert space as thoroughly as the classical mechanics spread phase space
volumes, then one would expect parameter counting arguments of the
sort I gave above to be relevant. In fact many initial wave functions
do *not* spread throughout phase space. This phenomenon has been the
object of intense research for several years and comes under the rubric,
'quantum localization.' For wave functions that get caught in a relatively
small region of Hilbert space, with a judicious choice of P_0 and P_1 (in our
notation), one will be able to solve the two-time boundary value problem,
despite what one might have thought would be rather dim chances—based
on parameter counting. For non-ergodic classical mechanics one expects
similar effects.

Exercise 5.3.5.   One formalism for two-time boundary conditions uses the density
matrix. Define an ensemble and associated density matrix ρ by the conditions: $\operatorname{Tr}\rho = 1$,
$\operatorname{Tr}\rho P = p \operatorname{Tr} P$, $\operatorname{Tr}\rho_T Q = q \operatorname{Tr} Q$, where P and Q are projections, $\rho_T = U\rho U^\dagger$ with
$U = \exp(-iHT/\hbar)$, and p and q are the input data.   (1) Show that in a maximum
entropy specification of ρ we get the usual form, $\rho = \exp[\lambda + \mu P + \nu U^\dagger QU]$, despite the
general lack of commutativity of P, Q, and H. Find equations for λ, μ, and ν. (2) The
important physical quantity turns out to be $\operatorname{Tr} PU^\dagger QU$. For $n = \operatorname{Tr} P = \operatorname{Tr} Q < \infty$, let
$\cos\theta \equiv \operatorname{Tr} PU^\dagger QU/n$. The condition $\cos\theta \sim 0$ for large T is a kind of ergodicity. Find
restrictions on p and q for $\cos\theta \sim 0$.   (3) Again letting $\operatorname{Tr} P = \operatorname{Tr} Q$, suppose $p = q$.
Find restrictions on p for general $\cos\theta$.

Appendix for this section (at the end of the chapter):
5.3.A   Relative path dynamics and wave packet localization

5.4 Notes and sources

Section 5.0. Maupertuis's variational principle of classical mechanics was published in 1747. Requiring that a particle's path minimize a functional based on
its time and place of *arrival* led to interpretations of the action principle in terms
of divine teleology. In those days that was OK, but what did cause trouble was

accusations of plagiarism from earlier work by Leibniz. At one point Maupertuis had led an expedition to Lapland to check the oblateness of the earth, only to be mocked by Voltaire—once this Leibniz business had erupted—as an 'earth flattener.' See [SOMMERFELD].

Section 5.1. A good place to learn about Lagrangian manifolds is [LITTLE-JOHN]. Arguments for path multiplicity in chaotic dynamics can be found in [SCHULMAN 94a]. Reprints, with commentary, of many fundamental papers on Hamiltonian dynamics are in [MACKAY]. Expositions of Morse theory (mentioned in Appendix A to Section 5.1) are in [MILNOR] and [POSTNIKOV]. Numerical methods for two-time boundary value problems with ordinary differential equations are in [FOX].

The two-time boundary value problem for the Kac ring model was studied in [SCHULMAN 76, 77], where the result in Exercise 5.1.1 was derived. The work on temporally displaced interactions mentioned at the end of the section is [SCHULMAN 95]. This paper also discusses causality in an asymmetric linear theory.

Section 5.2. Stochastic two-time boundary value problems were considered by [SCHRÖDINGER 32] in the context of trying to understand certain ideas of Eddington. From this developed the theory of Bernstein processes. For the Ehrenfest urn model, [COCKE] considered imposing future conditioning. Two-time boundary conditions for model and for physical systems were studied in [SCHULMAN 73, 76, 77] and several other articles. Much of the material in Section 5.2 is from [SCHULMAN 91a]. For the assertions (following Eq. (5.2.12)) on the projection operator, 'P_1,' see [GAVEAU 97].

Section 5.3. Two-time boundary condition problems for wave packets were studied in [SCHULMAN 88]. The semiclassical result in Appendix A to Section 5.3 for a slowly varying potential is based on [SCHULMAN 89]. For background on the path integral (used in that Appendix), see [SCHULMAN 81]. Many path integral developments since my 1981 book can be found in [CERDEIRA]. For relativistic two-time localization see [SCHULMAN 92]. The Hilbert space view is given in [SCHULMAN 94b]. As remarked, the analysis there extends [SCHULMAN 91b]. Quantum jumps were reported in [BERGQUIST]. Quantum localization and quantum dynamical suppression of classical chaos is treated in [BOHIGAS], especially Section 6.

The Heisenberg picture, with the possibility of a time dependent Hamiltonian, is given in [GOTTFRIED], Section 28. The Jacobi equation or the equation of geodesic deviation is discussed in [SCHULMAN 81], Section 12. See also Section 5.9 of [ADLER]. The generalized uncertainty relation in Exercise 5.3.4 is proved in [MERZBACHER].

References

ADLER 75   R. Adler, M. Bazin & M. Schiffer (1975) *Introduction to General Relativity*, 2nd ed., McGraw-Hill, New York.

BERGQUIST 86   J. C. Bergquist, R. G. Hulet, W. M. Itano & D. J. Wineland (1986) Observation of Quantum Jumps in a Single Atom, *Phys. Rev. Lett.* **57**, 1699.

BOHIGAS
93
O. Bohigas, S. Tomsovic & D. Ullmo (1993) Manifestations of classical phase space structures in quantum mechanics, *Phys. Rep.* **223**, 43.

CERDEIRA
93
H. Cerdeira, S. Lundqvist, D. Mugnai, A. Ranfagni, V. Sa-yakanit & L. S. Schulman (1993) *Lectures on Path Integration: Trieste 1991*, World Sci., Singapore. (Workshop and conf. at ICTP, Trieste.)

COCKE
67
W. J. Cocke (1967) Statistical Time Symmetry and Two-Time Boundary Conditions in Physics and Cosmology, *Phys. Rev.* **160**, 1165.

FOX
67
L. Fox (1957) *The Numerical Solution of Two-Point Boundary Problems In Ordinary Differential Equations*, Clarendon, Oxford. (Reprinted, 1990, by Dover, New York.)

GAVEAU
97
B. Gaveau and L. S. Schulman (1997) Theory of non-equilibrium first order phase transitions for stochastic dynamics, preprint.

GOTTFRIED
66
K. Gottfried (1966) *Quantum Mechanics*, Benjamin, New York.

LITTLEJOHN
92
R. G. Littlejohn (1992) The Van Vleck Formula, Maslov Theory, and Phase Space Geometry, *J. Stat. Phys.* **68**, 7.

MACKAY
87
R. S. MacKay & J. D. Meiss (1987) *Hamiltonian dynamical systems: a reprint selection*, Adam Hilger, Bristol.

MERZBACHER
78
E. Merzbacher (1970) *Quantum Mechanics*, 2nd ed., Wiley, New York.

MILNOR
63
J. Milnor (1963) *Morse Theory*, Princeton Univ. Press, Princeton.

POSTINKOV
67
M. M. Postnikov (1967) *The Variational Theory of Geodesics*, Saunders, Philadelphia. Russian ed., Nanka Press, Moscow, 1965.

SCHRÖDINGER
32
E. Schrödinger (1932) Sur la theorie relativiste de l'electron et l'interpretation de la mecanique quantique, *Ann. Inst. H. Poincaré* **II**, 269.

SCHULMAN
73
L. S. Schulman (1973) Correlating Arrows of Time, *Phys. Rev.* D **7**, 2868.

SCHULMAN
76
L. S. Schulman (1976) Normal and Reversed Causality in a Model System, *Phys. Lett.* A **57**, 305.

SCHULMAN
77
L. S. Schulman (1977) Illustration of Reversed Causality with Remarks on Experiment, *J. Stat. Phys.* **16**, 217.

SCHULMAN
81
L.S. Schulman (1981) *Techniques and Applications of Path Integration*, Wiley, New York.

SCHULMAN
88
L. S. Schulman (1988) Two Time Localization, *Ann. Phys.* **183**, 320.

SCHULMAN
89
L. S. Schulman (1989) Semiclassical Two-Time Localization and the Consequences of Restrictions on Wave Packet Spreading, in *Path Integrals from meV to MeV*, ed. V. Sa-yakanit *et al.*, World Sci., Singapore.

SCHULMAN
91a
L. S. Schulman (1991) Models for Intermediate Time Dynamics with Two-Time Boundary Conditions, *Physica* A **177**, 373.

SCHULMAN
91b
L. S. Schulman, C. R. Doering & B. Gaveau (1991) Linear decay in multi-level quantum systems, *J. Phys.* A **24**, 2053.

SCHULMAN
92
L. S. Schulman (1992) Relativistic Two-time Localization, *J. Phys.* A **25**, 3007.

SCHULMAN
94a
L. S. Schulman (1994) Accuracy of the semiclassical approximation for the time dependent propagator, *J. Phys.* A **27**, 1703.

SCHULMAN
94b
L. S. Schulman (1994) Particle detection via special states, in *The interpretation of quantum theory: where do we stand?*, ed. L. Accardi, Encic. Ital., Rome. Conf. at Columbia Univ., New York, 1992.

SCHULMAN
95
L. S. Schulman (1995) Time displaced interactions: classical dynamics and path integral quantization, *J. Math. Phys.* **36**, 2546.

SOMMERFELD   A. Sommerfeld (1964) *Mechanics*, Vol. 1, *Lectures in Theoretical Physics*,
64   Academic, New York. (Trans. 4th German ed. by M. O. Stern.)

Notes on the exercises

Exercise 5.1.4 As for Exercise 5.1.3, this is done in [SCHULMAN 77]. The small set of
balls has exponentially decaying color. For the cat map, the decay *is* influenced by the
future condition, much as in Fig. 4.3.4.

Exercise 5.2.1 $\Pr(X_{t+1} = x \mid X_t = y \,\&\, X_T = \beta) = \Pr(X_{t+1} = x \,\&\, X_t = y \,\&\, X_T = \beta)/\Pr(X_t = y \,\&\, X_T = \beta)$

Exercise 5.2.2 $\widetilde{W}_{xy} = W_{xy} \exp\left(-(x-y)\left(\frac{x+y}{2} - \xi\right)/2D(T-t)\right)$

Exercise 5.3.3 The time 0 wave function is

$$\phi_0(x) = \left[\frac{m\omega \cot(\omega T/2)}{\hbar\pi}\right]^{1/4} \exp\left[\frac{-m\omega}{2\hbar}\cot\left(\frac{\omega T}{2}\right)(x-x_0)^2 + i\frac{p_0 x}{\hbar}\right]$$

where $T = t_2 - t_1$ and x_0 and p_0 are the (*c*-number) time-0 position and momentum that
give x_k at t_k $(k = 1, 2)$ for the corresponding classical oscillator.

Appendix 5.0.A: Counting states
in a cubic centimeter of air

Quantum mechanics imposes a scale for the fine graining of phase space, namely $2\pi\hbar$. In
a $2n$-dimensional phase space each region of volume $(2\pi\hbar)^n$ gives rise to an independent
Hilbert space direction (or basis vector). With this scale we estimate the number of
microstates available to the atoms in a cubic centimeter of air. Actually, I'll estimate a
smaller number (by a few million powers of ten), the states associated with a monatomic
ideal gas of atomic weight 30, at room temperature and at atmospheric pressure. Let
N_1 be the number of particles in the given volume. Let $\bar{p}$ be a characteristic momentum
scale for the particles.

Exercise 5.0.A.1.   Show that $N_1 \sim 2.4 \times 10^{19}$, and $\bar{p} \sim 1.5 \times 10^{-18}$ g cm/s.

For each dimension (of three) and for each particle there are on the order of $\Delta x \Delta p/2\pi\hbar$
microstates. Taking $\Delta p \sim \bar{p}$ gives about 2.3×10^8 microstates per particle. The big
numbers come from exponentiating this to the number of degrees of freedom. The number
of microstates is thus

$$(10^{8.4})^{(3 \times 2.4 \times 10^{19})} \sim 10^{(10^{20.8})}$$

Suppose we put N_1 particles in a box 0.25 cm on a side and remove the walls, allowing the particles to enter a 1 cm box.[17] The initial constraint reduces the number
of microstates. 'Δx' is cut by a factor 4, so that the number of microstates is cut by
$4^{3N_1} \sim 4^{(10^{20})}$.

Now another scenario: imagine an initial condition in which the particles were in a
region 0.5 cm on a side. This cuts the phase space by a factor of about $2^{(10^{20})}$. A demand
that the gas spontaneously collect at a later time in a region 0.5 cm on a side would again

---

[17] The momentum range is assumed to be the same as for the previous calculation.

introduce the same factor. The overall factor is therefore $4^{(10^{20})}$. This is the same as for the smaller volume (0.25 cm on a side) demanded previously as an initial condition.[18]

Appendix 5.1.A: Multiplicity of solutions for classical mechanics

Because classical mechanics is derivable from a variational principle there is a topological theory associated with the two-time boundary value problem. You start with the action, which is a function on the space of paths, and in that space look at neighborhoods of paths where the action has an extremum—the classical paths. In the notes I give references for this approach, known as Morse theory. Here I collect a few examples.

Consider the Lagrangian $\frac{1}{2}\dot{x}^2 - V(x)$ with $V(x) = \frac{1}{2}x^\ell$ and $x \in \mathbb{R}$. Require that $x(0) = a$, $x(T) = b$. For $\ell = 0, 1$ there is a unique solution. For $\ell = 2$ the solution is $x(t) = [b\sin t + a\sin(T - t)]/\sin T$. This is fine, unless $\sin T = 0$. In that case there is either no solution or a continuous infinity of solutions, if $b = a\cos T$. If $\ell > 2$, the walls of the potential are sufficiently stiff that the harder you throw the particle the faster it returns. This implies that there is a countable infinity of solutions. Of special interest is a phenomenon that occurs even with a wall on only one side, for example the potential $V(x) = \exp(-\Omega x)$. If T is small there are two solutions to the boundary value problem, one direct, one with a bounce off the wall. If T is sufficiently large, there is *no* solution. There is a critical point for the variational action at the particular T between these two regimes (and for which there is one solution). This is associated with focusing and in higher spatial dimension leads to caustics (familiar from lens aberration).

Exercise 5.1.A.1. Show that the solution of the boundary value problem for $V(x) = \exp(-\Omega x)$, with $b = a$, is $x(t) = x_0 + (2/\Omega)\log\cosh(\Omega kt/2)$ with k, x_0, a and T related by $k^2 = 2\exp(-\Omega x_0)$ and $k\exp(\Omega a/2) = \sqrt{2}\cosh(\Omega kT/2)$. Show that the solution becomes unique for sufficiently large T and then ceases to be real. Relate this to the motion of the Lagrange manifold as illustrated in Fig. 5.1.1.

A phenomenon of particular interest commonly occurs in more than one dimension. A chaotic system—even with an upper bound on permitted energies—can yield many solutions, with the number growing exponentially in T. One can offer analytical examples using a compactified space of constant negative curvature, so as to allow closed paths. For physical applications a potential model is more realistic, but almost by definition chaos and analytic tractability are mutually exclusive. In any case the argument in the text, in which the growth described by the Lyapunov exponent is applied to the Lagrangian manifold, is adequate to show the exponential growth of solutions.

Appendix 5.3.A: Relative path dynamics and wave packet localization

The following is included for its path integral content and for reference in Section 9.4. Often, a process, a scattering or a chemical reaction, is well described by letting its major actors follow classical paths. But you still want quantum corrections. The technique below

---

[18] If the time between the two constraints is short there may be an additional reduction in the number of states, analogous to that implicit in the shorter times illustrated in Fig. 4.3.4.

starts from a situation in which a single classical path should be a good approximation, and in that context performs a quantum calculation. The quantum calculation is the problem of minimizing the spreads of a wave packet that is required to be localized around two spatial positions at two different times.

As in the main text, we optimize wave packet localization, with localization defined in terms of square-spread. We take the initial time to be 0, the initial position a, the final time T, and the final position b. Consider a Hamiltonian

$$H = \frac{1}{2m}p^2 + V \qquad (5.3.A.1)$$

with V sufficiently slowly varying to allow a semiclassical approximation. We examine the propagator for this Hamiltonian in the neighborhood of the localization positions given above. We first suppose that there is a unique classical path from a to b for the indicated times. Denote the path

$$\bar{x}(s), \qquad 0 \le s \le T \qquad \left(\bar{x}(0) = a, \quad \bar{x}(T) = b\right) \qquad (5.3.A.2)$$

We want the propagator for functions that move more or less along $\bar{x}(\cdot)$. We know from WKB analysis that such motion will be the small $\hbar$ behavior of smooth wave packets initially localized near $a = \bar{x}(0)$ with central momentum $\bar{p}(0) = m\dot{\bar{x}}(0)$. The propagator from the neighborhood of a (at $t = 0$) to the neighborhood of b (at $t = T$) is

$$G\left(b + \xi, T; a + \eta\right) = \sum e^{iS[x(\cdot)]/\hbar} = \sum e^{iS[\bar{x}+y]/\hbar}$$

$$= e^{iS[\bar{x}]/\hbar} \sum_{\substack{y(0)=\eta \\ y(T)=\xi}} \exp\left\{\frac{i}{\hbar}\left[\frac{\partial L}{\partial \dot{x}}y\Big|_0^T + \int \left[\frac{m}{2}\dot{y}^2 - \frac{1}{2}\frac{\partial^2 V}{\partial x^2}\Big|_{\bar{x}} y^2 + O(y^3)\right]\right]\right\}$$

$$(5.3.A.3))$$

where the sum, Σ, is the path integral. Note that we have expanded about $\bar{x}(\cdot)$, *not* about the classical path connecting $a + \eta$ to $b + \xi$. That is why the $\partial L/\partial \dot{x}$ appears. However, there is no other $O(y)$ term because $\bar{x}(\cdot)$ is a classical path. Next discard the $O(y^3)$ terms to get the WKB propagator. We write the truncated Eq. (5.3.A.3) as

$$G_{WKB}\left(\bar{x}(T) + \xi, T; \bar{x}(0) + \eta\right) =$$

$$\exp\left\{\frac{i}{\hbar}\left[S\left(b, T; a\right) + \bar{p}(T)\xi - \bar{p}(0)\eta\right]\right\} \sum_{\substack{y(T)=\xi \\ y(0)=\eta}} \exp\left\{\frac{i}{2\hbar}\int_0^T \left[m\dot{y}^2 - V''\Big|_{\bar{x}} y^2\right] dt\right\}$$

$$(5.3.A.4)$$

where the notation $S(b, T; a)$ $(= S[\bar{x}])$ shows the initial and final points along which the action is evaluated for the classical path. Define

$$\phi(u, t) = \exp\left[-\frac{i}{\hbar}\left(S\left(\bar{x}(t), t; \bar{x}(0)\right) + \bar{p}(t)u\right)\right]\psi\left(\bar{x}(t) + u, t\right) \qquad (5.3.A.5)$$

From Eq. (5.3.A.4) and the role of G_{WKB} as a propagator it follows that the function ϕ can be treated as a wave function propagated by the kernel

$$\widetilde{G}(u, t; \eta) \equiv \sum_{\substack{y(0)=\eta \\ y(t)=u}} \exp\left\{\frac{i}{\hbar}\int_0^t ds \left[\frac{m\dot{y}^2}{2} - \frac{1}{2}V''\Big|_{\bar{x}(\cdot)} y^2\right]\right\} \qquad (5.3.A.6)$$

In the present notation, the functional W of Eq. (5.3.2) whose minimization we desire takes the form

$$W = \int d\xi |\phi(\xi, T)|^2 \xi^2 + \int d\eta |\phi(\eta, 0)|^2 \eta^2 \qquad (5.3.A.7)$$

The time evolution given by Eq. (5.3.A.6) is the same as that induced by the (generally) time dependent Hamiltonian H_{SC}

$$H_{SC} = \frac{p^2}{2m} + \frac{1}{2}m\omega^2\xi^2 \tag{5.3.A.8}$$

where now $p = -i\hbar\partial/\partial\xi$ and

$$\omega^2(t) = \frac{\partial^2 V}{\partial x^2}\bigg|_{x=\bar{x}(t)} \tag{5.3.A.9}$$

The time dependence of H_{SC} does not preclude the method of Section 5.3. Using the quadratic nature of H_{SC} we get an explicit solution for the Heisenberg picture operators and then minimize W. Note that we use the Heisenberg picture with respect to H_{SC}, *not* the original H of Eq. (5.3.A.1). The time dependence of H_{SC} makes for a slight complication but is no bar to the use of the Heisenberg picture.

The smallest possible value of W—in the semiclassical approximation—is therefore the lowest eigenvalue of the operator

$$\widehat{W}_{SC} \equiv \hat{x}_{SC}^2(T) + \hat{x}_{SC}^2(0) \tag{5.3.A.10}$$

with

$$\hat{x}_{SC}(t) = U^\dagger(t,t')\,\hat{x}U(t,t') \quad \text{and} \quad U(t,t') = Te^{-i\int_{t'}^{t} H_{SC}(s)ds/\hbar} \tag{5.3.A.11}$$

Note the time-ordered product ('T') in Eq. (5.3.A.11), since H_{SC} need not commute with itself at different times. Implicit in $\hat{x}_{SC}(t)$ is the reference time t'. The momentum operator $\hat{p}_{SC}(t)$ is defined as in Eq. (5.3.A.11). For the operator equations of motion the Heisenberg picture Hamiltonian must also be specified. Call this object $\widehat{H}_{SC}$. The operator equations of motion are

$$\frac{d\hat{x}_{SC}(t)}{dt} = \frac{i}{\hbar}\left[\widehat{H}_{SC}(t), \hat{x}_{SC}(t)\right] = \frac{\hat{p}_{SC}(t)}{m}, \quad \frac{d\hat{p}_{SC}(t)}{dt} = -m\omega^2(t)\hat{x}_{SC}(t) \tag{5.3.A.12}$$

where use is made of the continued validity of $[\hat{p}_{SC}(t), \hat{x}_{SC}(t)] = \hbar/i$ under the unitary transformation Eq. (5.3.A.11). Eq. (5.3.A.12) is identical to the classical equations of motion and $\hat{x}_{SC}(t)$ can be written

$$\hat{x}_{SC}(t) = \hat{x}f(t) + \hat{p}g(t)/m \tag{5.3.A.13}$$

where $\hat{x}$ and $\hat{p}$ are the time t' operators and f and g are c-number functions satisfying the equation $\ddot{y} + \omega^2(t)y = 0$, with the time t' conditions

$$\begin{aligned} f(t') &= 1 & \dot{f}(t') &= 0 \\ g(t') &= 0 & \dot{g}(t') &= 1 \end{aligned}$$

The important feature of Eq. (5.3.A.13) is that $\hat{x}_{SC}(t)$ is a linear function of $\hat{x}$ and $\hat{p}$, and therefore $\widehat{W}$ of Eq. (5.3.A.10) is a harmonic oscillator whose eigenvalues can be found in a straightforward way. Note that from Eq. (5.3.A.13) it follows that

$$\left[\hat{x}_{SC}(0), \hat{x}_{SC}(T)\right] = i\frac{\hbar}{m}\left(f(0)g(T) - g(0)f(T)\right) \equiv \pm i\hbar\Omega \tag{5.3.A.14}$$

so that Ω depends implicitly on t' and the sign ($\pm$) is selected to make Ω nonnegative. It follows from Exercise 5.3.1 that the lowest eigenvalue W_0 of $\widehat{W}_{SC}$ is

$$W_0 = \hbar\Omega = \frac{\hbar}{m}\left|f(0)g(T) - g(0)f(T)\right| \tag{5.3.A.15}$$

If H_{SC} were the true Hamiltonian, Eq. (5.3.A.15) would be exact. Therefore if we look upon the right-hand side of Eq. (5.3.A.15) as the two-time localization bound for the true H_{SC} dynamics then it is clear that it cannot depend on the fiducial time t' used in defining the Heisenberg picture. It is easy to verify that Ω, as defined in Eq. (5.3.A.14), is independent of t'. Therefore, taking $t' = 0$ it follows that

$$W_0 = \frac{\hbar}{m}|g(T)| \qquad (5.3.A.16)$$

where $g(t)$ satisfies $\ddot{g} + \omega^2(t)g = 0$, $g(0) = 0$, $\dot{g}(0) = 1$. This agrees with our previous result for the case of ω time independent.

Eq. (5.3.A.16) bears out remarks in the main text on quantum dynamics on spaces of negative curvature and other instances of quantum chaos. The local, classical dynamics generated by H_{SC} is that of the Jacobi equation or the equation of geodesic deviation. For a space of constant negative curvature the matrix that plays a role analogous to $V''(\bar{x}(\cdot))$ will have a negative eigenvalue so that in the direction of the corresponding eigenvector wave packet spreading will grow rapidly. (Note that if $\omega^2 < 0$, $g(t) = |\omega|^{-1}\sinh(|\omega|t)$.)

In general the semiclassical result given in Eq. (5.3.A.16) is asymptotic for $\hbar \to 0$, but not uniform for $T \to \infty$. Thus for unbounded motion in which $V(x) \to 0$ for $|x| \to \infty$ we expect from Eq. (5.3.A.16) that 'Δx' $= \sqrt{W_0} \sim \sqrt{T}$. However, deviations from semiclassical behavior allow small pieces to break off from the main wave packet and to become separated by a distance that increases linearly with time. Because the size of the breakaway depends on $\hbar$, this does not change fixed T asymptotics; it does however prevent uniformity.

6

Quantum measurements: cats, clouds and everything else

Quantum measurement theory addresses several problems. All of them arise from applying a microscopically valid theory to the macroscopic domain. The most famous is the Schrödinger cat example in which the ordinary use of quantum rules suggests a superposition of macroscopically different states, something we do not seem to experience. Another problem is the Einstein-Podolsky-Rosen (EPR) paradox in which a fundamental quantum concept, entanglement, creates subtle and non-classical correlations among remote sets of measurements. Although it seems mere word play to observe that such an apparent micro-macro conflict ought to be viewed as a problem in statistical mechanics—by virtue of the way that discipline is defined—until recently the importance of this observation was seldom recognized.

The founders, Bohr, Schrödinger, Heisenberg, Einstein did not emphasize this direction, but over the years the realization that measurement necessarily involves macroscopic objects—objects with potentially mischievous degrees of freedom of their own—began to be felt. For me this was brought home by the now classic fourth chapter of Gottfried's text on quantum mechanics. He takes up the following problem. The density matrix ρ_0 for a normalized pure state is $\rho_0 = |\psi\rangle\langle\psi|$. It follows that $\mathrm{Tr}\,\rho_0 = \mathrm{Tr}\,\rho_0^2 = 1$. After a non-trivial measurement, the density matrix ρ is supposed to be diagonal in the basis defined by the measured observable, with $\mathrm{Tr}\,\rho = 1$ but $\mathrm{Tr}\,\rho^2 < 1$. The problem is that under pure unitary evolution the trace of ρ^2 shouldn't change: ρ_0 becomes $\rho = \exp(-iHt/\hbar)\rho_0\exp(iHt/\hbar)$, from which it follows that $\mathrm{Tr}\,\rho^2$ is still unity. Gottfried's answer is that for the true ρ we still have $\mathrm{Tr}\,\rho^2 = 1$, but that *for all practical purposes* ρ can be replaced by an effective $\hat\rho$ which is ρ with its off-diagonal matrix elements discarded. Remember, the basis corresponds to the measured observable. The reason that $\rho - \hat\rho$ can be discarded is that it cannot be measured. This is because in practice the relevant basis is local in coordinate space and since we cannot build operators to measure non-local observables the off-diagonal terms have no effect.[1] That is, no operator A that an experimentalist can realize can give a non-zero answer for $\text{Tr}[A(\rho - \widehat{\rho})]$. These concepts, 'for all practical purposes' and 'realizable apparatus' are the stuff of statistical mechanics. They are the concepts that entered when we derived the master equation in Chapter 2, and indeed discarding off-diagonal terms is what needed to be done there too. The justification at that point was based on coarse graining. But coarse graining is exactly the issue of what is practically realizable and what is not. In this way the arguments of Gottfried connect to those of statistical mechanics. This reinforces the view that the so-called time's arrow of quantum measurement is the same as that of thermodynamics.

In retrospect, with the mystery already present in statistical mechanical irreversibility, it's a wonder that it was found necessary to invent another mystery, wave packet collapse. The fashion has swung away from the latter terminology, but that doesn't mean the underlying problems are solved. In this book I argue that the solution to the quantum measurement problem lies in statistical mechanics, and indeed this is now becoming the conventional wisdom, but I will also argue that reconciling the conflict between quantum mechanical micro and macro predictions will require a comprehensive change in the foundations of statistical mechanics. The goal of the first portion of the book was to make that comprehensive change doubly palatable: first, that the change we will propose is logically reasonable; second, that until now it would have been experimentally unnoticed.

In this chapter we focus on the problem of superpositions of macroscopically different states, so called 'grotesque' states. Schrödinger's dramatic version of this micro-macro conflict involves a superposition of two states of a cat, living and dead. We will employ a different and not-quite-so-gedanken experiment to illustrate the problem. Every quantum measurement theory must deal with this in one way or another, ranging from the concrete idea of nonlinear deviations from unitary evolution, to momentary and undefined suspensions of those dynamics, to hidden variables, to proliferation of macroscopic states with significant support ('many worlds'). The list goes on. My goal in this book is to explain how I deal with the problem, and other ideas will be mentioned only insofar

---

[1] At one point Gottfried gets queasy about his locality requirement and puts in the word 'essentially' together with a footnote about superconductors and superfluids providing a complication, followed by another footnote about the immense number of degrees of freedom in a measurement apparatus, followed by yet another footnote arguing that superconductors would not act as pointers. (But who am I to complain about footnotes.)

as they are related or are to be compared. The theories sort themselves into two major categories on the basis of the following criterion: are their physical predictions the same or different from those of the Copenhagen interpretation? To the physicist this is obviously an important question, but, as might be expected in a field so heavily influenced by philosophy, it isn't always clear what the answer is, nor even what the Copenhagen prediction is. As will be seen below, the theory proposed here *is* experimentally distinguishable from the traditional 'interpretation,' although at the time of writing I do not have a convenient experimental test. This will be discussed in later chapters.

We now turn to a specific experimental context for a realization of the grotesque state problem of quantum measurement theory and for a statement of our own resolution of this problem.

6.1 A grotesque state of a cloud chamber

In this section we provide an explicit realization of a superposition of different states of a macroscopic system, states that are different at the macroscopic level. It is the existence of such states (called 'grotesque') that is the first problem of quantum measurement theory. We will then, in the next section, go into considerable quantum detail on the dynamics of our model of the cloud chamber. With the details of that calculation we will be in a position to make explicit our resolution of the quantum measurement problem.

The system we model, the grotesque state we produce, is a particle detector operating on the principle of a cloud chamber. This device, similar to a bubble chamber, relies on the amplifying effect of metastability to convert a microscopic stimulus into a macroscopic response. For example the passage of a cosmic ray induces a line of droplets marking the particle's path. This comes about in the following way. In the cloud chamber a gas is maintained at a temperature and pressure close to the liquid-gas phase transition. By slight compression, the stable state becomes the liquid rather than the gas, and the gas is now supercooled. But a supercooled vapor does not become liquid immediately. Sufficiently large droplets must form to act as nuclei for further condensation and it can take a while for such 'critical droplets' to form spontaneously. When a charged particle passes through the gas it ionizes atoms of the gas. This local perturbation facilitates the formation of a critical droplet. Once this droplet is present, the dynamics of the gas are such that it grows rapidly, amplifying the initial microscopic cause. In practice a line of droplets is formed and photographed soon after the compression that drove this system to metastability. Typically the system is decompressed and allowed to go through many detection cycles. Fig. 6.1.1 is a schematic diagram of three stages in the production of two droplets marking the path of a particle through the chamber.

In any small volume of supercooled gas through which the charged particle passes there may or may not be an ionization due to that passage. The scattering and (possible) ionization are describable by pure unitary time evolution within a fully quantum mechanical framework. If one were to calculate the results of such a scattering there would be two significant components to the wave function, one in which the atom was ionized and one in which it was not. The squares of these amplitudes would give the probabilities of droplet formation and non formation within that volume;[2] by appropriate further calculation one would obtain the efficiency of the detector. One generally stops doing quantum mechanics—and begins using classical mechanics—somewhere between the scattering process and the photograph. What I will do below is model the entire process, not for the cloud chamber, which has a rich variety of physical processes, but for a caricature, which nevertheless has the ability to take a microscopic event and promote it to an irreversibly registered macroscopic situation.

For the cloud chamber the initial state is of the following form:

$$\Psi_0 = \psi(\xi_1, \ldots, \xi_N)\phi(x) \qquad (6.1.1)$$

where $\xi_1, \ldots, \xi_N$ are the coordinates of the individual gas atoms and x is the coordinate of the impinging particle (call it 'X') to be detected. The

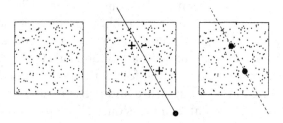

Fig. 6.1.1. Three successive stages in the operation of a cloud chamber. After compression (leftmost figure) the gas is supercooled, but in the absence of disturbance would not form liquid for a long time. In the second figure is shown a disturbance in the form of a particle that traverses the chamber, ionizing two atoms in its wake. In the third figure, the disturbing particle has long passed and droplets have formed at the sites of the ionization. One deduces the path of the particle (dashed line) from the droplets.

---

[2] This assumes that coherent quantum effects are negligible once ionization occurs. Otherwise the quantum calculation should be extended to larger scales.

gas atoms have some (extremely) complicated wave function ψ, and X can be thought of as being in a relatively simple wave packet state, for example, $\exp[-(\mathbf{x} - \mathbf{x}_0)^2/2\sigma^2 + i\mathbf{k} \cdot \mathbf{x}]$. The upper portion of Fig. 6.1.2 is a macroscopic view[3] of Ψ.

Now a few milliseconds go by. The particle has passed through the chamber, hardly deflected because of its high velocity. In its wake though it has left a disturbance, and in some places that disturbance has caused the formation of droplets of liquid (by the ionization and metastability induced amplification discussed above). The wave function of the system

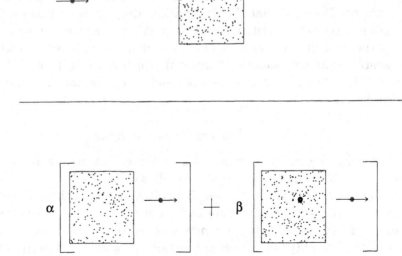

Fig. 6.1.2. The upper portion of the figure represents the wave function (Eq. (6.1.1)) before the particle X enters the cloud chamber. The lower portion represents the time-T superposition of two macroscopically different states (Eq. (6.1.2)). In one component, with amplitude α, no bubble has formed; in the other, with amplitude β, there is a bubble. For both components the rapidly moving particle X has passed through and out of the chamber.

---

[3] In Eq. (6.1.1) we omit the coordinates of the walls of the chamber, although in principle they could be included.

is now (the time is T, the total Hamiltonian H)

$$\Psi_T = \exp(-iHT/\hbar)\Psi_0$$
$$= \alpha\Psi_{\text{no droplet}}(\xi_1,\ldots,\xi_N,x) + \beta\Psi_{\text{droplet}}(\xi_1,\ldots,\xi_N,x) \tag{6.1.2}$$

where the 'droplet' and 'no droplet' subscripts refer to wave functions of the entire $N+1$ particle system. These have macroscopic states that respectively correspond to having or not having a droplet. The indicated states also incorporate any other changes in the system that occur in the course of the time interval T (for example, atoms in the initial state that were about to collide and were distant from the path of X will appear as having scattered in both pieces of the wave function in Eq. (6.1.2)). The coefficients α and β are in principle computable from the dynamics, and for normalized wave functions their magnitudes squared provide the first step in computing the efficiency of this detector. It also may happen that to a good approximation $\Psi_{\text{no droplet}}(\xi_1,\ldots,\xi_N,x)$ will factor to a product of gas wave function and X-wave function, but that is not important for the moment. Note too that the designation 'droplet' is a gross oversimplification of the actual situation and we could more accurately write this part of the wave function as a sum over the many possible wave functions representing different places and times at which droplet(s) form. In the lower portion of Fig. 6.1.2 we show a schematic representation of the time T wave function.

The problem and some solutions

The state Ψ_T is grotesque. It is not something that any of us has ever experienced. Nevertheless, its existence follows directly from the rules of quantum mechanics. When the founders of quantum mechanics realized where their microscopic formalism had landed them they looked for avenues of retreat. The Copenhagen maneuver is to back off from the naive attribution of 'reality' to the wave function: A wave function describes an ensemble and a measurement is the selection of one of the possibilities contained in that ensemble. The probability of a particular selection is the magnitude squared of the wave function for that outcome. This explanation, and finely nuanced variations, have generated uneasiness since they were offered. In this book I will not provide an exposition of the sources of the unease, nor of the subtle elaborations that developed for its amelioration. At this point the Copenhagen interpretation has passed every experimental test, and for many physicists intellectual dissatisfaction alone is not sufficient cause either to abandon a theory or to search for a more intellectually satisfying one with the same predictions.

Nevertheless, other interpretations as well as variant physical theories have been proposed. Again, this book is not devoted to an exposition of those ideas, although I will characterize some of them in the next few sentences. By variant physical theory, I mean that the time evolution is not quite given by the first line of Eq. (6.1.2), but rather there are small nonlinear effects. For interactions of a few microscopic particles these effects are negligible, but when a macroscopic measurement takes place the non linear effects become important and the system evolves to only one or the other of the components in Eq. (6.1.2), that is, one or the other of the pictures in the lower portion of Fig. 6.1.2. For these theories it is clear that one is talking about physics (rather than being marked by the stigma of philosophy) and they naturally lend themselves to experimental test, although for not all such proposals have the numbers led to feasible experiments. Another way to handle the selection of one component of the wave function comes under the rubric 'hidden variables.' Besides the wave function, it is postulated that there are other variables associated with the system. These are not variables for other particles in the environment, but are additional variables, with their own dynamics, for particles already under consideration. The idea is that when a measurement occurs the result of that measurement depends on the current value of the 'hidden' variable. The proponents of these theories say that they get the same experimental predictions as the Copenhagen interpretation so the virtue claimed is intellectual satisfaction. The last class of theories that I mention says, 'Believe what you see.' There is no dynamics but unitary dynamics and indeed you never discard pieces of the wave function. What happens is that in the course of time the wave function becomes a complicated object with support on many macroscopically different states. Nevertheless, this does not contradict experience! To see how this can be, consider what happens when you look at the state described by the lower portion of Fig. 6.1.2. Prior to your looking, the combined wave function of you (θ) and the cloud chamber was a tensor product of the following form:

$$\Psi_T \otimes \theta_{\text{you}}$$
$$= \{\alpha\Psi_{\text{no droplet}}(\xi_1, \ldots, \xi_N, x) + \beta\Psi_{\text{droplet}}(\xi_1, \ldots, \xi_N, x)\} \otimes \theta_{\text{you}}$$

$$(6.1.3)$$

After you look the wave function becomes

$$\alpha\Psi_{\text{no droplet}}(\xi_1, \ldots, \xi_N, x) \otimes \theta_{\text{you, not having seen a droplet}}$$
$$+ \beta\Psi_{\text{droplet}}(\xi_1, \ldots, \xi_N, x) \otimes \theta_{\text{you, having seen a droplet}}$$

$$(6.1.4)$$

(to avoid excess notation I have not indicated the changes in Ψ while you look). One piece of the wave function has support on a state in which you looked and didn't see a droplet. Another component has you with a memory of having seen a droplet. No version of you has seen the superposition of macroscopically different states. This is Everett's 'many worlds' interpretation. It is popular in quantum cosmology, although its advocates have found formulations that are less suggestive of science fiction.

A single macroscopic state using 'special states'

I come at last to a statement of how I propose to get only a single macroscopic state from the quantum evolution indicated *on the first line* of Eq. (6.1.2). The idea is that what we control when we do an experiment is the macroscopic state. Among the many microscopic states that are consistent with that macroscopic state are some that behave in unusual ways. With reference to Section 2.6, for some—rare—microscopic states an isolated ice cube in a glass of water will spontaneously grow and the water become warmer. I claim that for some (initial) microscopic states of the cloud chamber, you do not get the superposition of two different macrostates, but only one of them. For some of these special microscopic initial states you get—by pure unitary evolution— no droplet, that is a wave function of the type on the lower left of Fig. 6.1.2. For other special states you get—by pure unitary evolution— a droplet, that is a wave function of the type on the lower right of Fig. 6.1.2.

In my experience, the statement I have just made is often misunderstood. I do *not* deny the superposition principle, in fact I am fanatically conservative in maintaining the formal structure of quantum mechanics. To make this clear, in the next section I will provide a specific and explicit example of what I mean by a special state for a measurement. You will see that from most initial conditions the system goes into a superposition of macroscopically different states, but that for some particular—special— initial conditions this does not happen and only a single macroscopic state emerges, by pure unitary quantum evolution alone. You will also see that the same wealth of degrees of freedom that allows 'measurement,' also provides the possibility of finding special states. Statistical mechanics is *both* part of the problem and part of the solution.

The explicit construction in the coming section, important though it may be for the explanation, leaves unfinished the actual solution offered for the quantum measurement problem. That will involve claims to the effect that in every performed experiment the microscopic initial conditions are among those special states and not among the more numerous other states. But this point will be taken up after I give the example of special states for a model apparatus.

Note: *The construction of the model and the associated calculations constitute Section 6.2, and take several pages. To assist the reader,* *at the end of the section I have included a summary of the model and its features. If you prefer to move quickly to the general theory in Section 6.3, that summary should suffice.*

6.2 Model apparatus for a quantum measurement

In a cloud chamber a single microscopic particle initiates a series of events that culminate in a macroscopic change in the apparatus. The essence of the apparatus is its ability to amplify—in this case because it is poised for departure from a metastable state—and proceed to a macroscopically different state, from which it will not spontaneously return. This last feature irreversibly registers the measurement.

We make a quantum mechanical model of this process. For the system exhibiting metastability I use a three-dimensional array of quantum spins. These interact with ferromagnetic short range forces. In the presence of a magnetic field they align along that field. In Fig. 6.2.1 I show a two-dimensional slice of an array in which a magnetic field points up. For convenience I assume the temperature is so low that even for a fairly large array no spin is reversed.[4] To produce a metastable state we make a sudden reversal of the magnetic field. The upward pointing spins are no longer the lowest state of the system, but their transition to the stable state is slow and the system is metastable.[5] It simplifies the calculation to take the field just above the threshold for a single reversal to be energetically favorable.

With the field reversed the system is like the cloud chamber just after compression: the slightest disturbance will cause a transition to the stable phase. Here is how the array of spins functions as a detector. The particle to be detected, called X, interacts with the spins in such a way that it tends to flip them. When X traverses the array it can cause a spin to turn over. See Fig. 6.2.2. The neighbors of that spin are now in a more favorable situation for their own flips. With no spins flipped, six bonds must be broken for a flip (on a cubic lattice). With a flipped neighbor, the cost of a flip is only five broken bonds. Subsequent flips are yet

---

[4] The temperature should be low enough for the correlation length to exceed the system size. Note too that the macroscopic notion of temperature will only be invoked at a late stage in the calculation, and for the exhibition special states, not at all.

[5] The concepts of 'thermodynamic limit' and macroscopic metastability are known to be in conflict, but our low temperature and finite system make that theoretical problem irrelevant.

Fig. 6.2.1. Two-dimensional slice from an array of spins. The upward pointing field (indicated by the large arrow on the left) orients the spins in the up direction and the temperature of the system is low enough so that opposite pointing spins are unlikely.

more enhanced and the system rapidly passes to its stable state.[6,7] The apparatus thus takes the microscopic interaction that causes a single spin flip and amplifies it to the level of less sensitive detectability.

It is straightforward to propose a Hamiltonian for the interactions affecting the spin only. Take

$$H_s = -J \sum_{\langle \ell\ell' \rangle} \sigma_{\ell z}\sigma_{\ell' z} + h \sum_{\ell} \sigma_{\ell z} \qquad (6.2.1)$$

where ℓ and ℓ' range over the spins and the notation $\langle \ell\ell' \rangle$ in the first sum means summation over nearest neighbor pairs on the lattice. Throughout this section we take $\hbar = 1$. The spin-spin coupling constant is J, and for ferromagnetism it is positive. The external field is h. The operators $\sigma_{\ell z}$

---

[6] These last sentences are effectively an explanation of the concept of critical droplet.

[7] In an actual cloud chamber the condensation is interrupted by decompression, for another cycle of particle detection.

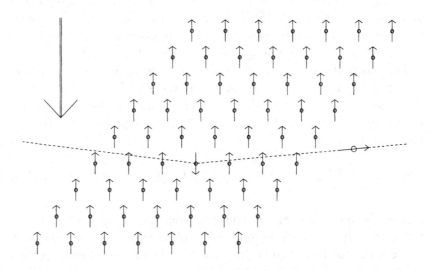

Fig. 6.2.2. The magnetic field for the array pictured in the previous figure has been reversed, so that its up state is metastable, not stable. A particle has passed through the array, flipping one spin and being scattered slightly in the process. Subsequently the neighbors of the flipped spin will turn over because the reversed bond allows them to follow more easily their tendency to align with the field.

are *quantum* spin operators. Explicitly

$$I = \begin{pmatrix} 1 & 0 \\ 0 & 1 \end{pmatrix}, \qquad \sigma_z = \begin{pmatrix} 1 & 0 \\ 0 & -1 \end{pmatrix}$$

$$\sigma_{\ell z} = I \otimes I \otimes \cdots \otimes \sigma_z \otimes \cdots \otimes I \qquad \left(\sigma_z \text{ is in the } \ell^{\text{th}} \text{ position}\right)$$

(6.2.2)

There is, however, more to the system than the spins alone. If one were actually to build such a system the flipped spins would give up their energy to the lattice that holds them in place. Furthermore, in a real system, that lattice is the thermal reservoir for maintaining the system's temperature. In different physical systems this lattice-spin interaction can take different forms. For convenience we assume that for our system the combined lattice and lattice-spin Hamiltonian is

$$H_p = \sum_{k=1}^{N} \omega_k b_k^\dagger b_k + \sum_{\ell} \sum_{k=1}^{N} \left(\gamma_k e^{i\phi_k(\ell)} \sigma_{\ell+} b_k + \gamma_k^* e^{-i\phi_k(\ell)} \sigma_{\ell-} b_k^\dagger \right) \quad (6.2.3)$$

The operators b_k and $b_k^\dagger$ are the phonon annihilation and creation operators for the kth mode (out of N) of the lattice. The mode's frequency is ω_k. The phonons couple to the spins with coupling constants γ_k, together with a phase, $e^{i\phi_k(\ell)}$, that varies with the position of the spin on the lattice. The operator $\sigma_{\ell+}$ is the raising operator for the ℓth spin and it and $\sigma_{\ell-}$ are defined analogously to $\sigma_{\ell z}$ (cf. Eq. (6.2.2)).[8]

The last portion of the model to be specified concerns the particle to be detected, X, in particular the way it interacts with the spins. We assume that the effect of X is to induce a strong coupling between the spin and the phonons. The Hamiltonian for X and its coupling to the detector are taken to be

$$H_d = \frac{p^2}{2m} + \sum_\ell v(x - x_\ell) \sum_{k=1}^N \left(c_k e^{i\phi_k(\ell)} \sigma_{\ell+} b_k + c_k^* e^{-i\phi_k(\ell)} \sigma_{\ell-} b_k^\dagger \right) \quad (6.2.4)$$

Again, the quantities c_k are coupling constants and the function $v(\cdot)$ is non-zero only when its argument[9] is close to zero. We take v dimensionless and of order unity for typical encounters. The coefficients c_k are assumed to be much larger than the usual lattice-spin coupling, i.e., $|c_k| \gg |\gamma_k|$ for the important range of k. The mass of X is m and in principle the function $\phi_k(\ell)$ could have been different from that Eq. (6.2.3).

The total Hamiltonian for the system is the sum of the three pieces we have displayed, $H_{\text{total}} = H_s + H_p + H_d$. The model defined thereby does not have an exact solution. But the point is not to provide exact solutions, only to arrive at the physical behavior, to show that notwithstanding an entirely quantum treatment, we have a system that behaves like a measuring apparatus. This is established in the following way. We first describe the response of the system qualitatively and then indicate what specific calculations are needed to justify the statements. The quantitative results will then be given. Having done this, we return to our primary goal, the exhibiting of 'special states,' particular initial states for which the detection event definitely takes place or definitely does not.

Since $v(\cdot)$ is short ranged, the impinging particle effectively interacts with only one spin. With this spin there is significant interaction for only a short time. As a result of this interaction there is some amplitude for flipping the spin, some amplitude for not flipping. The wave function of the array (cum particle X) now has two components, not yet characterizable as macroscopically distinct. In either case X moves away from

---

[8] H_p has the property that the number of up spins plus the total phonon number is conserved. In other models this number can vary. This conservation property also holds for H_d below.

[9] The expression '$v(\cdot)$' represents the function with its argument unspecified. This is standard mathematical notation; it indicates that v is a function, but avoids superfluous dummy arguments. Unfortunately, *Physical Review* has given me a hard time about using this notation, so I explain it here for those whose reading habits have deprived them of its benefits.

the array and the next stage in the history of the array takes place by means of the coupling due to the coefficients γ_k. For the component that has not flipped, the system is more or less where it was before, and it is unlikely to flip (we later examine this quantitatively). For the wave function component with the (single) flipped spin, the γ_k-coupling can either unflip the spin or can cause neighboring spins to flip. The energetic gain for flipping the neighbors is significant and for normal densities of phonon states the extra energy translates into much larger phase space for the transition. Therefore the most likely thing to happen (by far) is the flipping of the neighbors of the spin that was hit by X. Once this occurs the tendency becomes overwhelming and the system rapidly turns over completely. Since unflipping was unlikely, even after a single spin flip, the probability of detection will be the square of the amplitude for flipping after the first interaction of a single spin with X. This together with geometrical factors related to the range of $v(\cdot)$ will give the efficiency of this detector.

To justify the foregoing scenario we need the following information. 1) Flip probability for a single scattering event. 2) Probability of flip in the absence of any external particle (i.e., via the coefficients γ_k alone). This cannot be too large; otherwise the detector will be compromised by false positives. 3) Rates for multiple flip events. It must be shown that under the circumstances described the likelihood of two or more spins flipping either spontaneously (via γ_k) or because of X, are less likely than the single flip processes I described. If this were false it would not mean there was anything wrong with the detector, but it would invalidate our calculation. In the next few paragraphs we compute the foregoing quantities, although, not in the order listed.

Spontaneous spin flip: false positives and ground state stability

The rate for spontaneous transition is computed using the formalism developed for decay in Section 2.5. Because our Hamiltonian H_p only allows creation or annihilation of a single phonon in association with a spin flip, one can identify the matrix elements that correspond to the Hamiltonian given in Eq. (2.5.7). That form was

$$(2.5.7) \qquad H = \begin{pmatrix} h & C^\dagger \\ C & \Omega \end{pmatrix} \qquad \psi = \begin{pmatrix} x \\ Y \end{pmatrix}$$

where the first component ('x,' to be called component number '0' for the present calculation) is the initial state and the other components ('Y') are the decay products, the possible final states. For the calculation at hand, we take as the initial state $|+, n_1, n_2, \ldots\rangle$; here the integers n_k are the eigenvalues of $b_k^\dagger b_k$ and the '$+$' represents the state of the particular spin (from the array) on which we now focus. The diagonal matrix element of the Hamiltonian for this initial state has the value $H_{00} = -6J + h + \sum_k n_k \omega_k$. This state couples to all states of the form $|-, n_1, n_2, \ldots, (n_{k'}+1), \ldots\rangle$, $k' = 1, 2, \ldots$. These states have diagonal matrix elements $H_{00} + 12J - 2h + \omega_{k'}$. The coupling matrix elements, due to the terms $\gamma_k e^{i\phi_k(\ell)} \sigma_{\ell+} b_k$ and their adjoints, give rise to $\gamma_k e^{i\phi_k(\ell)} \sqrt{n_k + 1}$ in the first row of the Hamiltonian (to the right of the '00' diagonal term), and to its adjoint in the first column. In Section 2.5 we calculated the transition probability for the decay. This is the suitable quantity to study since the interaction with the bath is a continuing initiator of spin flip. The flip rate is given by the 'Golden Rule,' as found in Eq. (2.5.18). To use that formula we note that the energy deposited in the phonon mode is just so as to match initial and final energies, namely, $\Delta E \equiv 2h - 12J$, which we here take to be slightly positive. Let $\bar{k}$ be the phonon index for which $\omega_{\bar{k}} = \Delta E$, then the 'Golden Rule' takes the form

$$\Gamma = 2\pi \langle n_{\bar{k}} \rangle \rho(\bar{k}) |\gamma_{\bar{k}}^{\mathcal{N}}| \tag{6.2.5}$$

where the $\mathcal{N}$ superscript on γ_k recalls the factors introduced in taking the continuum limit, as discussed in Section 2.5. There is one volume-like factor in the matrix element and an inverse of such a factor in using the density of states (ρ). The combined effect is that for present purposes this aspect can be ignored.[10] The factor $\langle n_{\bar{k}} \rangle$ is the expected number of phonons in the initial state, and for inverse temperature $\beta = 1/k_B T$ is given by $[\exp(\beta \Delta E) - 1]^{-1}$. Unless $\beta \Delta E$ is so small as to get us into Bose-Einstein condensation issues, $\langle n_{\bar{k}} \rangle$ will be of order unity. Under the assumed physical conditions, *the important factor in Eq.* (6.2.5) *is* $\rho(\bar{k})$, *the density of states*. For low energy phonons this is proportional to the square of the frequency; thus $\Gamma \sim \Delta E^2 \ll 1$. It follows that if the detector contains M spins it will be unlikely to have any spontaneous flips provided $M \Delta E^2 |\gamma_{\bar{k}}^{\mathcal{N}}|^2 \ll 1$. This means that our detector is not 'macroscopic' in the sense of operating at *any* possible size, but is limited to dimensions that could be called mesoscopic. Once the signal is registered on our detector it can be recorded by a second level detector, as the droplets in the cloud chamber are photographed before the next decompression/compression cycle.

The same kind of estimates for the transition from spin-up to spin-down can be made to check that the state in Fig. 6.2.1, in which *no* spins oppose the field, is attainable. But we can also do this by a simpler Arrhenius rate estimate. For an array of M spins the expected number pointing in a direction opposite to the field

---

[10] When evaluating physical transition rates one would calculate per-unit-volume matrix elements as well as a per-unit-volume density of states.

is $M \exp(-\beta[\text{energy cost of a reversed spin}]) = M \exp[-\beta(12J + 2h)]$. With the mesoscopic size scales envisioned above this can be much smaller than one.

Multiple spin flips, their unlikelihood

Since our estimates involve one-flip-at-a-time transitions, we should check that it is unlikely for two spins to flip simultaneously, both for the scattering case and for the spontaneous case. This calculation was performed on a similar model (reference below) and the result is that instead of factors $|\gamma_k|^2$ or $|c_k|^2$, one gets $|\gamma_k|^4$ or $|\gamma_k c_k|^2$. Thus the condition for considering single-flips only is $|\gamma_k|^2 \ll |c_k|^2$, which we already assume. The collective soft mode in which all spins rotate at once has no energy barrier, but can be inhibited by fixing the direction of a small subset of the spins.

Scattering-induced spin flips

For convenience we assume that X traverses the detector rapidly, in particular on a time scale such that little takes place due to the usual spin-phonon interaction (mediated by 'γ_k') or among the phonons. Also we take the kinetic energy of X to be large compared to the interaction with the detector; thus the bend in the trajectory in Fig. 6.2.2 is highly exaggerated. Finally, the range of the potential v is such that only a single spin is involved in the initial stage of detection. Let that spin be labeled '0' and suppose (without loss of generality) that $\phi_k(0) = 0$. Under these circumstances we can confine attention to this spin, for which the result of the interaction will be

$$\psi_0 \to \exp\left[-i\int dt v(x(t) - x_0) \sum_{k=1}^{N}\left(c_k\sigma_{0+}b_k + c_k^*\sigma_{0-}b_k^\dagger\right)\right]\psi_0 \quad (6.2.6)$$

where ψ_0 is the initial spin wave function of spin 0, which is $\binom{1}{0}$ (i.e., 'up'). The function $x(t)$ is the path of X, which we assume unaffected by the interaction. We can now write $T \equiv \int dt v(x(t) - x_0)$, a quantity which is an effective contact time (since v is dimensionless) and incorporates the information on the closeness of the approach and other details of the trajectory. We are therefore interested in the unitary matrix

$$U = \exp\left[-iT\sum_{k=1}^{N}\left(c_k\sigma_{0+}b_k + c_k^*\sigma_{0-}b_k^\dagger\right)\right] \quad (6.2.7)$$

We henceforth drop the subscript '0' referring to the spin number. To obtain an explicit form for U, define a new set of boson operators. Let

$$\beta_1 = \sum_k \frac{c_k}{\tilde{c}} b_k \qquad \text{where } \tilde{c} \equiv \sqrt{\sum_k |c_k|^2} \qquad (6.2.8)$$

(The '1' is the phonon-state label.)   The other boson operators, $\beta_2, \ldots, \beta_N$, are a suitable set built from the b_k $(k = 1, \ldots, N)$ that are independent of β_1. In terms of the βs

$$U = \exp\left(-i\tilde{c}T(\beta_1\sigma_+ + \beta_1^\dagger\sigma_-)\right) \qquad (6.2.9)$$

It is an exercise in spin matrices to evaluate U. With the definition

$$\hat{\nu}_1 = \beta_1^\dagger\beta_1 \qquad \text{(an operator)} \qquad (6.2.10)$$

we find[11]

$$
\begin{aligned}
U &\equiv \exp(-i\tilde{c}T(\beta_1\sigma_+ + \beta_1^\dagger\sigma_-)) \\
&= \sigma_+\sigma_- \cos\left(\tilde{c}T\sqrt{\hat{\nu}_1 + 1}\right) + \sigma_-\sigma_+ \cos\left(\tilde{c}T\sqrt{\hat{\nu}_1}\right) \\
&\quad - i\left\{ \frac{\sin\left(\tilde{c}T\sqrt{\hat{\nu}_1 + 1}\right)}{\sqrt{\hat{\nu}_1 + 1}} \beta_1\sigma_+ + \frac{\sin\left(\tilde{c}T\sqrt{\hat{\nu}_1}\right)}{\sqrt{\hat{\nu}_1}} \beta_1^\dagger\sigma_- \right\}
\end{aligned}
\qquad (6.2.11)
$$

We use this form of U to give the wave function at time t. Because of the spin index and the two different phonon bases that we use, there is a certain notational overload. From the form of U it is obvious that the eigenvalues of $\hat{\nu}_k \equiv \beta_k^\dagger\beta_k$, for $k \neq 1$, are of no importance for the action of U. Therefore in looking at the evolution of the spin-0 wave function and the phonon wave functions we specify only the spin state (of spin 0) and the number of phonons of type 1, i.e., the eigenvalue of $\hat{\nu}_1$. Let this eigenvalue be designated ν_1. Thus if the initial spin-0-cum-phonon wave function is $\psi(0) = |+, \nu_1\rangle$, by Eq. (6.2.11) it evolves to

$$
\begin{aligned}
\psi(t) &= \cos\left(\tilde{c}T\sqrt{\hat{\nu}_1 + 1}\right)|+, \nu_1\rangle - i\frac{\sin\left(\tilde{c}T\sqrt{\hat{\nu}_1}\right)}{\sqrt{\hat{\nu}_1}}\sqrt{\nu_1 + 1}|-, \nu_1 + 1\rangle \\
&= \cos\left(\tilde{c}T\sqrt{\nu_1 + 1}\right)|+, \nu_1\rangle - i\sin\left(\tilde{c}T\sqrt{\nu_1 + 1}\right)|-, \nu_1 + 1\rangle
\end{aligned}
\qquad (6.2.12)
$$

This is the state of spin 0 after X has passed.[12] If the remaining processes are nearly certain (when spin 0 is down, the whole thing turns over,

---

[11] In Eq. (6.2.11), the expression $(\sin x)/x$ stands for an operator power series.
[12] Recall that this passage is rapid on the time scale of the phonon modes—which is why Eq. (6.2.7) well approximates the propagator during the passage.

otherwise it does not) then Eq. (6.2.12) gives the detection and non-detection amplitudes. If for example there is a thermal distribution of phonons in the initial state, the probability of detection is

$$\Pr(\text{detection}) = \left\langle \sum_{\nu_1} \exp(-\beta\nu_1\omega_{\nu_1}) \sin^2\left(\tilde{c}T\sqrt{\nu_1+1}\right)\right\rangle_{\tilde{c}T} \qquad (6.2.13)$$

where the sum over ν_1 is the thermal average, $\omega_{\nu_1} = \langle\nu_1|\sum_k \omega_k b_k^\dagger b_k|\nu_1\rangle$, and the subscript $\tilde{c}T$ on the angular, averaging brackets indicates that one should average over expected particle-spin contact times (since T is built from the trajectory of X). Eq. (6.2.13) can be carried further analytically, but there is no point in the present context.

Operation of the detector

This completes the information needed to see how the detector operates. Immediately after X passes through the array of spins the wave function is of the form given in Eq. (6.2.12). The neighbors of spin 0 (the spin that X came close to) now flip due to their interaction with the phonons and encouraged by the external field. The rate for these flips is much greater than that for spontaneous flips, because the energy of the created phonon is well away from threshold, and phase space factors (that inhibited the original turnover) are much larger. In three dimensions the flipping of M spins will take on the order of $M^{1/3}$ flip times—for the favorable, alignment-producing flips. If the probability of a spontaneous flip is small in this interval, then the false positive problem does not arise.[13]

With M spins overturned, even for modest M (perhaps 1000 or fewer), we have entered the classical domain in two senses. 1) The overturning of the spins is irreversible. Poincaré recurrence times for the phonons to coherently drive the spins back to the up-state are enormous. 2) The disturbance is large enough so that subsequent evolution (e.g., a probe of the detector's magnetic moment) can be treated classically (although there would in principle be a quantum description).

We next look at the wave function of the detector. After the first spin has flipped, the detector wave function is given by Eq. (6.2.12) (or a superposition of such terms) crossed into the wave function of all the other spins and phonons. Continuing, the wave function of the entire system after a further time interval, $\overline{T}$, sufficiently long for a large number of

---

[13] Even if it does, the important issue is whether false positives occur during the *spread* of the overall turnover time.

spins to fall, is given by the action of $\exp(-iH\overline{T})$ on the full detector state just described. Thus

$$\Psi = e^{-iH\overline{T}}\left[\left\{\cos\left(\tilde{c}T\sqrt{\nu_1+1}\right)|+,\nu_1\rangle - i\sin\left(\tilde{c}T\sqrt{\nu_1+1}\right)|-,\nu_1+1\rangle\right\}\right.$$

$$\otimes\Big|\text{spins other than 0, all of which are up}\Big\rangle$$

$$\left.\otimes\Big|\text{phonons other than }\beta_1\text{, all of which are unchanged}\Big\rangle\right]$$

(6.2.14)

As discussed above, the action of $e^{-iH\overline{T}}$ leads to

$$\Psi = \cos\left(\tilde{c}T\sqrt{\nu_1+1}\right)\Big|\text{all spins up; phonons hardly changed}\Big\rangle$$

$$- i\sin\left(\tilde{c}T\sqrt{\nu_1+1}\right)\Big|\text{spins down; phonons somewhat excited}\Big\rangle$$

(6.2.15)

This is a grotesque state in the sense of our previous section and is of the kind displayed in Eq. (6.1.2).

Remark: In our discussion we appealed to phase transition notions for the operation of the detector. This need not be taken literally, especially as one might become anxious about finite size effects and correlations when the order parameter is continuous. All we need is a bottleneck, a process whose rate is slow compared to everything else. It is the bypassing of the bottleneck via the incoming particle that allows that particle to be detected.

Scattering-induced spin flips with SPECIAL initial conditions

Our long interlude and explicit microscopic construction of a quantum measurement apparatus has had one purpose: to exhibit certain peculiar initial conditions for that apparatus. We showed above that after the passage of the particle X the wave function of spin 0 and the phonons was

$$\psi(t) = \cos\left(\tilde{c}T\sqrt{\nu_1+1}\right)|+,\nu_1\rangle - i\sin\left(\tilde{c}T\sqrt{\nu_1+1}\right)|-,\nu_1+1\rangle \quad (6.2.16)$$

where the only important fact about the phonon state is its quantum number with respect to a particular mode, '1'; the other modes can be doing anything and are unaffected by X. (With the departure of X the energy is spread, since the mode '1' is not an eigenstate of the Hamiltonian.) Furthermore, we consider a situation where once this one spin is flipped the continuation of the process is inevitable, and had it not flipped, nothing would have happened.

Now T is a number that will vary widely, depending on the proximity of X to the spin; there may also be variation in the coupling constants, c_k. In addition, from run to run there can be a wide range of values of ν_1, representing thermal fluctuations in the lattice near the spin site 0. Suppose it should happen that $\theta \equiv \tilde{c}T\sqrt{\nu_1 + 1} \approx \pi/2$. Then (by Eq. (6.2.16)) the spin-up component of $\psi(t)$ has coefficient $\cos\theta$, which is close to zero. The spin-down component, by contrast, will have magnitude near 1. With these values of T and ν_1 the final apparatus state is not grotesque. On the other hand, for another scattering event it may happen that $\theta \approx \pi$, so that despite the rather close coupling of scatterer and detector, there will be hardly any affect on the spin array. One spin does a handstand, but recovers completely. Again the final apparatus state is not grotesque. As can immediately be seen, these are but two possibilities for *special* states, states for which the experiment gives a *definite* result and the system does *not* find itself in a grotesque state at the end of the experiment.

It follows that if it should happen that for a given passage of an X particle the initial state (of X and of the phonons) is an eigenstate of $\hat{\nu}_1$ with an eigenvalue such that θ ($\equiv \tilde{c}T\sqrt{\nu_1 + 1}$) is close to [integer] $\times \pi/2$, then the final state will *not* be a superposition of two different macroscopic states, but only a single such state.

Summary of Section 6.2

An explicit microscopic example of a detector is given. Its operation models the metastability-driven amplification of a cloud chamber. The model consists of a collection of ferromagnetic quantum spins in interaction with an underlying lattice by means of phonon-spin coupling. These spins detect the passage of an impinging particle, X, through a spin-particle-phonon coupling. The total Hamiltonian is the sum of those displayed in Eqs. (6.2.1), (6.2.3), and (6.2.4).

In our calculation we assume that X rapidly passes near a particular spin (number 0) and we compute explicitly the effect of X on that spin and on the phonons. In Eq. (6.2.15) we show the wave function at the end of the entire process. The only initial-state label appearing explicitly in the wave function is 'ν_1,' which is the eigenvalue of a particular phonon number operator defined in the course of our calculation. In general, this state is grotesque.

We then show that for particular initial conditions, including particular values of the label ν_1, hence particular phonon states, one or the other of the components of this potentially grotesque state is zero. Looking at Eq. (6.2.15), an example of such a condition is that $cT\sqrt{\nu_1 + 1}$ be a multiple of $\pi/2$, where T is an effective contact time for the passing particle and c is a coupling constant. Such states are 'special.'

6.3 Definite quantum measurements

In Section 6.1 we produced a grotesque state of a cloud chamber, that is, a wave function for a large system that has components corresponding to more than one macroscopic configuration. In Section 6.2 we made a microscopic model of a detector, 'constructed' of quantum spins, in which the same thing happened. A particle traverses the detector and the final state is a superposition of |*all spins still up and few phonons*⟩ and |*all spins down and many phonons*⟩. However, in the quantum model we had a detailed description of the initial phonon state as well as a parameter (T) describing the closeness and duration of the approach of the particle. We found that *for certain initial phonon states and certain values of T, one of the components of the macroscopic superposition was simply not there.* Its amplitude was zero. For other initial phonon states, the other outcome was not there. Initial conditions for which only one of the macroscopically possible outcomes evolves, will be called *special*. Abstractly, for the general measurement situation one might have[14]

$$\Psi(\text{final}) \equiv \exp(-iHt/\hbar)\Psi(\text{initial})$$
$$= \Psi(\text{outcome 1}) + \Psi(\text{outcome 2}) + \dots \qquad (6.3.1)$$

whereas for 'special' initial conditions you would have

$$\Psi(\text{final}) \equiv \exp(-iHt/\hbar)\Psi(\text{initial; special for outcome 1})$$
$$= \Psi(\text{outcome 1}) \qquad (6.3.2)$$

Nothing ever happens to the system except application of the operator $\exp(-iHt/\hbar)$, but the microscopic details of the initial state were such as to lead to only one of the outcomes. For this initial condition one gets a definite outcome, rather than a grotesque superposition.

We come now to our general proposal to explain the apparent non-existence of grotesque states. *Every time you do an experiment for which the outcome is potentially grotesque, the microscopic initial conditions are 'special' in the above sense; that is, they lead to only a single one of the possible microscopic outcomes, and they lead there by pure unitary quantum evolution.*

A few obvious remarks: The phrase 'you do an experiment' is loose speaking. What I mean is, anytime there is a situation where a system *could* (as discerned from its macroscopic state) evolve to a grotesque state, it doesn't. This 'situation' need not be called an experiment, nor involve a person. Next, I am not limiting my claim to cloud chambers, quantum spins, or cats. It is therefore implicit in the claim that 'special'

---

[14] Notation: in the following two equations the objects within the parentheses are labels, not arguments.

states exist for any system large enough for potential grotesqueness. Finally, there is the manifest implication that in doing an experiment it is impossible to control its precise microscopic state. You set up a macroscopic situation, but the microscopic state of the system will be one of those that avoids superpositions of macroscopically different states in the course of its subsequent time evolution.

This is not the end of the list of claims, but these are already sufficiently dramatic that they require a pause. The reason for the long presentation of Section 6.2 was not only to explain fully what I was doing, but also to show that there actually are 'special' states. Just as we know (by time reversal) that there are microscopic states in which raindrops rise from puddles and fly to the clouds, so we have rare states from which a system does not evolve to the usual grotesque superposition, but only to one of its possible macroscopic final states. Unlike the time reversed states, as far as I know the existence of 'special' states does not follow from a symmetry principle, and in coming chapters the matter of their existence will be explored.

The other claim is the more dramatic. First you can't control your initial state, and second it is always one of the presumably rare 'special' ones. For those who have read the earlier chapters of this book, this claim should not be surprising. The thermodynamic arrow of time is what gives us our strong prejudice on the arbitrariness of initial conditions and the specificity of final conditions. In a dynamical system for which the boundary conditions are naturally defined at more than one time, the selection of allowable states (from among all candidates macroscopically possible) will not favor initial conditions in this way.

Therefore we can already anticipate that the constraint on initial states will have something to do with the giving of conditions in the future, conflicting with our primitive intuition that only statements about the past influence the present state of a system.

There is another claim implicit in this proposal. Suppose your laboratory is wealthy enough to have two cloud chambers and before doing the experiment you have to decide between them, based perhaps on choosing the one with a gas leak or the one with an unreliable compressor. (Well, maybe the lab is not so wealthy after all!) The chamber that you opt to use will turn out to be the one with the right 'special' state. Or suppose you decide to aim the beam a little differently. Then it will be other gas molecules that are primed for perfect detection or perfect non-detection. It follows that the 'special' states that occur are coordinated with your decision. But since (according to this theory) nothing ever happens except pure, unitary quantum evolution, the precursors of these states were heading where they were going before you made your 'decision.' So your decision was not a decision, and your wave function and that of the detectors are correlated. Pursuing this line of reasoning to ever greater scales, it follows that my ideas can only be valid if there is a single wave function for the entire universe. This wave function has the precise correlations necessary to guarantee 'special' states at every juncture where they are needed.

Again, it is my hope that the edge on the foregoing assertion has been taken off by the earlier parts of this book in which I discussed the arrow of time and past and future boundary conditions. A future boundary condition can trivially generate long range correlations. For chaotic systems these correlations demand extreme precision. What the foregoing discussion implies is that if I propose to motivate the appearance of special[15] states by a future boundary condition, that boundary condition should involve the entire universe. So if I am right about this explanation of the quantum measurement problem, not only must we reexamine statistical mechanics, but cosmology plays a role as well.

So far this proposal has generated two research programs: first, one should show the existence of special states and then find boundary conditions justifying their selection. We turn now to a third issue, one which gives rise to another problem, perhaps technically the most difficult.

An initial-value characterization of the world we have just described would be that it is an enormous conspiracy, evolving (via $\exp(-iHt/\hbar)$) from one special microstate to another. With such overwhelming choreography, how do we get the usual quantum probabilities? Suppose I do a z-component-measuring Stern-Gerlach experiment, sending in 100 atoms with their spins at 30° to the z-axis. Why do I find about 75 of them pointed in one direction (up) and about 25 of them in the other? What gives rise to probabilities in this deterministic world?

The answer is a simple one, basically the same answer that is given for the use of probability in classical mechanics. Behind this answer—for either physical scheme—lie philosophical problems (that we will not address) related to the nature of probability in a deterministic world.[16]

In brief: the probability of a particular outcome is proportional to the number of special states that lead to that outcome.

To be less brief, I will recall the classical rule and also define what is meant by 'number' of special states.

In classical mechanics, when an initial macrostate has more than one possible macroscopic outcome (as in flipping a coin), the probability of one or another outcome is proportional to the volume of the region of

---

[15] Henceforth I will usually drop the quotation marks on the word special, although I will still be using it in the sense defined here.

[16] References are given in Section 6.5.

the initial microscopic phase space that leads to the particular outcome. Let me say this less abstractly. You have a coin-tossing machine with a chaotic mechanical device that controls the force applied to the coin. Within the phase space for the machine, some regions (considered as initial conditions) give rise to heads, some to tails. Each time you operate the machine you set up macroscopic initial conditions, but in doing so you cannot control which of the two phase space regions the system is in. By the postulates of statistical mechanics, the probabilities of heads and tails are the relative volumes of the two phase space regions.[17] (This theme was taken up in Chapter 2.)

To connect this rule to the present proposal, I offer an observation and a reminder. The observation is that if two special microscopic states lead to the same macroscopic outcome, then any superposition of them does as well. Thus for a particular experiment and for a particular macroscopic outcome of that experiment, the special states for that outcome form a subspace of Hilbert space. The reminder concerns the quantum-classical correspondence for statistical mechanics. When going to quantum mechanics, the notion of volume in phase space becomes the number of dimensions of the Hilbert space. The integral in phase space $[\langle A \rangle = \int d^N q\, d^N p\, \rho(q,p) A(q,p)]$ becomes the trace in Hilbert space $[\langle A \rangle = \mathrm{Tr}\, \rho A]$.

Here then is the way quantum probabilities are recovered in my theory. You set up a macroscopic state. Let it represent a situation where it might shortly become grotesque. The gross macroscopic description of the system fixes a particular Hilbert space. The usual rule would be to calculate expectations by taking a trace over this entire Hilbert space. I postulate that you *take the trace only over the subspaces of special states*. If the operator whose expectation you are examining is a projection for a particular outcome of the experiment, this implies that the probability of that outcome is proportional to the dimension of the Hilbert subspace of special states for that outcome.

In Fig. 6.3.1 I show a similar rule for calculating probabilities when you have a cryptic constraint in classical mechanics. Instead of averaging over the entire volume associated with a particular outcome, you would average only over those subvolumes that also satisfy the cryptic constraint. Similar matters were taken up in Chapter 4, where future conditioning provided such cryptic constraints. In the quantum scheme presented here, one averages not over the entire macroscopically defined Hilbert space but only over those portions that also satisfy the property of not leading to grotesque states.

---

[17] *Equal* volumes is another matter; for honest coin-tossing they should be.

To summarize: *the probability that an experiment has a particular macroscopic outcome is proportional to the dimension of the subspace of special states that lead to this outcome.*

This rule imposes yet another challenge for our theory. One must show that the probabilities computed in this way agree with those obtained by the Copenhagen interpretation. Or at least one must show this for those experiments for which that interpretation has been confirmed.

This completes the statement of my proposal. In the next section I will outline the theoretical directions and experiments that would be needed to show these ideas to be conceptually consistent and consistent with Nature. We close this section with a telegraphic statement of the assertions.

1) Every apparatus capable of performing a measurement, or more generally, situations sufficiently complex to be able to evolve to grotesque
   states, have special microscopic states that evolve under pure quantum
   evolution ($\psi \rightarrow \exp(-iHt/\hbar)\psi$), to non-grotesque states, generally
   one of the possible outcomes of the experiment (or situation).
2) In every actually performed experiment, in every natural occurrence
   that could lead to a grotesque state, the (initial) microscopic state of
   the system is one of the 'special' states.

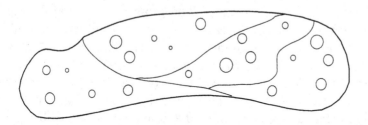

Fig. 6.3.1. Schematic diagram of a region of classical phase space. The entire region corresponds to a macroscopic characterization of a system. It is
divided into subregions which may be finer than coarse grains, each subregion
corresponding to a particular outcome of an experiment. The probability of an
outcome is proportional to the volume of the corresponding subregion. If there
are additional conditions that limit the microscopic state to the yet smaller subsets (indicated by small circles) within the subregions, then the probability of an
outcome is the total volume in the circular (in our diagram) regions within the
corresponding subregion. For the quantum case the overall Hilbert space consists
in general of 'special' subspaces for the various outcomes as well as subspaces
that give superpositions of outcomes.

3) The probabilities of various outcomes of an experiment are proportional to the sizes of the respective classes of 'special' states. 'Size' means Hilbert subspace dimension.

6.4 Issues to be addressed

Existence of special states in Chapter 7. There are several topics.
1) Explicit examples of special states.
2) Entanglement.
3) Physical situations where it looks difficult to find special states.
4) Physical situations where it looks possible and should be interesting.

Selection of special states in Chapter 8. Why should Nature select special microscopic states? From our earlier exposition on the arrow of time the reader should be willing to contemplate cryptic conditions on initial states, but why this particular cryptic condition—related to grotesqueness of wave functions?

Abundance of special states in Chapter 9. To recover the probabilities predicted by the Copenhagen interpretation the special states (or their subspaces) must come with the appropriate abundance. It turns out that, *modulo* points that still puzzle me, a Cauchy distribution of environmental disturbances gives the desired result. And it gives it universally. And it would not be possible to get any other probability relation, for example probabilities proportional to the fourth power of the (magnitude of the) wave function.

Experimental tests in Chapter 10. Several possibilities are considered, either based on better than usual control of microscopic degrees of freedom or a search for features related to the Cauchy noise (of Chapter 9) or on EPR type experiments.

In the indicated chapters, these items will be taken up in detail.

6.5 Notes and sources

Section 6.0. The textbook measurement theory discussion referred to is found in [GOTTFRIED]. The 'Schrödinger cat' gedanken experiment, in which a quantum experiment leads to a superposition of macroscopically different components, and one of these components arguably represents a thinking creature, is found in [SCHRÖDINGER]. The term 'grotesque' for a superposition of macroscopically different states was introduced in [GRIFFITHS 84].

Section 6.1. The Wilson cloud chamber has been an important experimental tool, for example in the discovery of positrons. It was most useful at what are now considered low energies, so that today you are more likely to encounter its relative, the bubble chamber (see [ARYA, HUGHES] or [PERKINS]). I prefer to use the cloud chamber as my metaphor since it is easier to describe an exothermic transition.

Reprints of many of the classic papers of quantum measurement theory appear in [WHEELER]. In recent years the literature has increased significantly. A sampling of conferences with presentations representing a broad spectrum of approaches to the problem is [MILLER, ACCARDI, GREENBERGER, BLACK, COLEMAN] and [HALLIWELL].

Theories stressing changes in dynamics that could lead to the apparently discontinuous results of quantum measurements ('collapse') are presented by [GHIRARDI 86, 89, PEARLE] and [WEINBERG], the latter disproved experimentally in [MAJUMDER]. For a general review of experimental work see [LAMOREAUX].

Hidden variable theories are mainly associated with the names Bohm and de Broglie although other theories that at first seemed different now turn out to be part of the same set of ideas. See [HILEY, GOLDSTEIN, NELSON] and [DELLA RICCIA].

The many worlds interpretation was proposed by Everett, whose thesis (under Wheeler) is reprinted in [DE WITT]. Articles by Hartle and others in [COLEMAN] present the 'post-Everett' version of this theory. See also [GRIFFITHS 84] and [ZUREK]. Gell-Mann has suggested (in the discussion in [SCHULMAN 94a]) that the theory presented in the present book is a what you would get from the theory given by him and Hartle ([GELL-MANN]) with a fixing of the density matrix to be particular pure states at distant past and future times. This is not the case, since with special states the entire initial state evolves to the entire final state, while in their picture the value of the relevant trace in their formalism (which is unity with perfect special states) would be phenomenally small.

Section 6.2. Versions of the model apparatus described in this section are given in [GAVEAU] and [SCHULMAN 91a], and these papers should be consulted for details of the calculation referred to in the text. Formally, the spin plus phonon part of the model Hamiltonian resembles extended versions of the Jaynes-Cummings model used in optics. See [SHORE]. For a discussion of macroscopic metastability see [SCHULMAN 90]. A sampling of recent papers that provide quantum treatments of various parts of the measurement process is [ALTENMULLER, BROYLES, HAAKE] and [NAKAZATO].

Section 6.3. The theory given here was first published in [SCHULMAN 84], with elaboration in [SCHULMAN 86]. Subsequent related works are [SCHULMAN 88a, 88b, 89a, 89b, 91b, 91c, 92a, 92b, 94a, 94b, 94c]. A review paper is [SCHULMAN 91a].

I will not provide philosophical references that deal with quantum measurement problems in general, as the great body of that work has little to do with the theory presented here. Of relevance though are references that deal with the foundations of probability and with matters of determinism and free will. One author, [O'CONNOR], aware of the indeterminism supposedly mandated for 'subatomic events,' questions how this could have bearing on the existence of free will. We physicists may be surprised to learn that for this philosopher the Gödel incompleteness theorem is more important than the so-called probabilistic nature of quantum mechanics. Another philosopher, [DENNETT], insofar as I can follow, also accepts the dogma of the intrinsic indeterminism of quantum mechanics. Nevertheless, he too finds this not as relevant as we physicists often believe, and feels that chaos (in the sense of dynamical sensitivity to small perturbation) can provide all the uncertainty one needs for free will. On page 87 of [SEARLE] is a critique of purported relations between free will and physical determinism with which I have particular sympathy (having expressed it in late night dormitory discussions, many years ago). I mention these things because I was once subjected to the fury of an outraged colleague from Munich. He yelled at me (this really happened): how could I advocate determinism? There could be no science in a deterministic world, since there could be no experiment—after all, the outcome of every experiment would be determined, so it's not an experiment. My point is that physicists, who have been known to make disparaging comments about philosophers (this too has really happened), are often poor philosophers themselves. In our function as physicists we should stick to just that—doing physics by the criteria that have served us over the years, some combination of theoretical consistency and conformity with experiment. Philosophy should be no more forbidden to us than gardening, but it shouldn't be the criterion for judging our physics. Well perhaps I exaggerate, since what just I called 'theoretical consistency' blends smoothly into intellectual satisfaction which can hardly be separated from philosophy.

Causality, probability, and explanation are interrelated ideas, and a philosophical work drawing them together is [SALMON]. Salmon does take up quantum mechanics, but again the issue of whether or not it is deterministic is given less importance than most physicists would have expected. A history of the origin of probability theory appears in [HACKING]. The contemporary scientist will discover there that many ideas now considered obvious did not emerge in European culture until recently (seventeenth century). From the epic *Mahábarata* there is indirect evidence of the earlier existence of a science of probability in India, and there are some probability calculations in the Talmud, but that's about it, notwithstanding the popularity of gambling in ancient and sophisticated cultures. In Europe, the Dutch, in providing an early form of annuity, did not address the actuarial problem and often lost money to the recipients of the annuities. A general collection of articles on the philosophy of science, including a few on the foundations of probability, is [FEIGL]. Thanks to Tom Reynolds and to Michael Harrison, I have become aware of a broader collection of works: [HINTIKKA], [RABINOVITCH], [SUPPES], [VON PLATO], and [WEATHERFORD].

References

ACCARDI 94	L. Accardi (ed.) (1994) *The interpretation of quantum theory: where do we stand?*, Encic. Ital., Rome. (Conf. proc., Columbia Univ., 1992.)
ALTENMULLER 94	T. P. Altenmuller & A. Schenzle (1994) Quantum Zeno effect in a double-well potential: A model of a physical measurement, *Phys. Rev.* A **49**, 2016.

ARYA
74
A. P. Arya (1974) *Elementary Modern Physics*, Addison-Wesley, Reading, Mass.

BLACK
92
T. D. Black, M. M. Nieto, H. S. Pilloff, M. O. Scully & R. M. Sinclair (eds.) (1992) *Foundations of Quantum Mechanics*, World Sci., Singapore.

BROYLES
93
A. A. Broyles (1993) Wave mechanics of particle detectors, *Phys. Rev. A*, **48**, 4252.

COLEMAN
91
S. Coleman, J. B. Hartle, T. Piran & S. Weinberg (1991) *Quantum Cosmology and Baby Universes*, World Sci., Singapore.

DELLA RICCIA
66
G. Della Riccia & N. Wiener (1966) Wave Mechanics in Classical Phase Space, Brownian Motion, and Quantum Theory, *J. Math. Phys.* **7**, 1372.

DENNETT
84
D. C. Dennett (1984) *Elbow Room: The Varieties of Free Will Worth Wanting*, MIT Press, Cambridge, Mass.

DE WITT
73
B. S. DeWitt & N. Graham (1973) *The Many-Worlds Interpretation of Quantum Mechanics*, Princeton Univ. Press, Princeton.

FEIGL
53
H. Feigl & M. Brodbeck (1953) *Readings in the Philosophy of Science*, Appleton Century Crofts, New York.

GAVEAU
90
B. Gaveau & L. S. Schulman (1990) Model Apparatus for Quantum Measurements, *J. Stat. Phys.* **58**, 1209.

GELL-MANN
94
M. Gell-Mann & J. B. Hartle (1994) Time Symmetry and Asymmetry in Quantum Mechanics and Quantum Cosmology, in [HALLIWELL].

GHIRARDI
86
G. C. Ghirardi, A. Rimini & T. Weber (1986) Unified dynamics for microscopic and macroscopic systems, *Phys. Rev. D* **34**, 470.

GHIRARDI
89
G. C. Ghirardi, P. Pearle & A. Rimini (1989) Markov Processes in Hilbert Space and Continuous Spontaneous Localization of Systems of Identical Particles, *Phys. Rev. A* **42**, 78.

GOLDSTEIN
87
S. Goldstein (1987) Stochastic Mechanics and Quantum Theory, *J. Stat. Phys.* **47**, 645.

GOTTFRIED
66
K. Gottfried (1966) *Quantum Mechanics*, Benjamin, New York.

GREENBERGER
86
D. M. Greenberger (1986) *New Techniques and Ideas in Quantum Measurement Theory*, New York Acad. Sci., New York (vol. 480, *Ann. New York Acad. Sci.*)

GRIFFITHS
84
R. B. Griffiths (1984) Consistent Histories and the Interpretation of Quantum Mechanics, *J. Stat. Phys.* **36**, 219.

HAAKE
87
F. Haake & D. F. Walls (1987) Overdamped and amplifying meters in the quantum theory of measurement, *Phys. Rev. A* **36**, 730.

HACKING
75
I. Hacking (1975) *The Emergence of Probability*, Cambridge, London.

HALLIWELL
94
J. Halliwell, J. Pérez-Mercader & W. H. Zurek (eds.) (1994) *Physical Origins of Time Asymmetry*, Cambridge Univ. Press, Cambridge.

HILEY
87
B. J. Hiley & F. D. Peat (1987) *Quantum implications: essays in honour of David Bohm*, Routledge, London.

HINTIKKA
81
J. Hintikka, D. Gruender & E. Agazzi (1981) *Probabilistic Thinking, Thermodynamics and the Interaction of the History and Philosophy of Science*, Reidel, Dordrecht.

HUGHES
91
I. S. Hughes (1991) *Elementary Particles*, 3rd ed., Cambridge Univ. Press, Cambridge.

LAMOREAUX
92
S. K. Lamoreaux (1992) A Review of the Experimental Tests of Quantum Mechanics, *Int. J. Mod. Phys. A* **7**, 6691.

MAJUMDER
90
P. K. Majumder, B. J. Venema, S. K. Lamoreaux, B. R. Heckel & E. N. Fortson (1990) Test of the Linearity of Quantum Mechanics in Optically Pumped ^{201}Hg, *Phys. Rev. Lett.* **65**, 2931.

MILLER 90   A. I. Miller (ed.) (1990) *Sixty-Two Years of Uncertainty: Historical, Philosophical and Physical Inquiries into the Foundations of Quantum Mechanics*, NATO ASI Ser. B: Phys. vol. 226, Plenum, New York.

NAKAZATO 93   H. Nakazato & S. Pascazio (1993) Macroscopic limit of a solvable dynamical model, *Phys. Rev.* A **48**, 1066.

NELSON 66   E. Nelson (1966) Derivation of the Schrödinger Equation from Newtonian Mechanics, *Phys. Rev.* **150**, 1079.

O'CONNOR 71   D. J. O'Connor (1971) *Free Will*, Doubleday, New York.

PEARLE 94   P. Pearle & E. Squires (1994) Bound State Excitation, Nucleon Decay Experiments, and Models of Wave Function Collapse, *Phys. Rev. Lett.* **73**, 1.

PERKINS 87   D. H. Perkins (1987) *Introduction to high energy physics*, 3rd ed., Addison-Wesley, Reading, Mass.

RABINOVITCH 73   N. L. Rabinovitch (1973) *Probability and Statistical Inference in Ancient and Medieval Jewish Literature*, Univ. Toronto Press, Toronto.

SALMON 84   W. C. Salmon (1984) *Scientific Explanation and the Causal Structure of the World*, Princeton Univ. Press, Princeton.

SCHRÖDINGER 35   E. Schrödinger (1935) Die gegenwärtige Situation in der Quantenmechanik, *Naturwiss.* **23**, 807. (A translation is reprinted in [WHEELER].)

SCHULMAN 84   L. S. Schulman (1984) Definite Measurements and Deterministic Quantum Evolution, *Phys. Lett.* A **102**, 396.

SCHULMAN 86   L. S. Schulman (1986) Deterministic Quantum Evolution through Modification of the Hypotheses of Statistical Mechanics, *J. Stat. Phys.* **42**, 689.

SCHULMAN 88a   L. S. Schulman (1988) Detection with Compulsory Nonabsorption, *Phys. Lett.* A **130**, 194.

SCHULMAN 88b   L. S. Schulman (1988) Two Time Localization, *Ann. Phys.* **183**, 320.

SCHULMAN 89a   L. S. Schulman (1989) Remote Two-Time Boundary Conditions and Special States in Quantum Mechanics, *Found. Phys. Lett.* **2**, 515.

SCHULMAN 89b   L. S. Schulman (1989) Semiclassical Two-Time Localization and the Consequences of Restrictions on Wave Packet Spreading, in *Path Integrals from meV to MeV*, ed. V. Sa-yakanit *et al.*, World Sci., Singapore.

SCHULMAN 90   L. S. Schulman (1990) System-size Effects in Metastability, in *Finite Size Scaling and Numerical Simulation of Statistical Systems*, ed. V. Privman, World Sci., Singapore.

SCHULMAN 91a   L. S. Schulman (1991) Definite Quantum Measurements, *Ann. Phys.* **212**, 315.

SCHULMAN 91b   L. S. Schulman, C. R. Doering & B. Gaveau (1991) Linear decay in multi-level quantum systems, *J. Phys.* A **24**, 2053.

SCHULMAN 91c   L. S. Schulman (1991) "Special" states in quantum measurement apparatus: Structural requirements for the recovery of standard probabilities, *Found. Phys.* **21**, 931.

SCHULMAN 92a   L. S. Schulman (1992) Relativistic Two-time Localization, *J. Phys.* A **25**, 3007.

SCHULMAN 92b   L. S. Schulman (1992) Definite Quantum Measurements, in [BLACK].

SCHULMAN 94a   L. S. Schulman (1994) Time Symmetric Cosmology and Definite Quantum Measurements, in [HALLIWELL].

SCHULMAN 94b   L. S. Schulman (1994) Special States in the Spin-Boson Model, *J. Stat. Phys.* **77**, 931.

SCHULMAN 94c   L. S. Schulman (1994) Particle detection via special states, in [ACCARDI].

SEARLE 84   J. Searle (1984) *Minds, Brains and Science*, Harvard Univ. Press, Cambridge, Mass.

SHORE 93   B. W. Shore & P. L. Knight (1993) The Jaynes-Cummings Model: A Topical Review, *J. Mod. Opt.* **40**, 1195.

SUPPES 84   P. Suppes (1984) *Probabilistic Metaphysics*, Basil Blackwell.

VON PLATO 94   J. von Plato (1994) *Creating Modern Probability: Its Mathematics, Physics and Philosophy in Historical Perspective*, Cambridge Univ. Press, New York.

WEATHERFORD 94   R. Weatherford (1994) *Philosophical Foundations of Probability Theory*, Routledge & Kegan Paul, London.

WEINBERG 89   S. Weinberg (1989) Precision Tests of Quantum Mechanics, *Phys. Rev. Lett.* **62**, 485.

WHEELER 83   J. A. Wheeler & W. H. Zurek (1983) *Quantum Theory and Measurement*, Princeton Univ. Press, Princeton.

ZUREK 91   W. H. Zurek (1991) Decoherence and the transition from quantum to classical, *Phys. Today* (October), 36.

7

Existence of special states

The existence of 'special' states can be established with ordinary quantum mechanics. We seek particular microscopic states of large systems that have the property that they evolve to only one or another macroscopic outcome, when other microstates of a more common sort (having the same initial macrostate) would have given grotesque states. Justifying the hypothesis that Nature chooses these special states as initial conditions is another matter. In this chapter we stick to the narrower issue of whether there exist states that *can* do the job, irrespective of whether they occur in practice.

In Section 7.1 we give several explicit examples. In Section 6.2 we exhibited an apparatus model and its special states. Here we look at the decay of an unstable quantum state, not as a single degree of freedom in a potential, but with elements of the environment taken into account as well. We also study another popular many-body system, the spin boson model. This has extensive physical applications and especially with respect to Josephson junctions has been used to address quantum measurement questions. A single degree of freedom in a potential, by the way, does *not* generally lend itself to 'specializing,' and an example is shown below.

In recent years, exotic non-local effects of quantum mechanics have been exhibited experimentally. Behind many of these lies *entanglement*, the property of a wave function of several variables that it does not factor into a product of functions of these variables separately. With special states you do not get macroscopic entanglement. You still get the EPR (Einstein-Podolsky-Rosen) effect and you still violate the Bell inequalities (as a quantum theory should). This will be explored more fully when we discuss possible experimental tests in Chapter 10. In the present chapter we will consider general constraints on the theory related to entanglement.

225

Following that we will discuss what appear to be problematic situations. There are cases where it is difficult to see where enough degrees of freedom can be found to allow special states. I will lay out the problems and possible solutions, none of which leaves me completely satisfied. It is my hope that readers of this book will be motivated to help resolve these questions.

By contrast there are several contemporary experiments, some dramatic, that offer the possibility of identifying the relevant special states. This would allow us to go beyond the models in Section 7.1, and will be discussed in Section 7.4.

7.1 Explicit examples of special states

General formalism

To analyze a measurement[1] you begin with the Hilbert space for the entire system. 'System' includes not only microscopic objects, but as much of the environment as is necessary for an accurate description.[2] Certain subspaces of this Hilbert space correspond to definite macroscopic states (cf. the connection of subspaces and coarse grains discussed earlier). In a measurement the system begins (at time 0, say) in one such subspace and can evolve (by some final time, T) to one of several (including, possibly, the original one). The measurement problem is that, in general, vectors in the original subspace evolve to a wave function with significant components in *several* of the possible final subspaces—a grotesque superposition of outcomes. Our search for special states is a search for vectors in the original subspace that evolve entirely to one or another of the possible outcome subspaces. We will confine the search to the interval $[0, T]$, assuming that outside that period the system is in contact with the larger world. We will also assume that if, within our model, the system at time T is not yet irreversibly committed to the subspace within which it finds itself, the contact with the world seals that commitment.

---

[1] The word 'measurement' suggests something done by a person. The term 'measurement situation' is often used to avoid this suggestion. I will use the less cumbersome single word, but intend it to have the broader and non-anthropomorphic meaning.

[2] Given the interconnectedness of the universe that is demanded by the theory of the last chapter, it may be surprising that any isolation, temporary or not, is possible, even theoretically. The reason isolation is possible is that once a system is large enough for a good description of the measurement and its special states, additional, relatively noninteracting, degrees of freedom enter, to a good approximation, as a Hilbert space cross product. As a result, the *relative* numbers of different classes of special states do not change.

We simplify this structure by considering only two possibilities: the system is still in its initial subspace or it is not. Having a special initial condition will then mean that at the time of observation the system is all in or all out. Richer situations are certainly of interest and will be discussed below.

Let the total Hilbert space be $\mathcal{H}$. The subspace in which the system is found at time 0 (initially) is designated $\mathcal{H}_0$ and projection operator for $\mathcal{H}_0$ is P. The orthogonal subspace to $\mathcal{H}_0$ is $\mathcal{H}_0^\perp$ and its projection operator is Q, so that $Q = 1 - P$. The total Hamiltonian is H and we conform to the notation used in Chapter 2 by giving the operators constructed from H the following names:

$$h = PHP, \qquad \Omega = QHQ, \qquad C = QHP \qquad (7.1.1)$$

This gives H and the total wave function forms[3] parallel to those of Chapter 2:

$$H = \begin{pmatrix} h & C^\dagger \\ C & \Omega \end{pmatrix} \qquad \psi = \begin{pmatrix} x \\ Y \end{pmatrix} \qquad (7.1.2)$$

There is nevertheless a major difference between the situation here and that studied in Chapter 2. Here, the space of initial states is itself multi-dimensional. At the least, P is a projection onto a coarse grain, so that the dimension of $\mathcal{H}_0$ is enormous.[4] However, we expect the dimension of $\mathcal{H}_0^\perp$ to be even more enormous. This is because the passage from one macroscopic state to another, and in particular an irreversible measurement process, activates or creates additional degrees of freedom, often with a quasicontinuum spectrum. Thus $\dim(\mathcal{H}_0^\perp) \gg \dim(\mathcal{H}_0) \gg 1$.

Since the initial state is in $\mathcal{H}_0$ it satisfies $P\psi_0 = \psi_0$. Under time evolution the state becomes $\psi_t = \exp(-iHt)\psi_0$ (we drop $\hbar$). The end of the period of isolation is called (time-) T. Our goal is to find an initial condition ψ_0 for which at T the state is entirely *in* $\mathcal{H}_0$ or entirely *out* of it. This would be a definite outcome and the associated ψ_0 a 'special' state. To be entirely in $\mathcal{H}_0$ means $p_T \equiv \|P\psi_T\|^2 = 1$, while being entirely out means p_T is zero. Since it is ψ_0 that needs to be special, it is convenient

---

[3] There is a slight abuse of notation in going from Eq. (7.1.1) to Eq. (7.1.2). For example, the operator 'C' that appears in Eq. (7.1.2) is not quite what is written in Eq. (7.1.1), but is its restriction to the spaces on which it can behave non-trivially.

[4] Coarse grains invariably involve many degrees of freedom. For example, for a single atom held in place by laser traction, the photons (with quasi-continuum spectrum) and the lasers must be included in the coarse grain as well as the 'single atom.'

to write p_T as

$$
\begin{aligned}
p_T &= \|P\psi_T\|^2 \\
&= \langle\psi_0|\exp(iHT)P\exp(-iHT)|\psi_0\rangle \qquad (7.1.3)\\
&= \langle\psi_0|A_T^\dagger A_T|\psi_0\rangle
\end{aligned}
$$

where

$$
A_T \equiv P\exp(-iHT)P \qquad (7.1.4)
$$

and we have used $P^2 = P$ and $P\psi_0 = \psi_0$. It is also convenient to define $B_T \equiv A_T^\dagger A_T$. Although A_T is neither unitary nor Hermitian, it is in a sense subunitary, a projection of the unitary $\exp(-iHT)$. The spectrum of the Hermitian B_T lies in the interval $[0,1]$. Correspondingly, 0 and 1 are the extreme values that can be taken by p_T. What we seek are wave functions ψ_0 for which B_T takes its extreme values. This points us in the direction of the diagonalization problem for B_T and the special states will be B_T's eigenfunctions for eigenvalues 0 and 1, if it has any.

Exercise 7.1.1.   Prove the above assertion about the spectrum of B_T.

This leads us to an issue that the reader may already have considered in the previous chapter. What if the special state is not perfectly special? That is, p_T is close to 0, but not 0, close to 1, but not 1. The answer lies in whatever physical considerations require special states. In the next chapter we argue that certain boundary conditions for the universe, at epochs close to a big bang and to a big crunch, can impose the avoidance of grotesque states, hence the occurrence of special states. If there is any play in the boundary conditions, there will be play in the constraint. It does seem to me that as the number of degrees of freedom grows the play in both should become small, but they need not be zero. Therefore in making our demand for special states, expressed perhaps as a demand on the spectrum of B_T, we allow small deviations from perfection.

This then is the framework for seeking special states: identify the operator B_T for the particular problem, and look for spectrum near zero and one.

Remark: The formulation just given is a special case of one of the two-time quantum boundary value problems discussed in Chapter 5. The arbitrary subspaces P_0 and P_1 of Section 5.3 are here replaced by P and $1 - P$.

Decay of a metastable quantum system

Given the parallel formalism above and in Section 2.5, all that remains is to identify a Hamiltonian suitable for decay. We are thinking in terms of an atom in an excited state, within an environment of other atoms, stray photons of all frequencies, the occasional cosmic ray, even neutrinos. Perhaps it would be better to think in time-dependent terms, collisions and scatterings, but for present purposes we imagine a box around the whole thing so that it makes sense to look at a total Hamiltonian. Time dependent phenomena then translate into particular initial conditions. The Hamiltonian h (of Eq. (7.1.2)) therefore describes the basic excited atom plus all sorts of junk from the environment. This spreads the available levels in a quasirandom way, so that one might contemplate one of the well known random spectra for h. For example, in spectra satisfying the Wigner conjecture the probability that a pair of nearest levels is separated by x is proportional to $x \exp(-\text{const} \cdot x^2)$. For the special state examples to be given, it doesn't much matter what spectrum is taken, so we will use equal spacing. For the spectrum of Ω the same considerations apply; now one is looking at the spread around the decayed atom plus decay products. Without loss of generality both h and Ω are taken to be diagonal, so that our remaining choices lie in the coupling, C.

Analysis of this model follows the steps from Eq. (2.5.8) to Eq. (2.5.10). With $P\psi$ described by (the now multidimensional) x we have

$$\dot{x}(t) = -ihx(t) - C^\dagger e^{-i\Omega t} \int_0^t e^{i\Omega s} C x(s) ds \qquad (7.1.5)$$

For the special state example that we now give, we take a form of the interaction C that is *not* general. Nevertheless, some of its structure should extend more generally (this question is considered in the reference cited), and it does provide a striking example of 'specialness.' We assume that C can depend on the state in $\mathcal{H}_0$ that is interacting, but not on the state in $\mathcal{H}_0^\perp$. That is, $C_{k\ell}$ only depends on ℓ. If the dimension of $\mathcal{H}_0^\perp$ is $N_0^\perp$, then the way to write C that allows the limit of large $N_0^\perp$ is $C_{k\ell} = c_\ell^* / \sqrt{N_0^\perp}$. Eq. (7.1.5) becomes

$$\dot{x}(t) = -ihx(t) - c \int_0^t \frac{1}{N_0^\perp} \text{Tr}\, e^{-i\Omega(t-s)} c^\dagger x(s) ds \qquad (7.1.6)$$

In the foregoing, c has a vector character (in $\mathcal{H}_0$), so that $c^\dagger x$ is a scalar. If Ω is assumed to have finely spaced levels and to extend over a wide range, the trace gives essentially a δ-function. The only delicate point is that the endpoint of integration comes at the argument of the δ-function. A bit of care shows that this gives half the contribution one would ordinarily get and the result of the integration is $\pi x(t)$ instead of $2\pi x(t)$. To make the dynamics in $\mathcal{H}_0$ more transparent we define a normalized form of the interaction and write Eq. (7.1.6), after the s integration as

$$\dot{x}(t) = -ihx(t) - \alpha \eta \eta^\dagger x(t) \qquad (7.1.7)$$

where $\eta = c\sqrt{\pi/\alpha}$ and the normalization $\eta^\dagger \eta = 1$ fixes the number α. These manipulations allow our dynamics to be conducted entirely in $\mathcal{H}_0$. Besides our assumptions on C, the broad spectrum of Ω meant that memory effects in $\mathcal{H}_0^\perp$ were not important (cf. Section 2.5). Notice, that as an operator within $\mathcal{H}_0$, $\eta\eta^\dagger$ is a one-dimensional projection.

The vector 'x' above (at time T) is exactly what the operator A_T gives when applied to the initial state. (Recall: $A_T \equiv P\exp(-iHT)P$.) That is, you start in $\mathcal{H}_0$, evolve a time T, and then see how much is still in $\mathcal{H}_0$. From Eq. (7.1.7) it therefore follows that

$$A_T = e^{-QT} \qquad \text{with} \quad Q = ih + \alpha\eta\eta^\dagger \qquad (7.1.8)$$

We can now address our original question, the eigenvalues of $B_T = A_T^\dagger A_T$. Fig. 7.1.1 shows the spectrum of B_T, evaluated numerically, for a variety of T values with $\mathcal{H}_0$ 41-dimensional and h having equally spaced levels. The value of α is $1/\pi$ and η is random.

It should be emphasized that Fig. 7.1.1 does not show the history of any particular initial state. The abscissa is time (T), but the ordinate gives the values of the eigenvalues of B_T. It happens that they vary smoothly, but for each T the eigenvector associated with the particular eigenvalue is different.

What the figure shows is that not only are there special states, but that for this model *most* of the states are special.[5] For example, when the measurement is to be taken at time 50, there are 31 dimensions of $\mathcal{H}_0$ for which the initial conditions will lead to no perceptible decay. There are five dimensions for which there is substantially full decay, and five dimensions somewhere in between.

If this excited atom (and its environment) were the unstable quantum system whose decay triggered the lethal apparatus in a Schrödinger cat experiment, and if the inspection of the chamber takes place at time 50 in these units, you[6] could do a nice job of saving the cat by using one of the near-1 eigenstates of B_T as an initial condition for the unstable system.

It is also true that although the curves in Fig. 7.1.1 are not the history of any particular states, the actual time dependence of $\|\exp(-iHt)\psi_{\text{special}}\|^2$ for $0 \le t \le T$ does resemble those curves. One thus has an example of a quantum jump.

---

[5] More precisely, the span of the special subspaces accounts for all but a few of the dimensions of the Hilbert space. With random selection of coefficients a vector would be drawn from different special subspaces and not itself be special.

[6] 'You' would have to be a Maxwell demon, since vectors in $\mathcal{H}_0$ are *micro*-states.

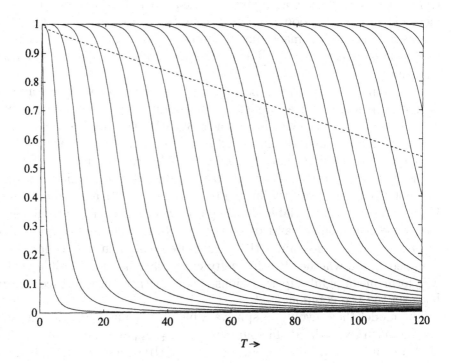

Fig. 7.1.1. Eigenvalues of the operator B_T for various T. The energy levels for the space of undecayed levels ($\mathcal{H}_0$) consists of 41 equally spaced levels. Each of these couples to the decay products in the same way although the overall strength of the coupling varies. Typically the spectrum consists of levels with eigenvalue (of B_T) close to one, levels close to zero, and four or five in between. The dashed line is the average of the levels.

The mechanism of the remarkable 'ticking' evidenced in the figure is partly understood and the reader is referred to the article cited below for mathematical details. One gets soliton-like behavior from a linear system, complete with a gaillard and pavane in the complex plane for the roots of a relevant polynomial. The result does not depend on our elimination of $\mathcal{H}_0^\perp$ above and numerical work in which those degrees of freedom are retained show the same features that have been displayed. One also need not be so finicky in the selection of a Hamiltonian although the ticking behavior is not typical of most quantum systems (or at least the associated linear decay, evident in Fig. 7.1.1, is not).

Spin boson model

Many physical systems are well described by the spin boson model, including some, such as the Josephson junction, that are large enough for transitions among the spin states to be considered measurements. The fact that one has a good microscopic model of a quantum device that stands near the micro/macro interface has led to its use for the posing of foundational questions as well as for applications to dissipative quantum tunneling.

The 'spin' of the model is usually a projection from a more complex degree of freedom. For example, a superconducting ring with a Josephson junction forms a device known as a SQUID. The total magnetic flux threading the ring behaves like a collective coordinate for the many-body electron wave function. This dynamical variable can be thought of as responding to a potential. When the states and excitations available within this potential are effectively limited to two, the collective coordinate acts like a two-state spin variable. This spin is more hefty than a single electron magnetic moment and often is characterized as *mesoscopic*. The tests of quantum mechanics that are contemplated in this context are designed to see whether for this hefty coordinate its two possible values will form meaningful superpositions, i.e., are there interference effects between the two components.[7]

The model thus involves a spin variable and the associated operators can be taken to be the Pauli spin matrices, σ_x, σ_y, etc. The spin variable is also coupled to a boson field. For the SQUID these are the phonons associated with the resistor in the Josephson junction and they provide ohmic dissipation. These boson modes will be labelled $k = 1, \ldots$, and the associated annihilation operators a_k. The Hamiltonian for the model is

$$H = \varepsilon \left(\frac{1+\sigma_z}{2}\right) + \Delta\sigma_x + \sum_k \omega_k a_k^\dagger a_k + \sigma_z \sum_k \beta_k(a_k^\dagger + a_k) + \sum_k \frac{\beta_k^2}{\omega_k} \quad (7.1.9)$$

where the bosons have frequencies ω_k, the spin boson coupling is proportional to (the real-valued) β_k, the up-down coupling due to the bosons is given by Δ and the unperturbed level energy difference between the spin levels is ε.

The formal structure for seeking special states in this context is exactly what we presented earlier in Eqs. (7.1.1–7.1.4). The space $\mathcal{H}_0$ consists of 'spin-up,' and 'bosons doing anything,' and its complement consists of 'spin-down,' and 'bosons doing anything.' If the initial state is 'up,' the projection operator 'P' defined earlier is simply $(1 + \sigma_z)/2$, where it is understood that there is a tensor product with the identity acting on all the boson modes.

---

[7] To digress: the limitations of such a test should also be appreciated. 'Heftiness' is not complexity—what it means here is that it is a collective coordinate for many degrees of freedom. Another feature not tested is the importance of spatial locality. For the SQUID, the value of the hefty coordinate does not affect the spatial position.

I will now report success in finding special states in this system. Unfortunately, I cannot report analytic control of the situation; all that can be presented is numerics. Based on numerical exploration, a limited degree of intuitive comprehension is possible. Nevertheless, the example is worth presenting, because it turns out that in this case the spectrum of B_T is all over the place (unlike the previous example). Since this is a more realistic model, it suggests that what goes on in Nature is that there are so many states around that some of them turn out to be special. This point was brought out by a referee of my journal article on this subject, who wondered why I was so surprised at the plethora of suitable eigenvalues of B_T. There are so many eigenvalues around; why shouldn't they be spread, getting close to 0 and 1.

Aside: The analytic machinery for this problem is considerable, especially functional integral techniques. For the spin variable one can use methods related to Kac's treatment of the telegrapher equation and which in this context involve things called 'blips.' Furthermore, coherent states for the boson degrees of freedom can be very powerful. In my hands, the special state problem, which is essentially a two-time boundary value problem in quantum mechanics, did not yield to these techniques. Perhaps a more limited program would be easier: use these methods to formulate a smaller numerical problem than the straightforward tensor products that I use below.

To study the spectrum of B_T numerically demands compromises, that is to say, truncations. You can't have an infinite number of boson modes and you can't allow each unlimited occupation. Because the Hilbert space is the tensor product of the spin and boson modes, the dimension climbs rapidly as more and larger modes are included. In practice three boson modes at most were used and cutoffs on the maximum boson excitation ranged from 2 to 50.

Remark: As in the previous example, we will be generating the spectrum of B_T for many different values of T. We will even connect the dots. This does not represent the history of a particular initial condition. It is simply convenient to generate all these data in one computation and the dotted lines are to orient the eye. Each point on the graph represents an *initial* state whose norm remaining in the Hilbert subspace of initial states is given by its ordinate, at the time given by its abscissa. It is also an eigenstate of B_T for the corresponding time.

Remark: The cutoff was implemented by disconnecting states with boson excitation number equal to or less than N_{cutoff} from those with more. In terms of the operators, the cutoff procedure replaced $a^\dagger$, etc., by finite dimensional square matrices.

For Fig. 7.1.2 only a single boson mode was used and it was restricted to (eigenvalues of) $a_k^\dagger a_k \leq 31$. As in the previous example the spectrum of B_T is evaluated for many different T. There is a marked abundance of special states. (Proximity to 0 and 1 is generally better than 10^{-3}.) They are present at almost all times evaluated, and if there is any tendency

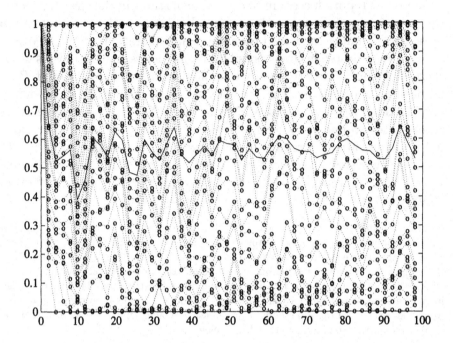

Fig. 7.1.2. Eigenvalues of B_T (ordinate) for various T (abscissa). The solid line is the average. Parameter values (for H of Eq. (7.1.9)) are $\omega = 0.9$, $\varepsilon = 1$, $\Delta = 0.6$, and $\beta = 0.2$. The boson cutoff is $N_{\text{cutoff}} \leq 31$.

at all it is for there to be more states near the extreme possibilities (0 and 1) than in the middle.[8] With increasing numbers of boson modes things can only get better, from the standpoint of finding special states. This is because you always have the option of not exciting the new mode. Numerically this is not guaranteed, since bringing in another boson mode means reducing the cutoff value to keep the matrices at practical sizes.

For the next figure (Fig. 7.1.3), three boson modes were used and it's more of the same. Although the boson cutoffs were severe (at most 4), there is still no difficulty producing states that drop completely at a specified time and those that drop not at all. Proximity to 0 and 1 is even better.

Remark: A point of interest is the degree to which these results are *generic*. This relates to parameter counting considerations discussed in Section 5.3. Briefly, special states persist

---

[8] There are many detailed questions one can ask here, for example, are the special states with eigenvalue 1 consequences of the cutoff? The answer is, some are, many aren't. These matters are taken up in the publication cited in the notes.

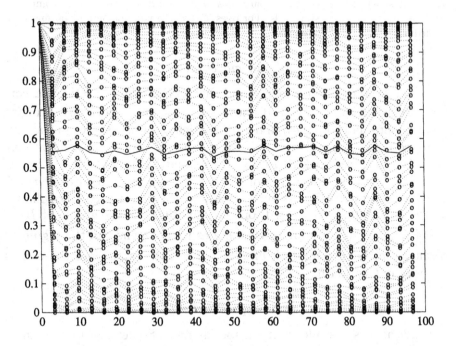

Fig. 7.1.3. Eigenvalues of B_T for various T. The solid line is the average. Parameter values (for H of Eq. (7.1.9)) are $\varepsilon = 1$ and $\Delta = 0.5$. The boson-number cutoffs are 4, 3 and 3. The associated frequencies ω are respectively 0.1, 0.9 and 1.1; $\beta = 0.2$ for all modes.

for this Hamiltonian even when they would be absent by parameter counting alone. Thus if we only allow excitations of from 0 to 16 for the single mode, $N_{\text{cutoff}} = 31$, case (what was shown in Fig. 7.1.2), there is still a plethora of special states, both for having and not having the transition. If the same exercise is done for a random Hamiltonian, making non-generic demands eliminates a class of special states.

What appears to be happening in this model is that the system tends to get stuck in small regions of the Hilbert space. In a sense this is always true, since unitary evolution does not change the dimension of a subspace. What is interesting here is that the continued evolution leaves the system in subspaces that are close to what you consider to be coarse grains. Therefore it is with respect to your physical characterization of spreading that the system remains relatively localized and special states are easy to find. Justification for this picture (for example by tracking evolution of individual states) is found in the cited article.

Aside: It should be clear that my numerical work was done with modest computing facilities, by *fin de siècle* standards. For questions where more or better information is needed it would be easy to improve.

Cloud model recast

The special states used in our 'cloud chamber' model of Section 6.2 conform to the structure outlined above (A_T, B_T, etc.). Formally, bringing in the coordinates of the impinging particle as well as those of all the spins could get complicated. If only one spin subject to an external force is considered—which is the approximation used there in producing special states—then $\mathcal{H}_0$ consists of spin-up crossed into all possible phonon states. $\mathcal{H}_0^{\perp}$ is then spin-down, again crossed into phonon states. If Eq. (6.2.11) is used, one can see that the eigenvalue of B_T is $\cos^2\left(\hat{c}T\sqrt{\hat{\nu}_1 + 1}\right)$. Finding special states then depends on the argument of the cosine being a multiple of $\pi/2$, which is the same requirement that we used in Chapter 6.

Idealized potential models

It is not always possible to find a special state. Sometimes the dynamics are just too simple. This does not contradict our basic thesis: on the contrary, it focuses on the complexity of quantum measurement situations and the fact that the role of statistical mechanics in this story is both as problem and as solution.

Consider a particle moving in one dimension in the presence of a fixed potential $V(x)$. Let the non-zero values of the potential be confined to $|x| \le R$ for some $R < \infty$. Suppose a particle is sent in from the left ($x < -R$) and scatters off this potential. In general there will be a reflected and a transmitted portion whose norms squared add to unity. Let's suppose that for the momentum range we now consider neither reflection nor transmission is 100%; otherwise there would be nothing to specialize.

If you wanted a single-degree-of-freedom special state you would try to have your particle either perfectly transmitted or perfectly reflected, rather than a bit of each. The specialness would be accomplished by sending in a carefully tailored wave packet. I'll now show that if the initial wave function is reasonably smooth, this can't be done.

It is sufficient to consider the following scattering states

$$\theta_k(x) = \begin{cases} e^{ikx} + \rho(k)e^{i\alpha(k)}e^{-ikx} & x < -R \\ \text{something complicated} & -R \le x \le R \\ \tau(k)e^{i\beta(k)}e^{ikx} & x > R \end{cases} \quad (7.1.10)$$

These are solutions to the time-independent Schrödinger equation. (By convention, $k \geq 0$.) The full time-dependent wave function is then

$$\psi(x,t) = \int dk \theta_k(x)\phi(k)e^{-ik^2t/2} \tag{7.1.11}$$

with $\phi(k)$ the momentum space wave function. ϕ must be such that $\psi(x,t) = 0$ for $x > 0$ and t large and negative. Similarly, for large positive times the integral in Eq. (7.1.10) should have no contribution from the $\exp(ikx)$ term for $x < 0$. This is part of the usual relation between time-dependent and time-independent scattering theory. These things are generally justified with the stationary phase approximation, not based on $\hbar \to 0$ (since this is a fully quantum result), but on large x and t. There does need to be some juggling to show that wave packet spreading does not get in the way.

Suppose we wanted to try for perfect reflection. Then we would require the integral

$$\psi(x,t) = \int dk \phi(k)\tau(k)e^{i\beta(k)}e^{ikx - ik^2t/2} \tag{7.1.12}$$

to be zero for $x > R$, for all times.[9] Pick a positive number v and write $x = vt + \xi$. Then for each t and for each ξ the following must vanish:

$$\psi(vt + \xi, t) = \int dk \phi(k)\tau(k)e^{i[\beta(k)+k\xi]} \exp\left[it\left(kv - \frac{k^2}{2}\right)\right] \tag{7.1.13}$$

Consider this in the limit of large t. Other than $t(kv - k^2/2)$, nothing depends on t. Therefore, unless one of the other terms is singular, the large t asymptotics will be given by the stationary point of $kv - k^2/2$. Now $\beta(k)$ and $\tau(k)$ are features of the quantum problem that is being solved. In general these will be smooth enough not to cause trouble. Also, by assumption, $\tau(k) \neq 0$ for the range of k examined. I will also impose smoothness on ϕ, thereby limiting the scope of the proof. With these assumptions, the integral can be evaluated by the stationary phase method, which in this case means that it is dominated by the region $k \sim v$. From Eq. (7.1.13)

$$\psi(vt + \xi, t) \sim \sqrt{\frac{2\pi}{it}}\phi(v)\tau(v)e^{i[\beta(v)+v\xi]}e^{itv^2/2} \tag{7.1.14}$$

---

[9] At this point one might be tempted to use the Paley-Wiener theorem or infinitely sharp tools of that sort, but I prefer not to demand exact vanishing of ψ in this domain, only that it be very small, smaller than what you would get, say, from wave packet spreading (for large t). My reasons are that special states are not supposed to make things *exactly* zero, and even in the usual story if one uses a Gaussian in k around a central k_0, there will still be leaks on the order of $\exp(-k_0^2|t|)$ for supposedly zero functions.

Therefore by examining ψ along a single line and going as far out on that line as is necessary, we find that the only way it could vanish is for $\phi(v)$ to vanish.[10] Since this is true for each v, the original wave function must have been identically zero.

Remark: Although this kind of potential model doesn't lead to special states, it's interesting that with just a bit more complexity you can get them. Consider the spin boson model, whose spin variable can be considered the coordinate in a double well potential in which only one state in each well need be considered. By introducing a 32-dimensional space—a single boson with cutoff '31'—we found above that special states for putting a particle in one or the other minimum were easy to produce.

Special states in Nature

We have just given several examples of special states. Is this what they really look like in Nature? In Chapter 9 we will take up the matter of the *abundance* of special states and from those results I believe that the states we have here exhibited are atypical. This section has still served the purpose of showing the richness of the microscopic degrees of freedom. However, it slights that richness to think that we have done anything but scratch the surface of possible special states.

7.2 Entanglement

An entangled wave function is one that is a function of more than one coordinate and which cannot be written as a product of separate functions of those coordinates, something of the form $a(x)b(y) + c(x)d(y)$ (where the functions are different and non-trivial). Generally one has in mind that the coordinates belong to different particles, but even for a single particle interesting effects can arise.

Two ways that this concept enters quantum measurement theory are as follows.

1) In the von Neumann description of a measurement, the initial state is of the form $\Omega_0(\xi) \sum a_n u_n(x)$, with Ω_0 the initial wave function of the apparatus, u_n states of the object on which the measurement is to be made, and a_n numerical coefficients for those (normalized) functions. After the measurement the wave function becomes $\sum \Omega_n(\xi) a_n \tilde{u}_n(x)$. The states Ω_n are macroscopically distinguishable. By the black magic of the Copenhagen interpretation this sum becomes a single one of its terms, and the probability of its becoming that is $|a_n|^2$. Prior to the 'black magic,' we had a grotesque, entangled state. Afterwards, it

---

[10] The $\sqrt{t}$ dropoff is ordinary wave packet spreading. 'Specializing' would imply reduction of the factors multiplying the $1/\sqrt{t}$.

is again a factored product. In the many worlds interpretation, the measurement does not disentangle, it's just that you don't notice, and for any one observer it appears to be disentangled (cf. Eq. (6.1.4)).

2) It is with microscopic entanglement that exotic effects of quantum mechanics occur. In the EPR experiment and its variations, a microscopically entangled state decays to physically separated pieces. Because the wave function presumably remains entangled, a measurement on one portion affects the other, remote portion, as well.

In the theory described in this book the first kind of entanglement is avoided by means of special states. The second kind, however, places additional demands on those states in that they must *disentangle* when necessary.

The condition of being entangled or not entangled is not conserved, so nothing mitigates against there being special states to take you from one condition to the other. On the contrary, this is a situation where one can use time reversal to find special disentangling states. Suppose one starts from a wave function of the form $\chi(x)\theta(y)$; under time evolution it becomes

$$\chi(x)\theta(y) \longrightarrow e^{-iHt}[\chi(x)\theta(y)] = \sum_{n,m} a_{nm}(t)u_n(x)v_m(y) \qquad (7.2.1)$$

where u_n and v_m are basis functions for the spaces associated with the variables x and y respectively. The sum in Eq. (7.2.1) is non-trivial in the sense that for $t > 0$ more than one coefficient a_{mn} is non-zero and the function no longer factors. Time evolution has taken us from a disentangled to an entangled state. In a time reversal invariant theory the opposite is clearly possible. Even without time reversal invariance, the argument still goes through—to find the initial state to disentangle for the physical Hamiltonian, do the calculation in Eq. (7.2.1) using its unphysical time reversal.

This shows that being or not being entangled can change in time. However, the demonstration has not necessarily produced the special state needed for disentanglement in any given physical situation, since the final states in Eq. (7.2.1) may be macroscopically distinct, and such a coherent recombination of a large number of degrees of freedom goes beyond what I am is claiming is possible.[11]

In certain instances disentangling may present particular problems. I will now discuss such an example and exhibit a special state for it, although in Nature I expect the actual special state to be different. Con-

---

[11] In fact part of our argument in the next chapter is based on the difficulty of coherently recombining grotesque states.

sider a pair of distinguishable spin-$\frac{1}{2}$ particles in a bound state of zero total angular momentum. Let u represent the spin wave function of a spin-$\frac{1}{2}$ particle with spin up (i.e., $\begin{pmatrix} 1 \\ 0 \end{pmatrix}$) and let v represent the corresponding spin down. Then if the spatial wave function of the pair of particles is spherically symmetric (call it f), the wave function will be

$$\psi(1,2) = \frac{1}{\sqrt{2}}[u(1)v(2) - v(1)u(2)]f(1,2) \qquad (7.2.2)$$

Suppose the pair of particles breaks apart and that we measure the z-component of spin for each of them individually at some distance from their original location, as in Bohm's version of the EPR experiment. Then at the end of the experiment the spin wave function will be either $u(1)v(2)$ or $v(1)u(2)$. Either of these is a superposition of spin-0 and spin-1, and therefore no longer a spin-0 eigenfunction of angular momentum. However, if the Hamiltonian by which these particles evolved is rotationally invariant you cannot go from a state of angular momentum 0 to one that includes a spin-1 component.

So far this is not a problem. As usual we are prepared to throw in all sorts of junk from the environment to clean up the angular momentum book-keeping. But because we are also going from an entangled to a disentangled state, it turns out not to be so easy, as we now show. The following equation (Eq. (7.2.3)) expresses a hope. We take as initial state, Ψ_{Initial}. It includes the particles of interest, but also has as much of the environment as desired, so long as its total angular momentum is zero. We act on it with a rotationally invariant Hamiltonian. What we want this to give us is another angular momentum zero state, but with the spin wave functions for particles 1 and 2 factored out. In effect we want to know whether it is possible to have an angular momentum zero state that has the factored form on the left hand side of the following equation:

$$u(1)v(2)\psi_{\text{Final}} = \exp(-iHT)\Psi_{\text{Initial}} \qquad (7.2.3)$$

where ψ_{Final} is taken to be almost anything (the 'almost' will be explained momentarily). It turns out that the factored form gets in the way of having zero angular momentum. To see this, write ψ_{Final} in an angular momentum expansion, and the 'almost' means that the expansion terminates at a finite value, designated J. Recall that $u(1)v(2) = [|00\rangle + |10\rangle]/\sqrt{2}$, where the labels in the kets refer to total and z-component of angular momentum. We thus demand that

$$\frac{1}{\sqrt{2}}[|00\rangle + |10\rangle] \sum_{j \leq J} F_j \qquad (7.2.4)$$

have total angular momentum zero (where each summand F_j has angular momentum j and has as many particle arguments as you wish, including

1 and 2). Consider the term $|10\rangle F_J$. By the usual angular momentum addition rules this has a component with angular momentum $J + 1$. No other term in the sum has so large an angular momentum, so that it cannot be canceled. Therefore it cannot be there, and the wave function must be zero.

Remark: Without the finite j restriction, a zero angular momentum state can be built with a factor $u(1)v(2)$. The norm of F_j for large j drops off like $[(\sqrt{3} - 1)/\sqrt{2}]^j$.

If one gives up the idea that the initial state is a total angular momentum zero eigenstate, then states can be found to accomplish the disentangling.[12] We introduce two additional particles besides those in Eq. (7.2.2) and suppose them to be disentangled with wave function (say) $u(3)v(4)$. Keeping only the wave function for the spins, we have

$$\psi_{\text{initial}}(1,2,3,4) = \frac{1}{\sqrt{2}}[u(1)v(2) - v(1)u(2)]u(3)v(4) \qquad (7.2.5)$$

A reasonable Hamiltonian would be something proportional to $\vec{s}_1 \cdot \vec{s}_3 + \vec{s}_2 \cdot \vec{s}_4$. Recall that, e.g., $s_{13}^2 \equiv (\vec{s}_1 + \vec{s}_3)^2 = \vec{s}_1^2 + \vec{s}_3^2 + 2\vec{s}_1 \cdot \vec{s}_3$ and that $\vec{s}_1^2$ and $\vec{s}_3^2$ are absolute constants. Therefore instead of $\vec{s}_1 \cdot \vec{s}_3$ we can use $(\vec{s}_1 + \vec{s}_3)^2$. It is now an exercise to show that

$$\begin{aligned}
\exp[i\pi(s_{13}^2 &+ s_{24}^2)/2]\frac{1}{\sqrt{2}}[u(1)v(2) - v(1)u(2)]u(3)v(4) \\
&= u(1)v(2)\frac{1}{\sqrt{2}}[u(3)v(4) - v(3)u(4)]
\end{aligned} \qquad (7.2.6)$$

It follows that the helper particles, numbers 3 and 4, are left in a spin-0 state, while particles 1 and 2 are disentangled and go off to their respective fates.

Exercise 7.2.1. Eq. (7.2.6) generalizes to 'helpers' in arbitrary spin-$\frac{1}{2}$ states. In particular show that if α and β are any two two-component spinors, then $\exp[i\pi(s_{13}^2 + s_{24}^2)/2][u(1)v(2) - v(1)u(2)]\alpha(3)\beta(4) = \alpha(1)\beta(2)[u(3)v(4) - v(3)u(4)]$.

As mentioned above, it is unlikely that this is what actually happens in the breakup of bound states. However, it does show that no general principles or symmetries get in the way. My guess is that photons and collisions with other particles are involved in an essential way. To see that photons can do the job, consider the following gedanken experiment. Take a helium nucleus with a single electron around it (i.e., a singly charged positive ion) and bring a muon with very small velocity into its neighborhood. To make it even more gedanken, start the muon with a wave

---

[12] It's not so difficult to give up this idea, since otherwise the initial state, presumably describing a macroscopic chunk of the world, would be rotationally invariant, hence featureless.

function resembling the $n = 3$ bound state it would have in the electric field of a single proton. For convenience imagine that the nucleus merely provides a fixed Coulomb potential, so that the initial wave function is simply an unentangled product of muon and electron wave functions. In short order, this time-zero state will form a bound state of a muonic helium atom with the emission of many photons. Eventually the muon will also decay, but we're not looking that far ahead. The initial conditions guarantee that the probability that the system does not end as a bound state is nil. The final state represents an entangled state of muon and electron wave functions, with many photons as well. In this argument, I am not concerned with whether it is possible to start a muon in the state I specified—all I wanted was to show the existence of a solution to a difficult angular momentum calculation. The photons are also entangled, but then that's always the case, since they are identical.

Remark: The purpose of the demonstration was to show that you could use photons to break up a bound state and reach a final unentangled state. This does not mean that you would need photons of just the right energies to do the job. In a real world breakup most of the energy could come from some major identifiable source, a collision or a single high energy photon. Our demonstration then shows that the angular momentum and disentanglement book-keeping could be finished by additional photons, but they could be of low energy.

Remark: What about the entanglement of identical particles? Does this help or hinder specialization? This is a problem of a mathematical nature that still needs to be resolved. My suspicion is that it makes finding special states easier, especially for bosons.

Remark: It is clear that the above discussion is relevant to the EPR experiment. When this is taken up in Chapter 10 we will argue that the specializing—which includes the disentangling—is most likely done at the source, the point at which the system breaks into constituents.

7.3 Impoverished environments

There are situations in which it is difficult to see where the special states could come from. Even if you put in as much of the environment as you can muster it still appears insufficient to provide definite outcomes. In this section I will mention a few examples and give my own speculations on what may be missing. Of course another possibility is that nothing is missing and that Nature simply doesn't do things the way I've suggested.

Scattering

Two particles scatter, go off in different directions and are detected at a distance from the collision by detectors occupying small solid angle. The textbook description of this leads you to expect a spherical wave emanating from the point of interaction, possibly modulated by a gentle angular dependence. My description says that the entire wave function of each particle has made its way to the respective detectors and has been absorbed there. What points them in the right direction?

I've started with this example because it is one for which many possible sources of specialization are possible. And because when they aren't possible it is no longer clear that the experiment has happened the way you think it did. First the textbook description usually employs center of mass coordinates, so in real scattering there is already directional and localized wave packet structure. As discussed in Section 7.1 (with respect to potential scattering), we expect that this is not sufficient for the whole job; nevertheless, it can help. Next there is the material within which the target is embedded or, in the case of colliding beams, other material in the immediate neighborhood of the collision. Bear in mind too that the focusing or other specializing does not take place only after the collision, but before as well.[13] Finally the typical scattering event is detected only after the particles have passed through solid pieces of matter and in particular the detection itself takes place in a system that interacts significantly with the particle—otherwise it would not detect.

If you remove enough material to invalidate the foregoing proposed sources (i.e., consider an experiment or situation where they are not present), then it may no longer be clear where or when the collision took place and your interpretation of the signal you receive can become ambiguous.

The scattering context will be explored more fully in Chapter 10 when we examine quantitatively whether it can be used for an experimental test of the present theory. It will be seen that coherent and incoherent scattering have enormously different degrees of effectiveness. For example you might wonder whether a neutrino created in the sun could interact sufficiently to be focused on a single nucleus of chlorine in a detector on planet Earth. It turns out that by the time the newborn neutrino wave function has expanded to 1 cubic centimeter it has interacted enough for the focusing to take place.

Before going on to the next example I would like to expand on a point mentioned above. The time at which the specializing takes place—i.e., the statistical deviation that we call a special state—can be before what one ordinarily identifies as the event of interest, during that event, after that event, or all of the above. If you watched a movie of a raindrop landing in a puddle, a really good movie with the ability to zoom in on molecular motion, and then you ran that movie backward, you would

---

[13] The 'specializing' does not represent some peculiar dynamics, just as measurement does not. I use the phrase to mean that the environment is in a statistically unusual state in which a collective blow can be delivered where needed.

see a specializing take place in which many molecules of the environment gave up their energy and momentum to relatively few molecules in such a way that the few broke away from the many and were propelled upwards into the air. This strange scenario, which in reverse happens every time it rains, helps one realize that having subtle molecular motions orient a single particle towards a detector should be relatively easy to achieve.

Nuclear decay

Consider the decay of a uranium nucleus, ^{238}U. This has a lifetime of 4.5×10^9 years, emitting an alpha particle to become ^{234}Th. Is it possible for the initial states of two different uranium nuclei to carry enough detailed structure so that one of them decays after 10 minutes, the other after 10^{10} years? Of course I am not proposing to do this with just the barrier penetration model of alpha decay, in which the many degrees of freedom other than the alpha are replaced by an effective potential and the alpha by a point in that potential. We already know (from Section 7.1) that such potential models offer little opportunity for specializing. But even using 238 separate particles, plus the occasional virtual pion, can there be such a richness of internal structure? On the other hand, if you include external influences, collisions with neighboring atoms, etc., can these relatively gentle forces reach into the nucleus so as to affect the decay?

The problem can be intensified by going to simpler systems and by considering multichannel decay as well. What makes one eta particle become two photons and another three pions? Does an eta still have too much internal structure? What about positronium? No one has found internal structure in an electron, yet an electron-positron pair in a relative s-state presumably annihilates at a random time based on the exponential decay law.[14] Now in Section 7.? our decay model showed that a bit of level mixing from the outside can be helpful, but was not sufficiently realistic nor quantitative to answer the present questions.

Furthermore, when the system finally does decay, the decay products head off in definite directions, giving rise to the same questions raised in connection with scattering. However, as for that situation, one can look to material in the neighborhood for sources of perturbation.

I don't have a firm answer for what specializes the time and nature of the decay. To discuss the possibilities I will anticipate results that we will obtain in Chapter 9. There we concentrate on the matrix element connecting two states of a system, decayed or undecayed, spin up or spin down, etc. Without probing in detail how the environment could contribute to this coupling, we ask, what must the distribution of the

---

[14] I don't know if this has ever been checked experimentally.

sizes of those contributions be in order to reproduce the usual quantum probabilities? The answer turns out to be the Cauchy distribution. This is a continuous distribution on the line. The probability of a Cauchy distributed random variable X taking a value between x and $x + dx$ is

$$C_a(x)dx = \frac{a/\pi}{(x - x_0)^2 + a^2}dx \qquad (7.3.1)$$

where a and x_0 are parameters characterizing the particular distribution. For those whose intuition has been developed on Gaussian random variables, the Cauchy distribution has many surprises. We'll go into this later, but for now I mention a most important feature. Suppose you are constantly being bombarded by kicks from a source whose output is Cauchy distributed (with non-trivial parameter a but with the parameter x_0 taking the value 0). I tell you in advance that the sum of those kicks, say 500 of them, is a large number, $10^9 a$. The question is, what is the least unlikely way for those 500 kicks to have their total add to $10^9 a$? Five hundred kicks of strength $10^9 a/500$? No, the answer is 499 small kicks (on the order of a) and one BIG kick, of size $10^9 a$. To give a contrasting example, the least unlikely way to have the total height of 100 humans add to 200 meters is to find a lot of tall humans—not to take 99 average people and one very big one.

This suggests that the actual decay events, when they take place, represent large, coordinated fluctuations. Although one can generally write off the environment because of the weakness of its coupling to the uranium nucleus, nevertheless when the moment of truth comes and the nucleus undergoes its transition to thorium plus alpha, it may be just those generally small outside influences that do it. And outside isn't always outside. One ordinarily thinks of the eta or ^{238}U nucleus as sitting in its ground state until the decay (so that my remark above about 238 separate particles was slightly misleading). But the ground state is not a true ground state; otherwise the particle would not decay. What we call the ground state is usually that of an idealized, truncated Hamiltonian, neglecting for example weak interactions. With all true interactions included, the state into which the system is born (such as the creation of the eta as a result of a strong interaction process) is an excited state of the full Hamiltonian and has a variety of excitations available.

These remarks on Cauchy distributed kicks are one category of possible response to the problem of impoverished environments. They can be characterized as, 'Just because something is small on the average, don't think it is always small.' (In fact Cauchy distributed things don't have an average.) Another class of responses has to do with marshalling forces you didn't notice. The world is full of tiny perturbations that you generally consider negligible, especially low energy photons and neutrinos, which because of zero mass (presumably) can be long ranging. Furthermore, from the missing mass problem we believe that in fact most of the universe has not been noticed. Perhaps it's this matter, these forces that do the specializing.

This last suggestion may very well be true, but if it is I would find it disappointing. Some, but not all, of the experimental tests that I will propose in Chapter 10 depend on the impoverishment of the environment, and then seeing whether your elimination of special state possibilities has changed the probabilities of outcomes. If there are unknown and thus uncontrollable specializing forces, it would be difficult to effect that impoverishment. This would not mean that there are no physical differences between this theory and others, only that this kind of experimental test will be more difficult.

7.4 Rich environments

The term 'quantum jump' began as a metaphor. One moment the atom is in its $2p$ state, the next in its $1s$ state. In 1930 you weren't supposed to ask what happened; in 1986 you could point to the moment at which a single atom made the transition. The reason you weren't supposed to ask was that all you calculated theoretically was the amplitude for remaining in the original state (the sort of thing we did in Section 2.5). It was then the rule of the Copenhagen interpretation that the norm of that amplitude squared gave the probability of having made the transition, but the transition itself was a kind of measurement. You couldn't ask too closely what happened, you only could use the amplitude to help evaluate the likelihood that Nature, in her arbitrariness, had selected one or another member of the ensemble.

Once you've narrowed the event to a few milliseconds (for transitions with lifetimes of 0.1 s and longer), and actually watched it happen, as several groups did in 1986 (references below), it becomes difficult to claim that another dynamical law takes over. Instead, this process too should yield to full quantum dynamical treatment.

What can such a treatment consist of? Here is more detail of the physical events. You have an atom (an ion of mercury, for example) in one of its excited states. The portion of its energy level structure needed to describe the experiment is shown in Fig. 7.4.1. The atom is bathed in a tremendous flux of photons, laser beams that keep it in place, excite it and sometimes drive it from one to another excited level. How lasers can keep an atom in place is described in references given below. The principal excitation furnished by the impinging photons is to drive

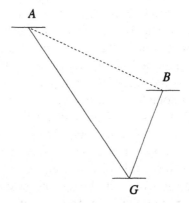

Fig. 7.4.1. Level scheme for quantum jumps. Level A is short lived, level B long lived and level G the ground state. A laser sends photons with the A-G frequency. The steady re-emission of these photons (as the ion cycles between A and G) is the sign of not being in the state B. Occasionally the atom drops or is kicked by another laser into B (suggested by the dashed line). The light from the A to G transition then ceases. When the ion drops from B to G, this light goes on again.

the atom to a short-lived (order of nanoseconds) excited state ('A' in the diagram). When the atom decays from A to G (the ground state), it emits a photon which can be detected in directions other than the one to which the laser beam is pointing. The atom goes through many, many of these excitation-decay cycles per second and a steady glow detected (away from the laser direction) at the frequency of the A-G energy spacing. Every once in a while the atom can decay (from A) to the level B. In the actual experiment this transition is enhanced by an additional laser beam at the right frequency. The level B is a lower energy long-lived metastable state with a lifetime on the order 0.1 s. While the atom is in B, the A-G cycling stops and the detected glow ceases. When the state B decays, the light resumes. In the experiment you see that one moment the atom is in the long-lived metastable state, B, and the next moment it has dropped. This is the so-called quantum jump. There should be no reason why each stage of this process should not be described by quantum mechanics. In such a calculation, the photons (and possibly other influences) cause a transition in which the evolution is not the gradual process described by the decay theory, but something much more rapid. This means that including all the electromagnetic fields, possible collisions, etc., now produce a 'jump,' rather as we produced a jump in Section 7.1 in our decay model. Except that here it's the real thing. Thus if you've managed to describe one of these jumps by quantum evolution alone you are exhibiting a special state.

As far as I know, the kind of analysis that I've just called for has not been performed.[15] Nevertheless, much of the mystery can be removed by distinguishing between the decay time and what might be called the transition time. Consider a barrier penetration situation with a high barrier. The lifetime of the state in the barrier may be long, but to the extent that one can associate a time with the duration of the passage through the barrier that passage time can be far shorter. Similarly, for transitions not describable in this way, there is nevertheless a second time scale, definable in terms of off-shell matrix elements involved in the transition. This is the time scale of the jump.

My reason for proposing to find the special states for this problem is not because it stands in contradiction to conventional views (although it did force more careful analysis of what those views ought to be). Rather it is because it is an experiment that focusses on a particular quantum event (the decay of the long-lived metastable state) and does so in a context where one can identify microscopic degrees of freedom that are involved in the transition.

Another rich environment, one mentioned earlier, is the SQUID. There too the quantum transition is relatively simple and the important features of the environment (the phonons) presumably known.

Finally, as more ambitious forays are made, both experimentally and theoretically, into the domain of systems in the microscopic and mesoscopic range, material will be available for the exhibition of special states. A recent example is work on Mott hopping. In this process, conductivity (under some circumstances) is viewed as a series of hops by a particle from one potential well to another. In one view of this, the hops represent a succession of 'wave function collapses' as the particle progresses. A recent proposal (referenced below) shows how by microscopic dynamics alone the wave function can be carried from well to well. This is in contrast to the picture the Copenhagen interpretation would suggest of one long series of amplitude leaks.

7.5 Notes and sources

Section 7.1. The decay example in the text appears in detail in [SCHULMAN 91a] and also in [SCHULMAN 91b]. The cloud model is in [GAVEAU].

Some of the wide ranging applications of the spin boson model as well as analyses of its properties can be found in [LEGGETT 87] and [WEISS]. Specula-

---

[15] See the notes below for reference to a many worlds calculation.

tions on foundational questions, stimulated by this model are in [LEGGETT 84]. Quite a few theoretical and experimental articles on this subject can be found in [GREENBERGER]. See especially [CHAKRAVARTY].

In a sense you can trivially get special states from the spin boson model since by appropriate adjustment of the boson vacuum the spin variable can be forced to one or the other state. This follows from the phase transition due to the effective long range coupling that comes out of the functional integral formulation. Physically it is not clear what this has to do with experiments since the immobility of the boson vacuum arises from the sluggish low frequency phonons. I don't know how one can prepare (in finite time) infinitely slow modes. This is a puzzle shared by S. Kivelson and me.

My article on special states in the spin boson model is [SCHULMAN 94]. The referee mentioned in the text turned out to be S. Goldstein. The telegrapher equation techniques, closely related to the 'blip' methods of those working with the spin boson model, are found in [KAC 56, 74].

In the text I use a one-dimensional model to show the difficulties of specializing single-particle-in-a-potential Hamiltonians. This lesson was not learned easily and is not universally appreciated. At one point I calculated the scattering of wave packets off the Coulomb potential. The time-dependent propagator for this potential is not known (notwithstanding claims in the literature to the contrary), but at the semiclassical level one can take advantage of a captivating formalism using the Kustaanheimo-Stiefel transformation (and on which those claims are based).[16] Unfortunately, even in the semiclassical approximation the unbound orbits are sufficiently unstable that not much could be done with wave packets (in the spirit of the two-time localization in a potential calculations of Chapter 5). In a similar vein there are authors who have discussed the measurement process in terms of one or a few degrees of freedom, failing to include the steps by which that microscopic 'measurement' reaches the macroscopic level. Examples are [VAN KAMPEN] and [LAMB]. My point is not that these works and the many works like them are wrong, but that they are incomplete. A variant of the potential model proof given in the text can be found in [SCHULMAN 88].

Section 7.2. The importance of entanglement was noted by Schrödinger in the early days of quantum mechanics. A history of the appreciation of this concept is found in [HORNE]. The general framework for quantum measurement mentioned at the beginning of this section is due to [VON NEUMANN].

Section 7.3. The barrier penetration model of alpha decay was introduced by Gamow and others. See [BLATT].

Section 7.4. Observations of 'quantum jumps' were reported in [NAGOURNEY], [BERGQUIST], and [SAUTER]. A general description of related experimental techniques is found in [WINELAND]. An explanation of how jumps can be seen, notwithstanding a long lifetime, is given in [BROYLES]. He gives a 'many worlds'

---

[16] The 'time' parameter introduced by the change of clock within the path integral turns out to be—*in the semiclassical approximation*—the 'eccentric anomaly' of Kepler problem fame.

view of the experiment and makes use of the fact that it is the shorter transition time scale that governs the jump, rather than the long lifetime of the metastable state. Regarding the time scale for quantum jumps, I've recently suggested, [SCHULMAN 96], that the minimal time scale is $\tau_J = \tau_Z^2/\tau_L$, where τ_L is the usual lifetime and τ_Z is the 'Zeno time,' i.e., $\hbar^2/\tau_Z^2 \equiv \langle\psi|(H - \langle\psi|H|\psi\rangle)^2|\psi\rangle$.
The microscopic model of Mott hopping mentioned in the text is [AZBEL].

References

AZBEL 94   M. Y. Azbel (1994) Oscillatory Wave Function Collapse in a Mesoscopic System, *Phys. Rev. Lett.* **73**, 138.

BERGQUIST 86   J. C. Bergquist, R. G. Hulet, W. M. Itano & D. J. Wineland (1986) Observation of Quantum Jumps in a Single Atom, *Phys. Rev. Lett.* **57**, 1699.

BLATT 52   J. M. Blatt & V. F. Weisskopf (1952) *Theoretical Nuclear Physics*, Wiley, New York.

BROYLES 92   A. A. Broyles (1992) Nature of quantum jumps, *Phys. Rev. A* **45**, 4925.

CHAKRAVARTY 86   S. Chakravarty (1986) Quantum Mechanics on a Macroscopic Scale, *Ann. New York Acad. Sci.* **480**, 25. In [GREENBERGER].

GAVEAU 90   B. Gaveau & L. S. Schulman (1990) Model Apparatus for Quantum Measurements, *J. Stat. Phys.* **58**, 1209.

GREENBERGER 86   D. M. Greenberger (1986) *New Techniques and Ideas in Quantum Measurement Theory*, New York Acad. Sci., New York (vol. 480, *Ann. New York Acad. Sci.*)

HORNE 90   M. Horne, A. Shimony & A. Zeilinger (1990) Down-conversion Photon Pairs: A New Chapter In the History of Quantum Mechanical Entanglement, in *Quantum Coherence*, ed. J. S. Anandan, World Sci., Singapore.

KAC 56   M. Kac (1956) *Some Stochastic Problems in Physics and Mathematics*. Lectures at the Field Research Lab., Socony Mobil Oil Co.

KAC 74   M. Kac (1974) A Stochastic Model Related to the Telegrapher's Equation, *Rocky Mount. J. Math.* 4, 497.

LAMB 86   W. E. J. Lamb (1986) Quantum Theory of Measurement, *Ann. New York Acad. Sci.* **480**, 407. In [GREENBERGER].

LEGGETT 84   A. J. Leggett (1984) Schrödinger's Cat and her Laboratory Cousins, *Contemp. Phys.* **25**, 583.

LEGGETT 87   A. J. Leggett, S. Chakravarty, A. T. Dorsey, M. P. A. Fisher, A. Garg & W. Zwerger (1987) Dynamics of the dissipative two-state system, *Rev. Mod. Phys.* **59**, 1.

NAGOURNEY 86   W. Nagourney, J. Sandberg & H. Dehmelt (1986) Shelved Optical Electron Amplifier: Observation of Quantum Jumps, *Phys. Rev. Lett.* **56**, 2797.

SAUTER 86   T. Sauter, W. Neuhauser, R. Blatt and P. E. Toschek (1986) Observation of Quantum Jumps, *Phys. Rev. Lett.* **57**, 1696.

SCHULMAN 88   L. S. Schulman (1988) Two Time Localization, *Ann. Phys.* **183**, 320.

SCHULMAN 91a   L. S. Schulman, C. R. Doering & B. Gaveau (1991) Linear decay in multi-level quantum systems, *J. Phys. A* **24**, 2053.

SCHULMAN 91b   L. S. Schulman (1991) Definite Quantum Measurements, *Ann. Phys.* **212**, 315.

SCHULMAN 94   L. S. Schulman (1994) Special States in the Spin-Boson Model, *J. Stat. Phys.* **77**, 931.

SCHULMAN 96   L. S. Schulman (1996) How quick is a quantum jump?, to appear in Proc. Adriatico Research Conf. (ICTP, Trieste) on 'Tunneling and its implications,' ed. D. Mugnai et al., World Scientific, Singapore.

VAN KAMPEN 88   N. G. van Kampen (1988) Ten Theorems about quantum mechanical measurements, *Physica* A **153**, 97.

VON NEUMANN 55   J. von Neumann (1955) *Mathematical Foundations of Quantum Mechanics*, Princeton Univ. Press, Princeton. (Trans. from German (1932: Springer, Berlin) by R. T. Beyer.)

WEISS 93   U. Weiss (1993) *Quantum Dissipative Systems*, World Sci., Singapore.

WINELAND 87   D. J. Wineland & W. M. Itano (1987) Laser cooling, *Phys. Today* (June), 34.

8

Selection of special states

Why should special states occur as initial conditions in every physical situation in which they are needed? Half this book has been devoted to making the points that initial conditions may not be as controllable as they seem; that there may be constraints on microscopic initial conditions; that this would not have been noticed; that such constraints can arise from two-time or future conditioning; that in our universe such future conditioning may well be present, although, as remarked, *cryptic*. In this chapter I will take up more detailed questions: what future conditions could give rise to the need for our 'special' states, and why should those particular future conditions be imposed.

Before going into this there is a point that needs to be made. Everything in the present chapter could be wrong and the thesis of Chapter 6 nevertheless correct. It is one thing to avoid grotesque states (and solve the quantum measurement problem) by means of special states and it is another provide a rationale for their occurrence. I say this not only to highlight the conceptual dependencies of the theses in this book, but also because there is a good deal of hand waving in the coming chapter and I don't want it to reflect unfavorably on the basic proposal. As pointed out earlier, the usual thermodynamic arrow of time can be phrased as follows: initial states are arbitrary, final states special. If you do not understand the origin of the thermodynamic arrow of time (and it's reasonable to say that Boltzmann didn't), then you don't know why there is a selection of a small subset of possible *final* states. But that doesn't mean you should stop using the second law of thermodynamics. Similarly, I am here claiming that there are restrictions on *initial* conditions. If my justification of this restriction is incorrect or incomplete, that does not mean that one must rule out the special initial condition explanation of the quantum measurement problem that I have offered.

The justification for the occurrence of special states that now follows should therefore be viewed as a *possibility*. Given that we are not used to thinking that our initial conditions are constrained and given the particular nature of the constraint demanded here, it is worth demonstrating that this constraint *can* emerge for plausible physical reasons. The reasons that I will offer are stated in terms of a cosmology with a big crunch. But I don't insist that this is what we're headed for. Consider the state of scientific knowledge in the late nineteenth century, and assume the (cosmo ⇒ thermo) thesis is correct. Could Boltzmann have guessed why there should be a restriction on final states?

Remark: Cross-reference: I have just alluded to two matters dealt with earlier: 1) fixing future conditions can cryptically constrain initial conditions; and 2) the thermodynamic arrow of time is equivalent to a restriction on final conditions. Topic 1 was discussed in Section 4.3, topic 2 in Section 2.6.

8.1 Conditions imposed by localization

Everything has its place, or at least every *thing* has a spatial locale. Notwithstanding wave packet spreading, every macroscopic object has a well defined position. In this section we examine the consequences of the demand that a system be well localized at two widely separated times. Our 'systems' are collections of particles and localization is measured in terms of the particles' wave function.[1] We will see that this demand can be difficult but not impossible to satisfy.

In Section 5.3 we derived a bound on two-time wave packet localization, with 'localization' measured in terms of square-spread. That is, for a given one-particle wave function ψ and a given spatial position x_0, the square-spread is defined by

$$S(x_0; \psi) \equiv \int_{\mathcal{M}} \psi^*(x)(x - x_0)^2 \psi(x) dx \qquad (8.1.1)$$

Here $x, x_0 \in \mathcal{M} \subset \mathbb{R}^n$. Suppose you wanted a wave packet to evolve from a time t_1 to a time t_2 in such a way that the sum of the spreads around a given pair of points x_1 and x_2 at the respective times is minimized. The Hamiltonian, H, is given, so the only question is to find the initial wave packet that does the job best. Actually, with H fixed it doesn't matter whether we look for an initial wave packet, a final wave packet or use any other time. For convenience we take a fiducial time, 0, and minimize the

---

[1] This localization demand applies to the entire wave function. By this definition, the many branches in a many world interpretation would (in general) represent a breakdown in localization.

following

$$W(\psi_0) = S(x_1; e^{-iHt_1/\hbar}\psi_0) + S(x_2; e^{-iHt_2/\hbar}\psi_0) \tag{8.1.2}$$

with ψ_0 the time-0 wave function. Let $H = p^2/2m + V(x)$ be time independent and let $T \equiv t_2 - t_1$. In Section 5.3 we found the following minimal values, W_0 for W

If	*Then*
$V(x) = 0$	$W_0 = \hbar T/m$
$V(x) = \frac{1}{2}\omega^2 x^2$	$W_0 = \hbar|\sin\omega T|/m\omega$
$V(x) = -\frac{1}{2}\omega^2 x^2$	$W_0 = \hbar|\sinh\omega T|/m\omega$
$V(x) = $ attractive, has bound state	$W_0 \lesssim \text{const}$
$V(x) = $ smooth and weak	$W_0 \sim T$, diffusive spread
$V(x) = \delta(x)$	$W_0 \sim T^2$

$$\tag{8.1.3}$$

These results are for a single particle in one dimension but they carry through more generally. If there is a bound state and all the amplitude is concentrated in it, then W_0 remains bounded, even as $T \to \infty$. If $V(x)$ goes to zero for $|x| \to \infty$ and doesn't have a bound state then by tailoring the wave function one can get spreading no worse than diffusive. The last case listed above, $\delta(x)$, was not considered in Section 5.3, but is a consequence of a result in Section 7.1. There we showed that with a simple potential model it is generally not possible to find a special state. Therefore if the initial position x_1 is on one side of the δ-function at 0, and the final point x_2 on the other, there will necessarily be a reflected wave at 0, and the two pieces of wave function will separate linearly in time.

Now consider N distinguishable particles, with N large. Let there be given $2N$ positions, $x_k^{(\ell)}$, with particle ℓ localized at $x_k^{(\ell)}$ at t_k, for $k = 1, 2$ and $\ell = 1, \ldots, N$. Generalizing Eq. (8.1.2), we seek to minimize

$$W(\psi_0) = \sum_{k=1,2} \int_{\mathbb{R}^{3N}} \left|\psi(x^{(1)}, x^{(2)}, \ldots, x^{(N)}, t_k)\right|^2 \sum_{\ell=1}^{N} \left(x^{(\ell)} - x_k^{(\ell)}\right)^2 \tag{8.1.4}$$

with $\psi(\cdot, t_k) \equiv \exp(-iHt_k)\psi_0$. We suppose the Hamiltonian to be of the form $\sum_\ell (p^{(\ell)})^2/2m^{(\ell)} + \sum_{\ell,\ell'} V\left(x^{(\ell)} - x^{(\ell')}\right)$, and take the two-particle potential to be the sort normally encountered for interatomic forces, repulsive at short range, attractive further out, followed by an approach to zero at greater distances.

Suppose the endpoint positions allowed the particles to remain well separated from one another throughout the long time interval $[t_1, t_2]$. If they remain separated, the value of W would be $3NT\hbar/m$ (assuming the particles all to have mass m). On the other hand, suppose the particles spent most of their time in large clumps. Then, roughly speaking, there would be *no* spreading, except in the clump-center-of-mass coordinates. With one stiff clump[2] this would make $W \sim 3T\hbar/Nm$. In this way the large number N goes from the numerator to the denominator.

For the applications that we have in mind, namely loose assemblies of particles at epochs separated by large time intervals, the initial and final positions will neither be already bound nor completely separated. If we want small W, we would need (initially) to have wave packets heading towards one another, with particles binding and occasionally unbinding when scattering. Cutting down on W would require that many particles spend a substantial part of the time interval in clumps. Towards the end of the interval there might be fragmenting scatterings if the boundary conditions required more separation than was typical during the intermediate times (the latter would depend on the overall constraint on W).

While the foregoing argument seems to me intuitively straightforward, the following two examples show more explicitly how a limit on the size of W can impose dynamical conditions during the time interval to which W refers.

First example: flat-bottomed oscillator

This example is a particle in one dimension (or the relative coordinate of a pair of particles). Our goal is to show that special initial conditions in which the particle exploits a potential can reduce spreading well below the free-particle diffusive value. Use the following Hamiltonian

$$H = \frac{p^2}{2m} + \frac{1}{2}m\omega^2 \theta\left(|x| - L\right)\left(|x| - L\right)^2 \tag{8.1.5}$$

(θ is the step function). The potential is shown in Fig. 8.1.1. To reduce the notational burden, let $x_1 = x_2 = 0$, $t_1 = 0$ and $t_2 = T$. If L is large compared to T, a wave packet that hangs around the origin the entire time, tailored to expand as little as possible (like the coherent state in the solution to Exercise 5.3.3, with $x_0 = p_0 = 0$) will give a spread, W, that is essentially T. But by matching the initial (and final) state to the potential a much smaller spread can be achieved. The trick is to take advantage of the harmonic potentials on the ends. For a classical particle no matter how rapidly the particle enters the potential it will spend one half period before being spewed out again. The time it takes to bounce

---

[2] Looser clumps and smaller T allow contributions from soft modes, but this is beside the point.

off either end is thus π/ω. This harmonic property is reflected in wave equation evolution by the fact that a wave packet sent into a half-parabola comes out again one half period later, essentially unchanged in shape, but with velocities reversed. Imagine then a small, high velocity wave packet sent (say) to the right, from the given initial point, 0. It will reach the region of potential on the right with little delay or spreading. Then, no matter how fast it was going, it comes out again a time π/ω later. During the time spent in the half parabola its shape is preserved—by the harmonic property—so that if it had small spatial spread on entry, it has small spatial spread on exit. The packet then zooms across the distance $2L$ and enters the potential on the other side. Again it spends the time π/ω in the potential and emerges, again ready to cover the large distance $2L$ at high speed.

   As just described, one can reduce the spread from the large value, T, to a small value, essentially the spread that occurs in the leftover time— what you get when you subtract from T all the intervals π/ω during which it does not spread at all. There is, however, a bit of checking that must be done before this picture can be accepted. Because of the discontinuity in the second derivative of the potential as the packet enters the harmonic region, there is some scattering. It must be checked that this is small, since amplitude scattered in this way acquires a spread of order L^2. In Appendix A to this section I include a scattering theory calculation that

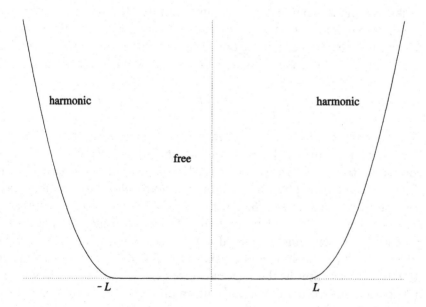

Fig. 8.1.1. Flat-bottomed potential. The potential is a separated parabola. It is zero in the middle and quadratic in $(|x| - L)$, when that quantity is positive.

constitutes the required checking. By the way, the Heisenberg uncertainty principle does not imply that the high momenta used in this example induce a large positional uncertainty. The central momentum is large, but the spread around it need not be. (Cf. Exercise 5.3.2.)

Second example: coordinated scattering

This example is slightly less artificial than the previous one. It illustrates the idea that dynamical scenarios involving bound states can cut down on wave packet spreading between widely separated epochs. Let A, B, C, and D be four different particles in a one-dimensional world. When distant from one another they do not interact, so that if they remain separate during the time interval $[0, T]$ the value of W, the (square) wave packet spread, is minimally $\hbar T \sum(1/m)$, where the sum runs over the four particles.

Suppose that B interacts only with A and only with short range attraction. Let the A-C and C-D interactions be infinite hard core repulsion. Take $m_C = m_D < m_A$ and $m_B \ll m_C$. The largest contribution to W is therefore from B. Our dynamical scenario is illustrated in Fig. 8.1.2 and is described below and in the figure caption.

Initially A, C and D all move toward B. Particles A and C meet at B with the right velocity relation to bring A to a halt. A and B now bind and their wave packet spreading is limited. C goes off to the right where it meets D whose velocity is tailored to send C back to the A&B pair. When C reaches A&B, they are split and all particles go to their final positions.

The foregoing description is valid under either quantum or classical dynamics. Classically, the A-C scattering includes the effect of A's speeding up as it approaches B, so that when A is left with zero kinetic energy *vis à vis* B the total A&B energy is negative (because of the attractive potential), and they are bound. In quantum mechanics the amplitude for the less likely outcome of a scattering is seldom zero, so that if one looks to a more realistic version of our scattering one must deal with amplitude that might not follow the neat lines of the figure. As for the flat-bottomed oscillator, such breakaway amplitude could increase W significantly.[3] To avoid this I refer to the results of Section 7.1. What we need is a special state for scattering, and we expect it to involve environmental degrees of freedom. As pointed out in Section 7.1, although pure potential scattering cannot be 'specialized,' with extra de-

---

[3] For the flat-bottomed oscillator this is treated in Appendix A to this section.

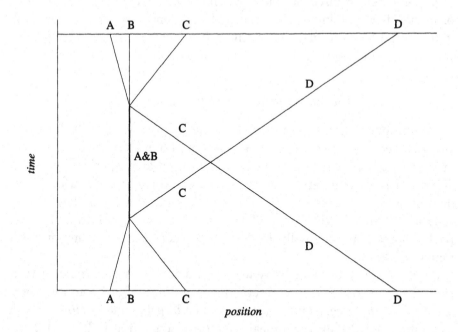

Fig. 8.1.2. Dynamical scenario for reducing wave function spreading. Particles A and C meet, bringing A to a halt and allowing A to bind to B. The impact causes C to reverse direction and eventually meet D. Particles C and D collide, sending C back to the A&B bound state, breaking it up. In this symmetric history, initial and final positions are the same.

grees of freedom much can be accomplished. For example, the spin boson model is based on tunneling between two wells in the presence of bosons. Even with only one boson mode 'specialized,' we showed that for a given final time all amplitude could tunnel or all amplitude could be prevented from tunneling—notwithstanding the fact that without tailoring some average amount of tunneling, neither 0 nor 1, would have taken place.

Assuming then that what can be accomplished for the spin boson model could be accomplished for the scattering in Fig. 8.1.2, the presence of a few boson degrees of freedom in the neighborhood of the collisions would complete the spread-minimization strategy.[4]

---

[4] I am assuming here that the boson wave function spreading is not itself a problem, either because they are tied down by other particles or because they are not part of the balance sheet, i.e., the localization constraint does not include their degrees of freedom.

Two-time boundary conditions with localization constraints

Suppose it is known about a system that some of its degrees of freedom are well localized at two widely separated times, for example that the quantity 'W' defined above is less than a number, W'. It is clear from the foregoing discussion that not every initial condition that looks localized at the initial time will manage to satisfy the two-time constraint. Rather, for solutions to the two-time boundary value problem there will be a tendency for clumpiness during the intermediate times. During collisions there will also be cooperative effects of local microscopic degrees of freedom to maintain the localization of individual particles throughout the time interval. Thus the way to achieve localization at the ends is to favor localized states during the intermediate times.

Remark: These features can be understood in terms of the stochastic dynamics models of Section 5.2. Suppose you demand unlikely initial and final states of a system. You can ask whether it will remain in those unlikely states throughout the interval or whether the least unlikely way to accomplish the goal is to wander throughout the state space and then collect again toward the end of the time interval at the unlikely state. In other words, one is balancing the unlikelihood of not spreading throughout the state space against the unlikelihood of collecting from all corners of that space back to the desired final state. For stochastic dynamics this question has a well defined answer, and depends on the transition rates between the states. Figs. 5.2.4 and 5.2.7 (in Section 5.2) illustrate possibilities. When the gathering together of the system to meet the final state is the most unlikely thing around, rather than do this, most histories will not show as much spreading—throughout the interval—as they would have without the future constraint.

Remark: 'Localization' in this section has been defined in terms of the square spread, W. Presumably qualitatively similar conclusions could be drawn with other definitions, but I have not explored this question.

Quantum measurement and delocalization

A quantum measurement is the ultimate wave function spreader. Consider the 10^{26} atoms in a Schrödinger cat. If you stick to pure quantum evolution, $\psi \rightarrow \exp(-iHt/\hbar)\psi$, which I believe you must, then after the experiment the wave functions of those 10^{26} atoms are spread by meters—with amplitude divided between a frisky pussy in the laboratory and its other world counterpart, discarded with or without (two additional worlds?) the honor due a martyr to Science. And after that it only gets worse.

Therefore, if one gives a two-time boundary value problem with particle localization at both ends, the wave functions that satisfy those boundary values will be those in which the big spreading events—quantum measurements inducing grotesque states—do not take place. The way they can be avoided is by means of what I have been calling 'special' states.[5] For the wave functions of the cat's 10^{26} atoms to localize 10^{11} years in the future is vastly less likely an eventuality than having a few hundred or a few million atoms and photons be correctly coordinated so as to prevent the split of the larger number at the time of the original experiment. In this context, 'less likely' means fewer solutions to the two-time boundary value problem for the Schrödinger equation (or Dirac equation, or ...). And 'fewer' means fewer dimensions in Hilbert space, as in our probability axiom, presented in Chapter 6.

Note: This (first part of the chapter) is the second half of our rationale for the occurrence of special states. The first half of that rationale (in the next section) is a proposal on why there should be spatial localization early and late in the history of the universe. To this we now turn.

Remark: This rationale is subject to all the weaknesses of our development, most significantly (in my opinion) the issue of identical particles. It should also be mentioned that the apparent arbitrariness of the localization criterion arises from not having a complete theory of what induces the two-time localization constraint. When the physics of that constraint is understood, the nature of the localization will be as well.

Remark: In our discussion of the existence of special states and in the quantum two-time boundary value problems of Section 5.3, there was an 'ϵ' floating around, a permitted deviation from perfection. It is at this point that we can see, at least in principle, how it arises. The need for special states comes from a presumably well-defined condition on the wave function. That condition is not a requirement of perfect point localization, but rather some quantitative measure of the early and late time wave functions of the matter in the universe. The value of the appropriate 'ϵ' will depend on that quantitative measure.

Appendix for this section (at the end of the chapter):
8.1.A   Wave packet scattering from a half parabola

8.2 Origins of spatial localization

> Now the earth was unformed and void
> (*tohu v'vohu*),
> and darkness was upon the face of the deep;
> —Genesis, Chapter 1

It is difficult to imagine a world without spatial localization, a *tohu v'vohu* without macroscopic variables. If, following the big bang, there was a period of this sort, at what stage did macroscopic variables emerge?[6] For

---

[5] It's the grotesque state that is avoided, not the measurement.

[6] Was this emergence late enough for the Riemannian arena of classical general relativity to be used or must one also wrestle with quantization of gravity? After the

the goals of this book we wish to establish a kind of wave function localization. However, in other studies of the quantum measurement problem,
spatial coordinates also have a preferred role. For example, as mentioned
in Chapter 6, Gottfried uses spatial separation to justify discarding portions of the density matrix, and we observed that this was further related
to spatial localization in statistical mechanical coarse graining. Similarly,
the 'pointer' states that are invoked in some work on quantum measurement theory turn out to be spatially defined, although as far as I know,
no explanation has been offered in that context for the special role of
space.

Perhaps the obvious answer is that space is special because particle
interactions are spatially local. At the level of classical gravity or non-
relativistic quantum mechanics I believe this is a valid justification, but
if one goes to the opposite extreme of questioning the foundations of
space and time, more caution is warranted. For example, in Section 3.3
we proposed an axiomatic basis for space-time based on a light cone
relation; that we could contemplate such a proposal is related to the
photon's mass being zero. This same mass-zero property is what gets in
the way of defining a position operator for the photon. For particles that
do have mass, spatial localization is natural, whether one is interested in
position operators for electrons or the range of forces mediated by massive
bosons.

In this section our goals are more specific. We want to justify spatial
wave function localization at an early time in the history of the universe.
With the views on cosmology and the arrow of time presented earlier in
the book, if we live in a world heading for a big crunch, such justification
would give the two-time localization needed for application of the results
of Section 8.1, above. As for our arrow of time arguments in Chapter 4,
we seek the simplest physical framework, and will try to avoid questions
of quantum gravity or uncharted areas of particle physics.

When matter is the source of the gravitational field the way to calculate
the strength of that source is to make the appropriate projection of the
density given by the full multiparticle wave function $|\psi(x_1,\ldots)|^2$.

Aside: In other approaches to quantum mechanics this is not obvious. For the Copenhagen view, ensembles obscure the issue, but since for distance scales at which one deals
with gravity all wave functions are 'collapsed,' the question is avoided. For the Everett
interpretation, which allows no evolution but the unitary evolution of pure quantum mechanics, in each branch, or world, the source of gravity is the wave function in that branch.
See the comments at the end of this section.

diatribe in Chapter 3, the reader may be surprised that I raise this question, but I
was not minimizing its importance, only despairing of finding reliable answers.

If at an early stage in the history of the universe the wave function has lost its translational invariance, then fine. If not, there comes a stage when instabilities due to gravitational forces cause more matter to gather in some places than in others. If there were no special states up to this point you might indeed have a wave function with many macroscopically different branches, in each of which the condensation could occur in different places. We know from current experience[7] firstly that there is such condensation and secondly that each branch does not feel the gravitational field of the wave function of other branches (if any exist). Since condensation is known to occur on a scale as small as thousands of kilometers this gives us wave functions bunched into regions no bigger than this. This is already enough localization for our purposes. Imagine a variant of a Schrödinger cat experiment in which a rocket does or does not take off from planet Earth, depending on whether or not there is a nuclear decay in a certain radioactive sample in Houston. Or perhaps the exact moment of explosion of a supernova is triggered by such an event. In any case, in a world of distinguishable particles, coherent recombination of these macroscopically different wave functions would not be possible. Therefore meeting a future boundary condition in which wave functions were again gathered into well defined planetesimals would require special states, and entail the whole story presented in Chapter 6. From this argument one could still have many branches, just no new ones since the initial branching. I can offer no specific objection to this last suggestion, but it would be a particularly inelegant way to run the universe.

There are other, more microscopic processes where localization occurs. Consider the creation of ^{12}C. This process is critical to the world as we know it.[8] If one assumes that ^{12}C is not part of the primordial creation process, then, as will now be explained, it necessarily forms fairly late, when there are mass inhomogeneities in the form of stars. The raw material for ^{12}C is ^{4}He, which does not need stars to form. However, the next step only proceeds under conditions of high density. Production of ^{12}C requires the formation (by a collision of two ^{4}He particles) of the unstable nucleus ^{8}Be, which must in turn be struck by another ^{4}He within the 3×10^{-16} s the ^{8}Be takes to come apart. This can only take place in the dense interior of a star.

Now consider the life and death of ^{12}C nuclei in our universe. Through nothing but the forces of gravity, they must have wave functions no larger than the innermost core of stars at their birth. (I expect one could do

[7] ... perhaps knowing *why*, perhaps not.
[8] Production of ^{12}C is a crucial step—a bottleneck—in the formation of heavier elements and in the establishing of the carbon cycle.

better, but for purposes of principle this is good enough.) In a time-symmetric cosmology those wave functions must coherently find their way back to these relatively small regions at the appropriate late epoch.

My guess, however, is that the localization sets in much closer to the ends. Presumably, nucleation of hadrons from the quark sea, since it is mediated by spatially local forces, will be sensitive to slight matter inhomogeneities leading to far greater localization than would follow from the discussion above. Whether there are structures in the range of distances from 10^{-33} cm to 10^{-15} cm, I would not venture to guess. As indicated, for present purposes, I believe much grosser arguments are sufficient. However, if one really did want to argue on grounds other than elegance against early, limited Everett branching, this range should be understood.

As indicated toward the end of the previous section, if the arguments we have given for localization could be made more precise, the nature and extent of this localization could be identified. That would be how we could tell whether it was square-spread or some other measure of localization that should be used and would fix the latitude available (the 'ϵ' of Eq. (5.3.12)) as well.

Finally, one should note that the localization limitations presented in this section may be quite different for fermions and for bosons, for massive and massless particles. In particular there may be little or no localization constraints for massless particles.

Aside: If experiment forced me to reject the quantum measurement theory presented in this book, I would find the many worlds interpretation the least unacceptable of the remaining options. For this reason it is worth mentioning why in that interpretation the moon in one 'world' does not influence the tides on another. First consider electromagnetism. Taking for simplicity distinguishable particles, one can ask whether particle number 15 feels the Coulomb force of particle number 17. Were you to look only at $\psi^*(\mathbf{r}_{15}, \mathbf{r}_{17})|\mathbf{r}_{15} - \mathbf{r}_{17}|^{-1}\psi(\mathbf{r}_{15}, \mathbf{r}_{17})$, it might appear to be irrelevant whether pieces of the wave function were on one or another branch. However, this would be missing the main point of the many world, measurement-induced branching process. After a measurement there is entanglement with the many degrees of freedom of the apparatus.[9] Thus the wave function used in the matrix element of the Coulomb potential should not be $\psi(\mathbf{r}_{15}, \mathbf{r}_{17})$, but $\sum_\alpha \psi_\alpha(\mathbf{X}, \mathbf{r}_{15}, \mathbf{r}_{17})$, with α running over the many world branches and $\mathbf{X}$ the apparatus coordinates. For this wave function, matrix elements with $\alpha \neq \alpha'$ will be substantially orthogonal and contribute nothing to the energy. The same arguments go through for gravity.

---

[9] As usual, the word 'measurement' implies only that there is significant macroscopic involvement (as discussed elsewhere), not that there is a human agent.

8.3 Further considerations

Although the argument in the previous section for particle wave function localization may be qualitatively correct, a quantitative sharpening would be useful—not only to add conviction, but to pin down the precise nature of the localization condition. This in turn should help in dealing with the question of how the identity of particles influences localization constraints.

This is also an opportunity to provide an answer to a question that may have occurred to some readers. We have phrased our justification for the selection of special states in terms of a two-time localization condition, a particular kind of two-time boundary condition. But, if quantum evolution is fully deterministic ($\psi \to \exp(-iHt/\hbar)\psi$, as usual), what difference does it make what sort of conditions you give on the wave function—a full specification of the wave function at any one instant determines it for all time.

The answer is Occam's razor. The physics is what it is. It's our estimate of plausibility that is influenced by the choice of wave function specification. Our theory consists not in modifying the laws of physics but in selecting the states on which those laws act. As such, I try to make that selection as plausible as possible. It is in the evaluation of that plausibility that each of us implicitly employs a principle of simplicity or reasonableness, in effect, Occam's razor. When our state selection is phrased in terms of initial conditions it may appear contrived or conspiratorial. As a two-time boundary value problem the state specification is no big deal. What then needs to be done is to justify two-time boundary value problems, but that was the subject of the first half of this book.

8.4 Notes and sources

The first form of the material in this chapter appeared in [SCHULMAN 88, 89]. Further developments are in [SCHULMAN 91]. Special states in the spin boson model are found in [SCHULMAN 94] and are developed in Section 7.1 above.

Pointer states are discussed in [ZUREK 82, 83]. An article on the special role of space is [SQUIRES].

Position operators for quantum mechanics can be problematic when special relativity is taken into account. Wigner was the pioneer in exploring this problem and an article by [WIGHTMAN], in an issue of *Reviews of Modern Physics* in honor of Wigner, provides an exposition. Even for massive particles things are far from obvious, but it is for massless particles that one is simply unable to provide the intuitive position notions that we bring from our non-relativistic experience.

The critical role of ^{12}C has been recognized for a long time. A nuclear physics book that discusses the process is [WONG]. One can also consult the rich and venerable text, [LEIGHTON].

I recently used an entanglement argument similar to that at the end of Section 8.2 to show that in the two-slit interference experiment one can destroy interference with *no* transverse momentum transfer. This should surprise no one (although it turns out there was controversy in the pages of *Nature* on this point). However, I did find an unexpected requirement for what it takes to *prevent* entanglement in situations where one is confident it should not occur. When a particle bounces elastically off a wall, we typically use the particle wave function with reflected momentum and ignore wall degrees of freedom. The justification is the macroscopic nature of the wall; but I found that there is also a requirement that the wall's position uncertainty not be too *small*. This has nothing to do with the quantum measurement theory of this book, and I refer the reader to [SCHULMAN 96] for details. A further consequence of this result is a quantitative lower bound (except with the use of special states) on decoherence in any such scattering.

References

LEIGHTON 59   R. B. Leighton (1959) *Principles of Modern Physics*, McGraw-Hill, New York.

SCHULMAN 88   L. S. Schulman (1988) Two Time Localization, *Ann. Phys.* **183**, 320.

SCHULMAN 89   L. S. Schulman (1989) Remote Two-Time Boundary Conditions and Special States in Quantum Mechanics, *Found. Phys. Lett.* **2**, 515.

SCHULMAN 91   L. S. Schulman (1991) Definite Quantum Measurements, *Ann. Phys.* **212**, 315.

SCHULMAN 94   L. S. Schulman (1994) Special States in the Spin-Boson Model, *J. Stat. Phys.* **77**, 931.

SCHULMAN 96   L. S. Schulman (1996) Destruction of interference by entanglement, *Phys. Lett. A* **211**, 75.

SQUIRES 90   E. J. Squires (1990) Why Is Position Special?, *Found. Phys. Lett.* **3**, 87.

WIGHTMAN 62   A. S. Wightman (1962) On the Localizability of Quantum Mechanical Systems, *Rev. Mod. Phys.* **34**, 845.

WONG 90   S. S. M. Wong (1990) *Introductory Nuclear Physics*, Prentice Hall, Englewood Cliffs.

ZUREK 82   W. H. Zurek (1982) Environment-induced superselection rules, *Phys. Rev. D* **26**, 1862.

ZUREK 83   W. H. Zurek (1983) Pointer Basis, and Inhibition of Quantum Tunneling by Environment-Induced Superselection, in *Proc. International Symp. Found. Quantum Mech.*, Tokyo.

Appendix 8.1.A: Wave packet
scattering from a half parabola

A wave packet comes from the left and scatters off the potential $V(x) = \frac{1}{2}m\omega^2\theta(x)x^2$. To justify the statements made in the main text we need to show that the packet bounces off this soft wall, substantially retaining its (reflected) shape. Also we need to calculate the above-barrier scattering from the discontinuity in V''.

The stationary (scattering) states of the Hamiltonian are (for $x < 0$)

$$u_k(x) = e^{ikx} + e^{-ikx+i\phi(k)} \tag{8.1.A.1}$$

The function $\phi(k)$ characterizes the scattering and will be given below. The general time-dependent solution is

$$\psi(x,t) = \int dk \left[e^{ikx} + e^{-ikx+i\phi(k)} \right] \exp(-ik^2t/2m)\widetilde{\psi}(k) \tag{8.1.A.2}$$

The packet to be used for the purposes of the main text is basically a coherent state with a large central momentum, and little spread about that momentum. Under these conditions we can use stationary phase arguments. Let the central momentum be $\bar{k}$, assumed positive. Thus $\widetilde{\psi}(k)$ is sharply peaked[10] at $\bar{k}$. A stationary phase argument applied to Eq. (8.1.A.2) shows that for large negative t, $\psi(x,t)$ is maximum at

$$\bar{x} = \hbar\bar{k}t/m , \qquad t \to -\infty \tag{8.1.A.3}$$

For large positive times it is the other term in the sum (in Eq. (8.1.A.2)) that contributes and the condition Eq. (8.1.A.3) is replaced by

$$\bar{x} = -\hbar\bar{k}t/m + \frac{\partial\phi(\bar{k})}{\partial k} , \qquad t \to \infty \tag{8.1.A.4}$$

So far this development has applied to bouncing a wave packet off any sort of wall. The harmonic nature of our wall is expressed through its particular phase shift $\phi(k)$. A straightforward calculation yields (for $V(x) = \frac{1}{2}m\omega^2\theta(x)x^2$)

$$\tan\left(\frac{\phi(k)}{2}\right) = \sqrt{\frac{2}{\nu+1/2}} \frac{\Gamma(\nu/2+1)}{\Gamma(\nu/2+1/2)} \tan\left(\frac{\pi\nu}{2}\right) \tag{8.1.A.5}$$

where

$$\nu = \frac{\hbar^2k^2/2m}{\hbar\omega} - \frac{1}{2} \tag{8.1.A.6}$$

For large ν, Stirling's formula applied to Eq. (8.1.A.5) yields

$$\tan\left(\frac{\phi(k)}{2}\right) \approx \sqrt{\frac{\nu+1}{\nu+1/2}} \tan\left(\frac{\pi\nu}{2}\right) \approx \left(1+\frac{1}{4\nu}\right)\tan\left(\frac{\pi\nu}{2}\right) \tag{8.1.A.7}$$

What we need for our calculation is $\partial\phi/\partial k$ whose approximate value follows from Eq. (8.1.A.7) and is given by

$$\frac{\partial\phi(k)}{\partial k} = \frac{\pi}{\omega}\frac{\hbar k}{m} , \tag{8.1.A.8}$$

plus $O(1/\nu)$ corrections (also derivable from Eq. (8.1.A.7)). To see that this is precisely the form needed to show harmonic behavior, go back to Eq. (8.1.A.4). Since $\partial\phi/\partial k$ is proportional to k it looks exactly like a shift in t, namely a delay by π/ω.

---

[10] With no loss of generality, the phase of ψ is also assumed to be stationary at $\bar{k}$.

Deviations from the perfect delay/scattering scenario arise from two sources: 1) wave packet distortion due to the $O(1/\nu)$ corrections to Eq. (8.1.A.8); and 2) above barrier scattering, here manifested through contributions to the integral beyond the stationary phase approximation.

As indicated, from Eq. (8.1.A.7) it follows that anharmonic contributions to $\partial\phi/\partial k$ will be of order $1/\nu$. Recall that the boundary value problem of the main text gave a time T for which the particle was supposed to get back to the origin as unspread as possible. Because most of the time is spent in the harmonic walls, there is only the leftover time for wave packet spreading. This leftover time is $\Delta T \equiv T - n\pi/\omega$, where n is the number of scatterings, the integer part of $T\omega/\pi$. Thus, the larger T and the larger L (half the distance between the soft walls) the larger ν, the less important this correction. Furthermore, since this is only a distortion of the wave, instead of the maximally slow spreading of the Gaussian it will spread a bit faster during ΔT.

Of more concern are the pieces of the wave packet broken off on each entry to and exit from the potential. Since these are separated from the main packet their contribution to W will be of order L^2 times their amplitude squared. We now estimate that amplitude squared. Non-stationary-phase contributions can be studied by going to the (almost[11]) exact form for large positive t

$$\psi(x,t) = \int dk \exp[-ikx + i\phi(k) - ik^2t/2m]\widetilde{\psi}(k) \qquad (8.1.A.9)$$

We have already performed a stationary phase approximation on this by looking at the region in k where $\widetilde{\psi}(k)$ is largest. Now we turn this around and distort the k contour to find a possibly complex k where $\partial\phi/\partial k$ vanishes. This will be the next largest source of potential trouble.[12] The vanishing of $\partial\phi/\partial k$ can be found using the same approximations (Eq. (8.1.A.7)) that we used above, so long as $|\nu|$ is large and $\arg\nu < \pi$. One finds that ν is given by

$$\mathrm{Im}\,\nu \sim \mathrm{const} + \frac{2}{\pi}\log\mathrm{Re}\,\nu\,, \qquad \mathrm{Re}\,\nu \sim \mathrm{integer} + \frac{1}{2} \qquad (8.1.A.10)$$

This gives values of ϕ that have a large imaginary component, which means that the amplitude for this contribution is small. In particular, at this complex k,

$$|e^{i\phi}| \sim \frac{1}{8\pi}\frac{1}{|\nu|^2} \qquad (8.1.A.11)$$

With $n \approx \omega T/\pi$ scatterings, the amount of norm squared removed from the main wave packet due to this contribution is

$$n\left[\frac{1}{8\pi}\frac{1}{|\nu|^2}\right]^2 \sim \frac{T}{k^8} \qquad (8.1.A.12)$$

Now k is essentially $2nL/\Delta T$, and ΔT is at most π/ω. Therefore k is at least of order TL. The contribution to W from this kind of scattering is thus $L^2T/(TL)^8 \sim L^{-6}T^{-7}$. This contribution is small for large T and L, even ignoring the fact that $\widetilde{\psi}(k)$ is small at the complex k value used in this calculation.

---

[11] The incoming wave packet has been dropped, a good, but not perfect, approximation.
[12] In principle one should balance this against $|\widetilde{\psi}(k)|$, but enough is enough.

9

Abundance of special states

Pure quantum evolution is deterministic, $\psi \to \exp(-iHt/\hbar)\psi$, but as for classical mechanics probability enters because a given macroscopic initial condition contains microscopic states that lead to different outcomes; the relative probability of those outcomes equals the relative abundance of the microscopic states for each outcome. This is the postulated basis for the recovery of the usual quantum probabilities, as discussed in Chapter 6. In this chapter we take up the question of whether the allowable microstates (the 'special' states) do indeed come with the correct abundance. To recap: 'special' states are microstates not leading to superpositions of macroscopically different states ('grotesque' states). For a given experiment and for each macroscopically distinct outcome of that experiment these states form a subspace. We wish to show that the dimension of that subspace is the relative probability of that outcome.

This is an ambitious goal, especially considering the effort needed to establish that there are *any* special states—the subject of Chapter 7. As remarked there, the special states exhibited are likely to be only a small and atypical fraction of all special states in the physical apparatus being modeled (e.g., the cloud chamber). In one example (the decay model) there is a remarkable matching of dimension and conventional probability, but I would not make too much of that.[1] What is especially challenging about the present task is that we seek a *universal* distribution. What I mean is that measurements can be performed with all sorts of apparatus, in all sorts of environments. In every one of these the 'number' of special states must reflect relatively simple properties of the system being measured. The Copenhagen interpretation says, look at the wave

---

[1] Such a matching must occur when the special states substantially exhaust the Hilbert space, as they do there. However, I do not expect such exhaustion to be general.

function of the system being measured. If there are two outcomes, with amplitudes α and β, then the respective probabilities are $|\alpha|^2$ and $|\beta|^2$. Simple. For us, you must count special states for each way of measuring this, and for every possible apparatus the count of states must come out the same, reflecting those amplitudes, α and β.

It is precisely this daunting generality that motivates the approach in the present chapter. We back off from attempting to describe or even identify all the microscopic degrees of freedom of the apparatus and environment. We will assume though that there is *something* out there that can kick, push and shove the degrees of freedom of the particular quantum system on which the measurement is to be performed. Rather than propose a model of the 'kicker,' we examine what general demands need be placed on the kicks so as to reproduce standard probabilities. It will turn out that there really is an answer to this question. Moreover, finding this answer is a non-trivial test of the theory. If Nature had used $|\psi|$ or perhaps $|\psi|^4$ for probability, there would be no answer. Except for having the wrong direction in time, you might say that our theory *predicts* the use of $|\psi|^2$ for probability. The kick distribution function that we find, the Cauchy distribution (details below), has a certain modern flavor to it, in the sense that things that might once have been considered mathematical playthings show up in Nature. An example of this is the occurrence of fractals, and indeed there is a relation between certain fractal sets and the Cauchy distribution.[2]

The derivation presented below is not a complete story. From the clean way things come together I feel that there is an essential rightness in it; however, there is an important assumption whose physical justification I do not fully understand. It is an assumption about the class of kicks involved in specializing and before getting down to cases we will articulate this assumption.

Allowable kicks

By a 'kick' in a system I mean an unusual time evolution. For the given macrostate most microstates do one thing; the system that we say is 'kicked' does something else. The time evolution is entirely according to the correct microscopic dynamics; it's just that, as usual, the microstate is rare, perhaps 'special' in the sense I have been using that term. With this description, the term 'kick' is not quite right, since it suggests a perturbation. However, from another point of view the terminology is more suitable. Our approach in this book has been to include the environment in the microstate. If, for a change, we focus on the small system being

---

[2] See the Appendix to this section for elaboration.

measured, then one could think of the rare environmental states as perturbations, 'kicks.' This is the perspective of the following pages. We imagine the small system subject to a succession of perturbations whose nature follows from the precise initial condition of the environment.

At the end of the measurement the system will be in one of several possible states, spin-up or spin-down atom, decayed or undecayed nucleus, captured or uncaptured photon. Often there are more than two options: three or more possible spin states, multiple decay channels, different grains of a photographic film that a particle strikes and activates. The collection of possible final states defines a set of microstates of the microscopic system on which the measurement is being performed. Call these states (of the microscopic system being measured) 'outcome' states. Our assumption about kicks is this: *the kicks move the microscopic system within two-dimensional real subspaces defined by basis vectors selected two-at-a-time from vectors proportional to the initial state and to the outcome states.*[3] In other words, for n-state discrimination, the measurement itself defines $n+1$ directions in Hilbert space (1 initial, n final). The kicks relevant for the specializing of that measurement allow pairwise connections of these directions.

I have ideas on why such a restriction is reasonable; however, I find them less compelling than the remarkable success of the (forthcoming) calculation and prefer to let that speak for itself. In principle, the assumption now being made should not be a separate axiom of the theory. For example, if the special state constraint follows from cosmological two-time localization, the kick restrictions should also follow from that condition.

Semiclassical and phase space considerations

There are other arguments that can be marshalled for the correctness of the special state count. In Section 9.4 two of these are presented. One of these is explicitly semiclassical and involves enough detailed calculation to allow a pinning down of its assumptions. The other does not have the latter virtue, but I find it most compelling in semiclassical terms. I don't know whether there is an underlying relation between the arguments of Section 9.4 and those of the rest of Chapter 9; in any case it doesn't hurt to approach the problem from more than one direction.

---

[3] The general implementation of this assumption is embodied in Eq. (9.2.1). See also the Remark at the end of Section 9.1.

9.1 Two-state discrimination

We begin with measurements that distinguish two possibilities. The metaphor will be a measurement of the z-component of spin in a Stern-Gerlach experiment, although the results apply generally. In such an experiment there is a source-preparation region from which we have reason to believe there emerges an atom with total spin 1/2 and with a spin state oriented at a particular angle with respect to the z-axis. Downstream from this region are collimators, magnets and finally detectors in which macroscopic detection occurs. For many repetitions, the spin-down versus spin-up ratio of detections is the tangent squared of the angle. Specifically, suppose we know enough about the preparation to assign (by conventional analysis) the following wave function to the initial atomic spin state

$$u_\theta = e^{i\theta\sigma_x/2} \begin{pmatrix} 1 \\ 0 \end{pmatrix} \qquad (9.1.1)$$

In the usual description, we would say that after passing through the magnets, a relative amount of atomic wave function proportional to $\cos(\theta/2)$ will arrive at the UP detectors and an amplitude proportional to $\sin(\theta/2)$ will arrive at the DOWN detectors. By the usual prescription, the DOWN and UP detectors will fire with probabilities in the ratio

$$\tan^2 \frac{\theta}{2} = \frac{\sin^2 \theta/2}{\cos^2 \theta/2}$$

According to the theory of this book, the entire wave function will evolve to one in which only the UP detector fires or it will evolve to one in which only the DOWN detector fires. In the present discussion we will not try to identify the particular degrees of freedom of the environment that yield this evolution. Nevertheless, there are general statements we can make about where those degrees of freedom are most likely to occur. What we need to invoke is an idea that was implicit in our stochastic two-time boundary value studies in Section 5.2. Suppose you need a rare outcome. Although the ways to accomplish this may be few relative to the universe of unconstrained possibilities, nevertheless among those ways that *are* allowed, some are more likely or more common than others.[4] Thus if we need rare special states in order to get a non-grotesque outcome, we should look for the *least unlikely* among them.

For the Stern-Gerlach experiment, the least unlikely way to get detection at only the UP or only the DOWN detector is to have the angle θ (in

---

[4] With a phase space understanding of likelihood, events that are 'likely' and states that are 'common,' are the same thing.

Eq. (9.1.1)) be modified even before the atom enters the region of the magnets. Such a change in the state affects only a single degree of freedom. If whatever 'special' things need to happen don't happen until the atom is passing the poles of the magnet, then the translational degrees of freedom of the atom would also have to be affected. I'm not saying that in some experiments you don't get this, just that in this one there would be far more microscopic perturbations of the environment that can successfully affect a two-dimensional Hilbert space (the spin variable) than could successfully act on the many dimensions of the particle's translational degrees of freedom.

Another matter of principle can also be mentioned here, although by this point in the book I could probably stop beating this drum. The fact that the 'specializing' takes place before one has reached the experiment proper is perfectly reasonable. If we are meeting a final boundary condition, the influence of that boundary condition extends through all time. For greatest effectiveness the temporal separation between the 'experiment proper' and the 'specializing' will not be too great; but there is no prejudice about the sign of that separation. (For readers who may have skipped parts of the earlier presentation, Chapter 5 should help in understanding our notion of causality.)

We begin then with the wave function u_θ of Eq. (9.1.1). Somewhere in its progress toward the field region it will be changed to either an UP state (u_θ with $\theta = 0$) or a DOWN state ($\theta = \pi$). Physically this can be the result of 'stray' magnetic fields, collisions with other atoms, collisions with passing photons, perhaps even with neutrinos. For the moment we are not asking. In Chapter 10, when we discuss experimental tests, this question becomes important, but for now our approach is phenomenological.

As discussed in the introductory section of this chapter, we will assume that whatever kicks are administered are such as to move the system within the real subspace defined by the final (outcome) states and by the initial state. For the state in Eq. (9.1.1) the wave function is not real, but by taking 'DOWN' to be $\begin{pmatrix} 0 \\ i \end{pmatrix}$ (i.e., a particular choice of phase) we can bring the parametrization to explicitly show only a single real dimension. Similarly for other initial states.[5] The 'kicks' for our situation are thus rotations of the angle θ. To analyze the perturbing rotations, we define a function $f(\psi)$ to be the number of ways the environment can change the angle θ of u_θ by an angle ψ. As usual, 'number of ways' means number of Hilbert space dimensions, although if these kicks can be described semiclassically, phase space volume might be a simpler description. In either case, 'number of ways' translates into probability.

---

[5] The general prescription is given at the end of this section.

At this stage there is a fundamental point to be made about the function $f(\psi)$. Presumably it is symmetric about $\psi = 0$. Similarly it should be largest for ψ near zero and smaller for larger values of $|\psi|$. The latter condition arises from the expectation that systems mostly stay where you put them. A natural first guess about the kicks to the system would be that they are small on the scale of the angle θ. However, we need θ to go all the way to 0 or π. Therefore there would need to be a lot of small kicks. By the central limit theorem, the distribution function associated with N kicks is $\sim \exp(-\psi_{\text{total}}^2/2N\sigma_f)$. If you try to use this functional form to reproduce the ratio of probabilities, $\tan^2(\theta/2)$, you will find there is no value of σ_f, the standard deviation for f, that does the job. A way to see the problem for small σ_f is to imagine, say, that $\theta = 30°$. Then getting ψ_{total} to achieve that value would demand an extremely unlikely Brownian walk. But the likelihood of the yet longer 60° walk would be suppressed by exponentially small factors.

With a traditional physics background in probability, the foregoing dilemma may seem insuperable. The key to its solution is to invoke a probability distribution that has appeared in the physics literature, but which until recently was overshadowed by the far more common Gaussian. The unjustified assumption in the foregoing paragraph was the existence of the quantity σ_f. Not all distribution functions have finite moments. Without second moments you can't get the central limit theorem and the sum of a large number of random variables does *not* tend to a Gaussian. We will see this in detail below. In Appendix A to this section we will explore the mathematics and a give a few examples.

We next take into account the periodic nature of the angle. As indicated, a kick by θ will bring the atom to a final UP state. A kick by $\pi - \theta$ will bring the atom to a final DOWN state. But adding 2π to either of these also accomplishes the respective result. This makes it useful to define a function

$$F(\theta) = \sum_{k=-\infty}^{\infty} f(\theta + 2k\pi) \tag{9.1.2}$$

It is $F(\theta)$ that gives all possible ways of getting u_θ to become UP, and it is $F(\pi - \theta)$ that gives us the ways of getting to DOWN. Our requirement on f (or F) is therefore[6]

$$\tan^2 \frac{\theta}{2} = \frac{F(\theta + \pi)}{F(\theta)} \tag{9.1.3}$$

---

[6] For cosmetic reasons Eq. (9.1.3) is written in terms of $F(\pi + \theta)$. This is justified by the symmetry of f about 0 and by the 2π translational invariance.

This functional equation (for f) has a solution. Even without knowing the exact solution one can make an intelligent guess from the physical requirements. Consider θ close to zero. Presumably the likelihood of finding UP should vastly dominate and little 'kicking' should be required. This implies that $F(\theta)$ would be dominated by the $k = 0$ term in Eq. (9.1.2), i.e., simply $f(\theta)$. Nothing interesting should be happening for $F(\theta + \pi)$, so that we find that $f(\theta) \sim 1/\theta^2$ for small θ. It turns out that this is an exact solution! That is, the following is an identity:

$$\tan^2 \frac{\theta}{2} = \frac{F(\theta + \pi)}{F(\theta)} \qquad \text{with} \qquad F(\theta) = \sum_{n=-\infty}^{\infty} \frac{1}{(\theta + 2n\pi)^2} \qquad (9.1.4)$$

From our physical requirements, this fixes the form of f to be $1/\theta^2$, down to θ as small as has been experimentally checked. We have thus both accomplished the goal of finding a distribution function to give—universally—the physical probabilities, and gone as far as possible, based on the physics, in characterizing that distribution.

One expects, however, that at some microscopic scale, as $\theta \to 0$, the function $f(\theta)$ ceases to diverge. Although we have no physical preference, it is convenient to take as a finite distribution function a well known, and in a sense canonical, mathematical distribution, the Cauchy distribution. Let me explain why I call it 'canonical.' (Appendix A has more information.) It plays a role, for functions with asymptotic $1/\theta^2$ behavior, analogous to that played by the Gaussian among functions for which the central limit theorem holds (having finite second moments). The general Cauchy distribution is parametrized by two quantities, and we use the notation

$$C_{a,x_0}(x) = \frac{a/\pi}{(x - x_0)^2 + a^2} \qquad (9.1.5)$$

What is special about the Cauchy distribution[7] is that it is *stable*. Roughly, 'stable' means that when you add two random variables with the same kind of distribution you get another one of the same kind, with its parameters related to those of the summands. Let me be more precise. The possible 'kinds' of stable distribution are characterized by a parameter, usually called α, with $0 < \alpha \leq 2$. For each α, the associated distributions will have additional parameters. For example, our Cauchy distribution has parameters a and x_0, or the normal distribution has parameters σ (standard deviation) and x_0 (average). The normal distribution is stable with $\alpha = 2$, and has the property that the sum of two normally distributed random variables is a normally distributed random variable

---

[7] ... among those with inverse square asymptotics.

whose standard deviation is the sum of that of its summands. For the Cauchy distribution it is the parameter 'a' that adds. What makes the stable distributions important is that they serve as a kind of basin of attraction for sums. That is, it is common to be interested in a random variable that is a sum of many smaller ones. Thus Brownian motion (in an appropriate limit) is the sum of many small, independent steps. Even if the distribution function for the individual steps is not normal, so long as it has a finite standard deviation,[8] the sum *is* normally distributed.[9] That's the central limit theorem. The other stable distributions do not have finite second moments, sometimes not even first moments. If you take a sum of many of these, you do not get Brownian motion. You get something known as a Lévy flight.[10] When summing a large number of these infinite-moment random variables, the random variable to which they tend can be one of the stable distributions. If the asymptotic behavior of the distribution is $1/x^2$, then they tend to the Cauchy distribution. That's why my choice of a Cauchy distribution is analogous to the use of a normal distribution for finite second moment cases.

Since the distributions we study are symmetric about zero, we will consider cases for which $x_0 = 0$, and drop x_0 from the notation. Let's check some of the assertions of the last paragraph for the Cauchy distribution. Suppose X_1 and X_2 are independent and are Cauchy distributed with parameters a_1 and a_2, respectively. Let S be their sum. To get the distribution function for S we should perform a convolution of those for X_1 and X_2. However, as is well known, the way to do this is to Fourier transform, and then the transform of the distribution function for the sum is the product of the transformed functions (sometimes one does a Laplace transform and works with the moment generating function—for present purposes it makes no difference). It is an elementary fact that

$$C_a(x) = \frac{a/\pi}{x^2 + a^2} = \frac{1}{2\pi} \int_{-\infty}^{\infty} dk \exp(-a|k| + ikx) \qquad (9.1.6)$$

Since $\exp(-a_1|k|) \cdot \exp(-a_2|k|) = \exp(-(a_1 + a_2)|k|)$, it follows (by what I said about Fourier transforms or moment generating functions) that the distribution function for S is *also* Cauchy, with parameter $a_1 + a_2$.

Aside: The family resemblance of the various stable distributions is easiest to see in terms of their Fourier transforms. The general form is $\exp(-a|k|^\alpha)$. Cauchy thus corresponds to $\alpha = 1$ and the normal distribution to $\alpha = 2$. For $\alpha < 2$, the dropoff in the distribution function is $1/x^{\alpha+1}$ for large x. Exactly at $\alpha = 1$ there are logarithmic corrections possible

---

[8] ... plus some other technical requirements.

[9] The sum is the total distance covered by the random walk.

[10] See Fig. 9.1A.1, in the Appendix.

(still leaving a stable distribution) and no shortage of exotica at all values of α. Only for $\alpha = 1$ and 2 is the distribution function itself simple.

For our purposes, the most important single feature of the Cauchy distribution is the way in which large fluctuations occur. Let S_N be a sum of N random variables, all Cauchy distributed with the same parameter, a. Suppose it is known that S_N attained a value A, where $A \gg Na$. In other words, there was a large fluctuation. That is an unlikely event, but how did it happen? What was the least unlikely way for this to have occurred? For the Cauchy distribution and for the normal distribution there are two entirely different answers. For the Cauchy distribution, the least unlikely way is for a single one of the summands to be large, essentially equal to A, and all the others small, near a. For the normal distribution the least unlikely way is for the deviation to be shared among all, approximately equally. There are easy estimates to see this. To calculate the probability that a particular variable, say the first summand in S_N, is nearly equal to A and all the others near zero, we note that

$$C_a(x) \sim \begin{cases} \dfrac{1}{a\pi} & \text{for } x \sim 0 \\[2mm] \dfrac{a}{\pi x^2} & \text{for } |x| \sim \infty \end{cases} \qquad (9.1.7)$$

Call P_{one} the probability that only number 1 is big, the others small. Call P_{all} the probability that all the random variables share the deviation, in particular each takes the large value A/N. These are given by

$$P_{\text{one}} = \frac{a}{\pi A^2} \left(\frac{1}{a\pi} \right)^{N-1}, \qquad P_{\text{all}} = \left(\frac{a}{\pi (A/N)^2} \right)^N \qquad (9.1.8)$$

The ratio is

$$\frac{P_{\text{one}}}{P_{\text{all}}} = \left(\frac{a}{A} \right)^2 \left(\frac{A}{Na} \right)^{2N} \qquad (9.1.9)$$

Thus as A grows P_{one} always wins. By contrast for the normal distribution (with standard deviation a^2) the corresponding quantities are (essentially)

$$P_{\text{one}} = \exp\left(-A^2/2a^2 \right), \qquad P_{\text{all}} = \exp\left[-N(A/N)^2/2a^2 \right]$$

$$\frac{P_{\text{one}}}{P_{\text{all}}} = \exp\left(-\frac{A^2}{2a^2} \left(1 - \frac{1}{N} \right) \right) \qquad (9.1.10)$$

In this case, large A makes P_{all} the winner. But it's a small victory, as the normal distribution is extremely intolerant of large deviations.

The exact nature of the neighborhood-of-zero cutoff for the Cauchy distribution did not play a role in our argument, and all that matters is the slow dropoff with large deviations. There are other interesting and (to the uneducated intuition) peculiar features of the long-tailed stable distributions. For example, you generally do *not* improve your estimate of the average of the distribution by repeating a measurement. In addition to the Appendix, I give references below.

Returning to our physical situation, we find that the specialization of the microscopic state is most likely to occur by a single, resounding whack. Our model of the source of the kicking is not sufficiently detailed to say whether this is a large energy fluctuation, a large field, or what. My guess would be that by the time our statistical arguments made sense, the perturbing forces could be treated semiclassically. This is another reason why, although I was happy to be able to exhibit special states in Chapter 7, I don't think they're the actual special states that occur in Nature.

Deviations from standard probabilities

Before going on to the problem of multi-state discrimination, I wish to take up the extent to which deviations of $f(\theta)$ from $1/\theta^2$ behavior (as $\theta \to 0$) could lead to experimentally distinguishable results, that is, DOWN-to-UP ratios different from the $\tan^2(\theta/2)$ of Eq. (9.1.3).

We suppose that f has the form of the Cauchy distribution with a parameter γ. The small θ dependence implied by this supposition has not been forced upon us by our physical demands; nevertheless, it is a convenient function and for the kind of estimate we seek the exact form should not much matter. Thus we take

$$f(\theta) = \frac{\gamma/\pi}{\theta^2 + \gamma^2} \tag{9.1.11}$$

To see the experimental consequences perform the sum

$$F(\theta) = \frac{\gamma}{\pi} \sum_{n=-\infty}^{\infty} \frac{1}{(\theta + 2n\pi)^2 + \gamma^2} \tag{9.1.12}$$

$$= \frac{1}{2\pi} \operatorname{Im}\left(\frac{1}{\tan\frac{1}{2}(\theta - i\gamma)}\right) = \frac{(\frac{1}{2\pi}) \tanh\frac{1}{2}\gamma}{\sin^2\frac{1}{2}\theta + \cos^2\frac{1}{2}\theta \tanh^2\frac{1}{2}\gamma}$$

For an initial state u_θ, the probability of observing UP is

$$\Pr(\text{UP}) = \frac{F(\theta)}{F(\theta) + F(\theta + \pi)} = \frac{\cos^2\frac{\theta}{2} + \sin^2\frac{\theta}{2} \tanh^2\frac{\gamma}{2}}{1 + \tanh^2\frac{\gamma}{2}} \tag{9.1.13}$$

For small γ this gives

$$\Pr(\text{UP}) \cong \cos^2\frac{\theta}{2} - \frac{\gamma^2}{4} \cos\theta \tag{9.1.14}$$

This is *not* the standard result. An atom going in with spin up ($\theta = 0$) has probability $\gamma^2/4$ of being misread as down. In Chapter 10 we will discuss whether this effect could be measurable.

Remark: As indicated earlier, each initial state and pair of outcome states define a two-dimensional real Hilbert space. We now give a general prescription for defining that space: let the initial state be ψ and the outcome states be ϕ_1 and ϕ_2. These are related by $\psi = \alpha\phi_1 + \beta\phi_2$ for complex numbers α and β whose norms squared add to unity. Now ψ can be redefined (by multiplying by a phase factor) so that α is real and equal to $\cos\theta$ for some θ. This will in general change the phase of β as well. Next, ϕ_2 is redefined so that (the new) β is real, in which case it will necessarily equal $\sin\theta$. (The sign ambiguity is irrelevant.)

> *Appendix for this section (at the end of the chapter):*
> 9.1.A Cauchy and other long-tailed distributions

9.2 Multi-state discrimination

For many experiments there are more than two possible outcomes. A Stern-Gerlach experiment for an atom with spin one will show three bands on the photographic film, three possible values for the z-component of spin. In some experiments, for example measurements of position, there appears to be continuum of possible outcomes, although presumably a finite granularity in the detectors imposes a finite limit for the dimension. In this section we extend the results of Section 9.1 to situations where any finite number of possible 'outcome' states is possible.

As for the two-state result, we impose conditions on the class of allowable 'kicks.' The only states defined by the experiment are the outcome states, which we assume to be orthogonal to one another, and the initial state. As before, we will not consider kicks outside the two-dimensional subspaces defined by pairwise combinations of outcomes and by the initial state (or by the state at any stage of the kicking process). By adjusting phases, the subspaces within which kicks are allowed can be taken to be real two-dimensional Hilbert spaces. The precise implementation of this condition will be evident in the forthcoming calculations.

As indicated in the introduction to this chapter, I am not sure why this works so well. In a sense it is the most frugal possible foray into Hilbert space. The physics of the experiment defines the outcome states and the initial state and we are only going to move along the edges of the polygon defined by this. In any case, we shall see that as for the case of two-state discrimination the mathematics is remarkably cooperative.

For the n-state discrimination problem, the initial wave function is an n-component vector, $\psi = (\psi_1, \ldots, \psi_n)$, expressed in the basis defined by the outcome states. We wish to show that the probability of observing the system to be in the state k ($1 \leq k \leq n$) is $|\psi_k|^2$.

In the previous section we showed how to fix the phases so that all coefficients in ψ are real; however, it is just as easy to proceed with the calculation without doing this. Although we will only use a single real parameter for each pairwise kick, the formal rotation calculations are more convenient if complex numbers are not banished. Let the basis vectors for ψ's n-dimensional space be denoted $\hat{e}_k$, with $(\hat{e}_k)_\ell = \delta_{k\ell}$. Introduce the notation $\psi_j \equiv \sqrt{r_j} \exp(i\phi_j)$. The kicks to be used act in the two-planes defined by the pairs $\psi_k \hat{e}_k$ and $\psi_j \hat{e}_j$. The kick that rotates $\hat{e}_k \psi_k + \hat{e}_\ell \psi_\ell$ into $\tilde{\psi}_\ell \hat{e}_\ell$ uses the generator

$$\sigma_{k\ell} = \omega \hat{e}_k \hat{e}_\ell^\dagger + \omega^* \hat{e}_\ell \hat{e}_k^\dagger \qquad (9.2.1)$$

with

$$\omega = i e^{i(\phi_k - \phi_\ell)} \qquad (9.2.2)$$

It is then easy to see that

$$e^{i\theta \sigma_{k\ell}}(\hat{e}_k \psi_k + \hat{e}_\ell \psi_\ell) = \hat{e}_\ell e^{i\phi_\ell} \sqrt{r_k + r_\ell} \qquad (9.2.3)$$

provided θ satisfies

$$\tan \theta = \sqrt{\frac{r_k}{r_\ell}}, \qquad 0 \leq \theta \leq \frac{\pi}{2} \qquad (9.2.4)$$

There are $n(n-1)/2$ allowable kick directions, as defined by Eq. (9.2.1). As the reader can imagine, I tried the other possibilities, and in fact will show below that a single kick cutting across the n complex dimensions would not work.[11] This raises a consistency question for our 'allowable-kick' ansatz. As for ordinary rotations, the Hilbert space rotations generated by the $\sigma_{k\ell}$ do not commute with one another, and if applied indiscriminately would take the wave function out of the relatively small space to which our ansatz confines us.

Once again, the properties of the Cauchy distribution solve the problem. As follows from the previous section, the evidence is that the cutoff parameter, 'γ,' of Eq. (9.1.11) is extremely small. From the asymptotic form of the Cauchy distribution, Eq. (9.1.7), this implies that the important kicks are rare. As further shown in the previous section, when you need a big deviation from the long-tailed Cauchy distribution, the *least unlikely* way is for it to be done by a single large deviation, rather than a sum of smaller ones. This means that overwhelmingly the rotation in Eq. (9.2.3) occurs in one shot. You don't go halfway in one k-ℓ plane, then rotate partially in another plane and rotate again in the original k-ℓ plane (which would send ψ all over the complex n-sphere). Rather, it

---

[11] ... in the sense of reproducing standard probabilities.

is vastly less unlikely that the rotation of Eq. (9.2.3) occurs in a single
'specializing' event, and then other planes come into play, one by one.
Thus for a wave function ψ with (say) all ψ_k non-zero, the state will
become special by means of $n-1$ big kicks (in the planes permitted by
our ansatz), plus a lot of irrelevant jittering.

We can now calculate the relative probabilities for ψ to go to the various
$\hat{e}_\ell$, $\ell = 1, \ldots, n$, and for definiteness compute the number of ways to reach
$\hat{e}_n$. To avoid too much abstraction, we will first work with $n = 3$ and
then proceed by induction. Our goal then is to kick the state

$$\psi = \begin{pmatrix} \psi_1 \\ \psi_2 \\ \psi_3 \end{pmatrix} \qquad \text{to the state} \qquad \hat{e}_n = \begin{pmatrix} 0 \\ 0 \\ 1 \end{pmatrix} \qquad (9.2.5)$$

with a possible phase difference. Because of the restriction on available
kicks this must be done in two steps. Let us suppose that we first send
$\psi_1\hat{e}_1$ to $\hat{e}_3$ and then afterward send $\psi_2\hat{e}_2$. To calculate the relative number
of ways to take the first step we fall back on our results of the previous
section. Our earlier result $F(\theta) \sim 1/\sin^2(\theta/2)$ becomes

$$G_{1\to3} = \frac{1}{r_1/(r_1 + r_3)} \qquad (9.2.6)$$

This is deduced as follows. The effective rotation angle (for $\psi_1\hat{e}_1 \to \hat{e}_3$)
depends on ψ_1 and ψ_3 alone, since ψ_2 is not involved. Therefore the term
'$r_1 + r_3$' is needed to renormalize the state $(\psi_1, 0, \psi_3)$. The cosine of the
angle of rotation that is now required is thus $|\psi_3|/\sqrt{|\psi_1|^2 + |\psi_3|^2}$. Its
(inverse) sine squared is what appears in Eq. (9.2.6). This formula of
course continues to make use of the distribution of kicks that we found
to be necessary in the previous section.

Next we rotate $\psi_2\hat{e}_2$ into $\hat{e}_3$. The third component of ψ subsequent to
our previous operation was $\sqrt{r_1 + r_3}e^{i\phi_3}$, so that the relative number of
ways to finish the job (i.e., to send $\psi_2\hat{e}_2$ into $\hat{e}_3$) is

$$G_{2\to3} = \frac{1}{r_2}$$

The weight associated with the combined operation is therefore

$$G_{1\to3,2\to3} = G_{1\to3}G_{2\to3} = \frac{r_1 + r_3}{r_1 r_2}$$

Had we operated in the opposite order the corresponding weight would
have been

$$G_{2\to3,1\to3} = \frac{r_2 + r_3}{r_1 r_2}$$

Note, by the way that distinct orders of operation do *not* require us to bring in new $\sigma_{k\ell}$s. This follows (from Eq. (9.2.3)) because the *phase* of the new non-zero state (after each step) is unchanged. As a result, even among the two-plane outcome state-basis rotations only half of them are needed.

This does not exhaust the routes to $\hat{e}_n$. We can first send $\psi_1 \hat{e}_1$ to $\hat{e}_2$ and then the two combined to $\hat{e}_3$. The weight for these steps is

$$G_{1\to 2,2\to 3} = \frac{1}{r_1/(r_1+r_2)} \frac{1}{r_1+r_2} = \frac{1}{r_1}$$

Similarly

$$G_{2\to 1,1\to 3} = \frac{1}{r_2}$$

All four terms combined give

$$G_3(r_1, r_2, r_3) = \frac{2}{r_1 r_2}$$

where the quantity $G_3(r_1, r_2, r_3)$ is defined to be the relative weight for reaching $\hat{e}_3$ from an initial state ψ, with $r_1 + r_2 + r_3 = 1$.

According to the probability postulate of Chapter 6, the general requirement to be satisfied by $G_\ell(r_1, \ldots, r_n)$ (going back to the n-state case) is

$$r_\ell = \frac{G_\ell(r_1, \ldots, r_n)}{\sum_k G_k(r_1, \ldots, r_n)} \tag{9.2.7}$$

Our putative solution G_3 above should thus satisfy

$$r_3 \stackrel{?}{=} \frac{\frac{2}{r_1 r_2}}{\frac{2}{r_1 r_2} + \frac{2}{r_2 r_3} + \frac{2}{r_1 r_3}} \tag{9.2.8}$$

The answer to Eq. (9.2.8)'s question (recalling $\sum r_j = 1$) is an emphatic yes.

It is interesting to consider Eq. (9.2.7) as a functional equation in its own right. If $G_\ell(r_1, \ldots, r_n)$ is to depend only on r_ℓ, then for $n > 2$ the only (measurable) solution is $G_\ell = r_\ell$. The significance of this result is that (for $n > 2$) it precludes the transition $\psi \to \hat{e}_\ell$ with a single kick. This *could* have been achieved by including rotations in planes other than those defined by the outcome state-basis, but even without knowing the propagator for Lévy flights on the complex n-sphere it is clear that that propagator should depend only on the invariant distance, which is a function of r_ℓ only (for $\psi \to \hat{e}_\ell$).

A solution of Eq. (9.2.7) is

$$G_\ell(r_1,\ldots,r_n) = \frac{\text{const}}{\prod_{j\neq\ell} r_j}$$

Clearly our detailed counting for $n = 3$ turned up such a solution.

We will now give an inductive argument to show that Cauchy distributed kicks in the outcome state-basis give the solution

$$G_\ell(r_1,\ldots,r_n) = \frac{(n-1)!}{\prod_{j\neq\ell} r_j} \tag{9.2.9}$$

Suppose Eq. (9.2.9) is true for $n-1$. We wish to send ψ to $\hat{e}_n$, where ψ has n nonzero components. The first step involves a particular component, say the jth ($j \neq n$). There are two choices, either it is sent to n or to some other component, say k.

Case I. $j \to n$. The first step has weight $[r_j/(r_j + r_n)]^{-1}$. For the subsequent operations we are back to the case of $n - 1$ components, but with r_n replaced by $r_j + r_n$. Bringing all other components to $\hat{e}_n$ thus has weight

$$\frac{(n-2)!}{\prod_{k\neq j,n} r_k}$$

to give combined weight

$$\frac{(r_j + r_n)(n-2)!}{\prod_{k\neq n} r_k}$$

Case II. $j \to k$. The first step has weight $[r_j/(r_j + r_k)]^{-1}$. For the next steps there are $n-1$ components and r_k has become $r_k + r_j$. Those steps have weight

$$\frac{(n-2)!}{(\prod_{\ell\neq j,k,n} r_\ell)(r_k + r_j)}$$

The combined weight for Case II is therefore

$$\frac{(n-1)!}{\prod_{\ell\neq k,n} r_\ell}$$

Bringing together Case I and Case II (for each k) and summing over possible initial steps j gives

$$
G_n(r_1, \ldots, r_n) \overset{?}{=} \sum_{j \neq n} (n-2)! \left\{ \frac{r_j + r_n}{\prod_{\ell \neq n} r_\ell} + \sum_{k \neq j, n} \frac{1}{\prod_{\ell \neq k, n} r_\ell} \right\}
$$

$$
= \frac{(n-2)!}{\prod_{\ell \neq n} r_\ell} \sum_{j \neq n} \left\{ r_j + r_n + \sum_{k \neq j, n} r_k \right\}
$$

$$
= \frac{(n-2)!(n-1)}{\prod_{\ell \neq n} r_\ell}
$$

which completes the induction.

Remark: Presumably, for (very) large n the result of this section extends to the interpretation of the position wave function $\psi(x)$, $x \in \mathbb{R}^3$.

Remark: Earlier in this section I indicated that it was only because we had a long-tailed kick distribution that we could consistently apply our 'allowable kick' ansatz (so as to avoid Hilbert space excursions forbidden under the ansatz). Should this 'perils of Pauline' rescue enhance the theory's credibility or have the opposite effect? This is not as frivolous a question as it sounds, and people have given a lot of thought to what it means to 'confirm' a theory. The reader will not be surprised that I find the meshing of apparently disparate components of the demonstration to be a positive indicator, rather than the opposite.

9.3 Probabilities and the wave function norm

Why does the square of the norm of the wave function give probability? That it does, presumably enjoys complete experimental verification, although I do not know if there has been systematic testing of this rule. For example, the small θ case discussed at the end of Section 9.1 would be difficult to check. In the Stern-Gerlach context, one would come up against another sort of deviation that has been theoretically predicted, so if a difference from the usual probabilities were found, the interpretation would be unclear. The deviation to which I refer was pointed out by Wigner, Araki and Yanase, who showed that when additive conservation laws apply to an experiment, perfect application of the probability rule cannot be made. Moreover, the scale of the deviation is governed by the ratio of the microscopic system to the macroscopic apparatus. (See the Appendix to Chapter 10.)

$|\psi|^2$ as the choice of mathematicians et alii

For the mathematical physicist the $|\psi|^2$ rule provides elegant coordination with Hilbert space concepts. The inner product is the fundamental construct, distinguishing Hilbert spaces from less particular objects, so that the most natural course is to associate this structure with the physical quantities described by the theory.

To Born, who received the Nobel Prize for the association of $|\psi|^2$ and probability, these mathematical considerations were not uppermost, and in fact his initial proposal was to use $|\psi|$ rather than $|\psi|^2$. In a note added in proof in his original publication he went over to the square.

The Everett theory is supposedly self-interpreting, and the approach used to obtain probabilities is to consider the many branches of the wave function consequent to the N-fold repetition of a particular experiment, for $N \rightarrow \infty$. One then defines an operator that gives the frequencies in *most* of the branches of the wave function. The proof is based on a theorem of Finkelstein and I give a reference below. I am not satisfied though about the 'self-interpreting' aspect, since to use that theorem one must accept the axiom that measurements can result only in the eigenvalues of Hermitian operators.

$|\psi|^2$ as the choice of the present theory

In Sections 9.1 and 9.2 we found that with an appropriate selection and distribution of kicks, the special states could recover the standard $|\psi|^2$ probabilities. What I now wish to show is that it could not recover any other power. Our theory thus requires the choice that Nature actually makes.

To see this, we go to the two-state discrimination case of Section 9.1 and suppose that a power different from two occurred in Nature. Our development of the relation between kick distribution and probability would go through as before. Suppose that the physical probabilities were given by a power β of the wave function norm. In place of Eq. (9.1.3) the equation to be satisfied by the distribution function would be

$$\left|\tan\frac{\theta}{2}\right|^\beta = \frac{F(\theta+\pi)}{F(\theta)} \quad \text{with} \quad F(\theta) = \sum_{k=-\infty}^{\infty} f(\theta+2k\pi) \quad (9.3.1)$$

For θ near zero, we again expect $F(\theta)$ to be dominated by the $n = 0$ term, implying that $f(\theta) \sim B\theta^{-\beta}$, with B a constant. Now although θ has just been described as 'small,' it is smallness on the macroscopic level that is meant. Even with small macroscopic θs, so long as we have not reached the level of any 'γ'-like cutoff (cf. the parameter in Eq. (9.1.11)), the asymptotic power law behavior holds. (In other words, the 'small' θ regime within which the $n = 0$ term in F dominates is still large, or asymptotic, with respect to the functional form of its distribution function.)

The argument given in Section 9.1 showing that a finite-second-moment distribution could not provide correct probabilities did not depend on the power β being two. Thus if there is any solution to Eq. (9.3.1), for any β, it will be one of the long-tailed distributions. Furthermore, from the perspective of the theory presented here the only allowable distributions should be the stable ones. This is because by modeling the environment in ever more detail, each kick could be microscopically decomposed and thought of as a sum of kicks. (This reflects the close relation of *stable* distributions and *infinitely divisible* distributions.) Thus if θ is large enough so that the microscopic details of the distribution are lost (and any macroscopic θ should suffice) then the relation ($\theta^{-\beta}$) that holds at small-but-macroscopic angle will hold at all angles. Therefore, for any macroscopic angle, we have, not merely similitude, but equality: $f(\theta) = B\theta^{-\beta}$. Our putative identity therefore takes the form

$$\left| \tan \frac{\theta}{2} \right|^{\beta} = \frac{F(\theta + \pi)}{F(\theta)} \qquad \text{with} \qquad F(\theta) = \sum_{n=-\infty}^{\infty} \frac{1}{|\theta + 2n\pi|^{\beta}} \qquad (9.3.2)$$

(B cancels). For $\beta \neq 2$, Eq. (9.3.2) is simply not true. First, for the case $\beta \leq 1$, the sum in Eq. (9.3.2) diverges and in any reasonable cutoff scheme the ratio (in Eq. (9.3.2)) gives 1, not $|\tan \theta/2|^{\beta}$. Next, for $\beta > 1$, we can see that Eq. (9.3.2) is false by expanding near $\theta = 0$. To lowest order, if Eq. (9.3.2) were true we would have

$$\left| \frac{\theta}{2} \right|^{\beta} = \frac{2\sum_{n=0}^{\infty}[(2n+1)\pi]^{-\beta}}{\theta^{-\beta}} \qquad (9.3.3)$$

The θ^{β} cancels. The right-hand side of Eq. (9.3.3) is related to the Riemann zeta function and were Eq. (9.3.3) valid we would have expressed that function as a simple combination of powers. But it is not valid, and equality in Eq. (9.3.3) holds *only* at $\beta = 2$.

We have thus shown that *from our point of view*, that is, when probabilities are obtained by counting special states, no power other than the second (for $|\psi|$) can satisfy the restrictions developed here.

9.4 Abundance at the semiclassical level

In this section we give two arguments with a semiclassical flavor. The approach is different from that of the previous sections. In one argument we carry over what was learned from the harmlessness (with respect to present-time dynamics) of remote future boundary conditions in our cat map studies (Chapter 4). The justification of the 'harmlessness' was this: the precise distribution imposed by the future condition looks random.

And a random distribution is the way to estimate the uniform distribution. Our claim will be that special states reproduce the results of *all* states, and thus reproduce the usual predictions which implicitly include an (ordinarily uninteresting) average over all apparatus states. The second argument uses an explicit form for the propagator of a wave function to provide a special state justification for the association of $|\psi(x)|^2 dx$ with the probability of finding a particle in an interval dx. Because of the assumptions made about the apparatus, this argument is related to the earlier one, but it seems to me to have additional content because at a crucial stage the particular form of the semiclassical propagator plays a role. (This 'particular' form is itself not a theorist's curlicue, but is an essential part of the usual correspondence principle.)

Random sampling of phase space

Thinking semiclassically, the subspaces of special states become volumes in the phase space of the system cum apparatus cum environment. If, as in the linear decay model (Section 7.1), these subspaces substantially exhaust the Hilbert space, then semiclassically they fill the phase space. More typically I expect them to occupy a small portion of the phase space, on the order of the numbers seen in other applications of two-time boundary conditions (on the scale of $\exp(-$Avogadro's number$)$, cf. Appendix A to Section 5.0). Suppose in this latter case that the occupied regions are effectively randomly distributed, much as the points for the future-conditioned cat map were effectively randomly distributed among all possible initial conditions in the two-time evolution shown in Fig. 4.3.2. Then for both these cases the density matrix constructed from special states alone will have the same expectations as the density matrix constructed from the entire Hilbert space or phase space. Thus, if $\Psi = \phi \otimes \omega$ represents the overall wave function, ϕ for the system being measured and ω for apparatus cum environment, then the density matrix would be $\rho = \mathcal{N}^{-1} \sum_\omega \Psi \Psi^\dagger$, with $\mathcal{N}$ taking care of normalization. (There is no sum over ϕ, since that is the same microscopic state in all realizations.) Let A be an observable being measured, for example something built of degrees of freedom of one grain on a photographic plate. Then you answer the question, 'Did the particle land in that grain?' by looking at the value of $\mathrm{Tr}\,\rho A$. If the density matrix for the special-states-only is $\tilde{\rho} = \widetilde{\mathcal{N}}^{-1} \sum_{\tilde{\omega}} \Psi \Psi^\dagger$ (where the sum is over $\{\tilde{\omega}\}$, the collection of special states), then being able to replace ρ by $\tilde{\rho}$ means that all experiments will return the same answer when only special states are used.

This argument is given in more detail in the reference cited in Section 9.5. However, I want to point out (as is done in there too) that the implication is weaker in the quantum context than in the classical (cat map) one. Recall that relaxation is a tenuous concept in quantum mechanics and certain kinds of conditioning can influence a system at times remote from that conditioning. Thus you have no guarantee that the distribution of the special states will be random among all states. Furthermore, the actual conditioning makes reference to the events of this particular measurement—it's with respect to this measurement that they are 'special.' Thus our argument is hardly a proof. However, I expect it to have validity at the semiclassical level. In the discussion below on probability distributions in space we will see how an argument along these lines is actually used. The special state conditioning and the properties of the measurement are in practice decoupled. Thus the fact that the conditioning is related to the requirement of having a definite outcome is uncorrelated with the particular outcome. In this way the objection raised a moment ago is defused.

Probability distributions in space

By making plausible assumptions on the ability of a source to spit out particular kinds of wave packets for a particle, and by assuming the validity of the semiclassical approximation, we can show that the quantity $|\psi(x)|^2 dx$ is the probability of finding a particle in an interval dx.

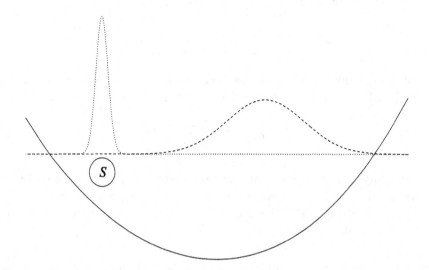

Fig. 9.4.1. A particle is emitted by a source (marked S), into a slowly varying potential (solid line). Its initial wave function (magnitude) is indicated by the dotted line. It has initial momentum to the right and its time-t wave function is shown by the dashed line.

Let there be a source, S, that emits particles in the neighborhood of a point[12] x_0 such that immediately upon emission they are subject to a slowly varying potential $V(x)$. This is shown Fig. 9.4.1. Call the initial wave function $\psi_0(x)$, and its spread $\Delta^2 \equiv \int dx(x - x_0)^2 |\psi_0(x)|^2$. Let the propagator be $G(x, t; y) \equiv \langle x| \exp(-iHt/\hbar)|y\rangle$. Then the time-$t$ wave function is $\psi(x, t) = \int G(x, t; y)\psi_0(y)dy$.

When V varies slowly on the appropriate scale (this scale can be found using standard WKB methods), the exact propagator can be replaced by its semiclassical approximation. For cases in which there is a single classical path (i.e., solution of the classical equations of motion in the given potential $V(x)$) from y at time 0 to x at time t, the semiclassical approximation for $G(x, t; y)$ is given by

$$G_{WKB}(x, t; y) = \sqrt{\frac{i}{2\pi\hbar}\frac{\partial^2 S}{\partial x \partial y}} \exp\left(\frac{iS}{\hbar}\right) \qquad (9.4.1)$$

where S is the classical action of the classical path, namely $\int L dt$ along the path, with L the Lagrangian. S can be considered a function of the points, x and y, and the time,[13] t, that define the boundary value problem; it is in that sense that we have the partial derivative of S with respect to the position variables that appears within the square root in Eq. (9.4.1).

Suppose that Δ is small on the appropriate scale. Then the time-t wave function is given by[14]

$$\psi(x, t) \approx \sqrt{\frac{i}{2\pi\hbar}\frac{\partial^2 S}{\partial x \partial x_0}} \exp\left(\frac{i}{\hbar}S(x, t; x_0)\right) \qquad (9.4.2)$$

The magnitude squared of this wave function near a point x for which Eq. (9.4.2) is a valid approximation is therefore

$$P(x) = \frac{1}{2\pi\hbar}\left|\frac{\partial^2 S}{\partial x \partial x_0}\right| \qquad (9.4.3)$$

According to the Copenhagen interpretation, this is the probability density for finding the particle near x. What we want to show is that this quantity coincides with the special state count for putting all the particle's wave function near x.

---

[12] For convenience this calculation is being made for a one-dimensional coordinate space, but everything goes through for higher dimension.

[13] We assume the Lagrangian to be time-independent.

[14] This formula breaks down far from x_0, as is clear from the normalization requirement. For large $|x - x_0|$ the phase variation in the propagator is great enough to see details below the level of Δ, vitiating the approximation leading to Eq. (9.4.2).

We must therefore investigate whether the available special states for having the entire wave function arrive near x come with the frequency given by Eq. (9.4.3). This requires assumptions on the properties of the source. Let us suppose that the wave function ψ_0 comes with a spread in momentum that is far from minimal, that is, it is much larger than $\hbar/\Delta$. Let us also suppose that for rare internal states, the emitter is nevertheless capable of producing near minimal packets, in particular wave functions that are close to coherent states with relative x and p spreads that minimize wave packet spreading in the potential $V(x)$ (roughly what we got in Appendix A to Section 5.3). To a good approximation such packets follow the classical path associated with their central momentum (and initial central position). These will be the special states of this system. As in earlier parts of this chapter, we do not try to pin down the dynamics within the apparatus that allows it to produce these special states, only assume that it *is* possible, and examine the consequences.

We thus have an emitter that, when it is not doing anything 'special,'[15] produces a wave function with spatial spread Δ and momentum spread much larger than $\hbar/\Delta$. Its 'special' states yield packets with narrower (minimal) momentum spread. It is reasonable to assume that whatever is the mechanism that allows the emitter to produce these narrower momentum-spread states, it is not sensitive to the precise value of that momentum within the range demanded by boundary conditions. We are appealing to the kind of semiclassical argument used earlier in this section. We say that any momentum in this band has the same number of ways to be produced (as measured in phase space or Hilbert space). It follows that the number of special states will be proportional to the number of classical states, which for this case will be the relative number of solutions of the two-time boundary value problem.[16] We thus need to calculate the number of states between (initial momentum) p_0 and $p_0 + dp$ that have position label x_0 and whose corresponding classical paths arrive, at time t, between x and $x + dx$. This quantity is proportional to $\partial p_0/\partial x$, where the functional dependence of p_0 on x is defined by satisfying the boundary value problem, just as the functional dependence of S above was defined through a boundary value problem. In fact these are the same functional dependencies. Recall that $S(x, t; x_0)$ is related, as a function of its arguments, to p_0, through $p_0 = -\partial S/\partial x_0$. It follows that

---

[15] More precisely: when it is not in an unusual microstate.

[16] Actually if one is using coherent states it is possible to make a count of state vectors. Although the coherent states are overcomplete, it is known that one can form a *complete* basis set from them by taking a subset of state labels on an appropriately spaced lattice, $|n\Delta x, m\Delta p\rangle$, and leaving out just one of these. By counting basis vectors in this way one can take care of the proportionality constant between the classical result and the Hilbert space dimension needed for my probability postulate.

the number of special states that arrive in the desired interval, $[x, x+dx]$, is proportional to the magnitude of $\partial^2 S/\partial x \partial x_0$. Comparing to Eq. (9.4.3) above, we have the desired proportionality between the number of special states and the magnitude squared of the wave function, agreeing with the usual quantum probability.

The foregoing argument is another perspective on getting probability to be the *square* of (the magnitude of) the wave function. You might have tried any power, but the classical mechanics, the correspondence principle, and the connection postulated in Chapter 6 between probability and special state count, tie it down to be the second power.

There is an interesting subtlety in the foregoing argument, and although I now call it 'interesting' I must own up to a short period of desperation when I thought this subtlety undermined my derivation of probabilities. The question is, what if there are two classical paths? Thus, if from y at time 0 to x at time t there are two solutions of the classical equations of motion, the propagator becomes

$$G_{WKB}(x,t;y) = \sqrt{\frac{i}{2\pi\hbar}\frac{\partial^2 S_1}{\partial x \partial y}}\,e^{iS_1/\hbar} + \sqrt{\frac{i}{2\pi\hbar}\frac{\partial^2 S_2}{\partial x \partial y}}\,e^{iS_2/\hbar} \qquad (9.4.4)$$

Again assume the initial wave packet to be well localized. Now the probability (the absolute value squared) will contain a cross term—quantum interference effects—no big deal. But the special state count, which went through so smoothly above, does *not* have interference terms. It would seem that I am up against the problems that strike the most naive attempts to interpret quantum phenomena in classical terms.

The solution lies in examining more closely what it would take for the interference—at the quantum level—to occur. Recall that two different paths that arrive at the same point must have *different* initial momenta. For example, to get two paths in the potential of Fig. 9.4.1, you would have a faster moving path bounce off the potential and meet a slower one on the way back. Will these interfere? If U is the exact evolution operator for the time t, then the interference term is given by $[U(t)|x_0p_1\rangle]^\dagger[U(t)|x_0p_2\rangle]$, where p_1 and p_2 are the initial momenta in question, and the labels can be taken as coherent state labels, selected from an appropriate ('von Neumann') lattice that is complete but not overcomplete. But by the unitarity of U, this inner product equals $\langle x_0p_1|x_0p_2\rangle$. For the complete (but not overcomplete) lattice of coherent states the inner product of different states is not quite zero. Nevertheless, as p_1 and p_2 separate, the overlap drops off rapidly. Thus, although in principle one must allow for quantum interference, it will only be significant for close values of the initial momentum. But for momenta so close that the wave functions overlap significantly the heuristic semiclassical arguments that I used for my state counting will not be valid. In other words, one can certainly get interference effects, but the special states I wanted to use when that was unimportant are no longer suitable. What special states would be valid in this case, I don't know, although I suspect one should look to the uniform semi-classical approximation that is valid in the neighborhood of caustics.

9.5 Notes and sources

Sections 9.0–9.3. The association of $|\psi|^2$ (and $|\psi|$) with probability is in [BORN].

The first report of the results in Sections 9.1–9.3 was in [SCHULMAN 91a]. The multi-state discrimination work is in [SCHULMAN 91b]. In these publications the assumption on the class of allowable kicks (Section 9.0) was explicitly applied to the multi-state case, but it was only after a helpful discussion with Bill Wootters that I became aware that the assumption was necessary for the two-state case as well (see additional discussion below).

Information on the Cauchy distribution can be found in the classic texts by [FELLER], especially the second volume. Here too can be found theorems on stable distributions and their domains of attraction. An excellent review of long-tailed distributions, with many physical applications, is found in [MONTROLL 87]. See also [DUDEWICZ]. As described in Appendix A to Section 9.1, a Lévy flight is a random walk in which successive steps are selected from one of the Lévy distributions. (See Fig. 9.1A.1.) Unlike Brownian motion (generated by Gaussian noise) the flight does not converge to a continuous curve. Nevertheless, these flights describe many physical phenomena. [MANDELBROT] uses them to describe the distribution of galaxies. On another scale, [SOLOMON] finds that tracer particles in a fluid rotating in an annular tank (with shear in the overall flow) follow Lévy flights as they move from one local attractor to another. That is, under appropriate conditions the flow consists of a chain of vortices sandwiched between unbounded jets. The tracer particle gets caught for a while in a vortex, gets free, and moves on to another vortex. The distance travelled between vortices is Lévy distributed. The video display of this phenomenon is impressive. Moreover, this experiment illustrates a general property of dynamical systems, since the flow of the tracer particle can be solved using the stream function in a way that is formally identical with the Hamilton equations of motion. A recent work aimed at explaining what is called 'the ubiquity of Lévy distributions in Nature' is [TSALLIS].

The relation of long-tailed distributions to other phenomena in physics is discussed in [MONTROLL 83, 87, SHLESINGER] and [WEST]. The 'catalogue' of fractal phenomena mentioned in the Appendix is in [AHARONY]. The subject of $1/f$ noise has an enormous literature. An early review is [HOOGE]. [PRESS] gives examples from astronomy as well as general background. Other physical examples and a variety of theoretical explanations, including self-organized-criticality, can be found in [ALERS, BAK, CHRISTENSEN, DUTTA] and [PROCACCIA]. Some years ago, [VOSS 75, 78] reported that music—all sorts of music, from Beethoven to the Beatles—shows a $1/f$ spectrum, when appropriately analyzed. Besides studying existing music, they produced random music with the $1/f$ spectrum. Apparently most listeners found this spectrum aesthetically superior to other patterns; see [GARDNER].

In my earlier publications the system was not from the start restricted to the two-dimensional real subspace defined by the outcome states, with phase for superpositions determined by the initial state. Rather, I argued that kicking along the direction defined by that phase was the least unlikely way to accomplish the specialization. It is, but not by all that much. Since it turns out that large kicks are possible, in fact necessary, a slightly larger kick along a different axis in the full complex Hilbert space (in this case $\mathbb{C}^2$) still gives a contribution of the same general size as that along the optimal axis. In a similar vein, for the multi-state (say N) case, going in any direction in $\mathbb{C}^N$ could be more effective than the one-subspace-at-a-time way we do it. Nevertheless, it appears that just because the N possible directions define a large Hilbert space ($\mathbb{C}^N$), it doesn't mean that all these Hilbert space directions are physically significant nor relevant for the measurement. (This is the essence of our assumption about kicks.) I should mention that I also am not happy defining the kick distribution in the following way: let θ be the angle that you need to cover in the two-real-dimensional subspace (which is well-defined, as in the text, no matter what you say about kicks). Allow all kicks, including those outside that subspace, but define the kick distribution function as the total probability for all those kicks. It is this that needs to be the Cauchy distribution. It follows that kicks about particular axes would have another distribution such that the Cauchy distribution would be the appropriate marginal distribution. One might expect that there would be lots of distributions whose marginals are Cauchy. One reason that I don't like this proposal is that with this point of view kicks of more than 2π should be lumped together with those shorter kicks to which they are equivalent—but then you would not have the Cauchy distribution, which is defined on the entire line. Also this proposal opens the possibility of making rotations in all sorts of Hilbert space directions. In a sense our Stern-Gerlach metaphor makes us want to allow y-axis rotations besides the x-axis rotations that seem absolutely required by the form of Eq. (9.1.1). But in general the Hilbert space could be sliced many ways and other directions allowed as well. This is why, although I don't fully understand the limitation to two-real-dimensional subspaces defined by the outcomes and by the initial state, nevertheless, they have a self-consistent frugality that other proposals do not.

Confirming—or not confirming—a hypothesis can be more subtle than one ordinarily supposes; see [SALMON].

Showing that the only allowed possibility for probabilities is $|\psi|^2$ is in [SCHUL-MAN 91b]. As indicated in Section 9.3, the source of this rule in the usual interpretation is not all that obvious. For the Everett interpretation, the way that probabilities emerge is discussed by [SMOLIN], where he uses the theorem of Finkelstein mentioned in the text.

The consequences of additive conservation laws for quantum measurements were found in [WIGNER] and studied by Araki and Yanase, among others. In an Appendix to Chapter 10 we discuss this theorem, and in Section 10.4 provide additional references.

Section 9.4. The arguments presented in this section were first given in [SCHULMAN 86]. Information on semiclassical approximation of the propagator can be found in [SCHULMAN 81]. This also gives the semiclassical propagator in the neighborhood of caustics. For more on coherent states, von Neumann lattices, etc., see [KLAUDER].

References

AHARONY 90
A. Aharony & J. Feder (1990) *Fractals in Physics: Essays in honour of B. B. Mandelbrot*, North Holland, Amsterdam. Reprinted from *Physica D*, Vol. 38, Nos. 1–3 (1989).

ALERS 91
G. B. Alers & M. B. Weissman (1991) Mechanical relaxations and $1/f$ noise in Bi, Nb, and Fe films, *Phys. Rev.* B **44**, 7192.

BAK 87
P. Bak, C. Tang & K. Wiesenfeld (1987) Self-Organized Criticality: An Explanation of $1/f$ Noise, *Phys. Rev. Lett.* **59**, 381.

BORN 26
M. Born (1926) Zur Quantenmechanik der Stossvorgänge, *Z. Phys.* **37**, 863. Trans. in [WHEELER], On the Quantum Mechanics of Collisions.

CHRISTENSEN 92
K. Christensen, Z. Olami & P. Bak (1992) Deterministic $1/f$ Noise in Nonconservative Models of Self-Organized Criticality, *Phys. Rev. Lett.* **68**, 2417.

DUDEWICZ 76
E. J. Dudewicz (1976) *Introduction to Statistics and Probability*, Holt, Rinehart & Winston, New York.

DUTTA 81
P. Dutta & P. M. Horn (1981) Low-frequency fluctuations in solids: $1/f$ noise, *Rev. Mod. Phys.* **53**, 497.

FELLER 50
W. Feller (1950) *An Introduction to Probability Theory and Its Applications*, 2nd ed., vol. 1, Wiley, New York.

FELLER 71
W. Feller (1971) *An Introduction to Probability Theory and Its Applications*, 2nd ed., vol. 2, Wiley, New York.

GARDNER 78
M. Gardner (1978) Mathematical Games, *Sci. Am.* **238**, April, 16.

HOOGE 76
F. N. Hooge (1976) $1/f$ Noise, *Physica* **83B**, 14.

KLAUDER 85
J. R. Klauder & Bo-S. Skagerstam (1985) *Coherent States, Applications in Physics and Mathematical Physics*, World Sci., Singapore.

MANDELBROT 83
B. B. Mandelbrot (1983) *The Fractal Geometry of Nature*, Freeman, New York.

MONTROLL 83
E. W. Montroll & M. F. Shlesinger (1983) Maximum Entropy Formalism, Fractals, Scaling Phenomena, and $1/f$ Noise: A Tale of Tails, *J. Stat. Phys.* **32**, 209.

MONTROLL 87
E. W. Montroll & B. J. West (1987) On an enriched collection of stochastic processes, in *Fluctuation phenomena*, 2nd ed., E. W. Montroll & J. L. Lebowitz, eds., North-Holland, Amsterdam.

PRESS 78
W. H. Press (1978) Flicker Noise in Astronomy and Elsewhere, *Comments Astrophys.* **7**, 103.

PROCACCIA 83
I. Procaccia & H. Schuster (1983) Functional renormalization-group theory of universal $1/f$ noise in dynamical systems, *Phys. Rev.* A **28**, 1210.

SALMON   W. C. Salmon (1973) Confirmation, *Sci. Am.* **228**, May, 75.
73

SCHULMAN   L. S. Schulman (1981) *Techniques and Applications of Path Integration*,
81   Wiley, New York.

SCHULMAN   L. S. Schulman (1986) Deterministic Quantum Evolution through Mod-
86   ification of the Hypotheses of Statistical Mechanics, *J. Stat. Phys.* **42**,
              689.

SCHULMAN   L. S. Schulman (1991) "Special" states in quantum measurement appa-
91a   ratus: Structural requirements for the recovery of standard probabilities,
              Found. Phys. **21**, 931.

SCHULMAN   L. S. Schulman (1991) Definite Quantum Measurements, *Ann. Phys.*
91b   **212**, 315.

SHLESINGER   M. F. Shlesinger (1989) Lévy Flights: Variations on a Theme, *Physica* D
89   **38**, 304.

SMOLIN   L. Smolin (1984) On quantum gravity and the many-worlds interpre-
84   tation of quantum mechanics, in *Quantum Theory of Gravity*, ed. S.
              Christensen, Adam-Hilger, Bristol.

SOLOMON   T. H. Solomon, E. R. Weeks & H. L. Swinney (1994) Chaotic advection
94   in a two-dimensional flow: Lévy flights and anomalous diffusion, *Physica*
              D **76**, 70.

TSALLIS   C. Tsallis, S. V. F. Levy, A. M. C. Souza & R. Maynard (1995) Statistical-
95   Mechanical Foundation of the Ubiquity of Lévy Distributions in Nature,
              Phys. Rev. Lett. **75**, 3589.

VOSS   R. F. Voss & J. Clarke (1975) "1/f noise" in music and speech, *Nature*
75   **258**, 317.

VOSS   R. F. Voss & J. Clarke (1978) "1/f noise" in music: Music from 1/f noise,
78   *J. Acoust. Soc. Am.* **63**, 258.

WEST   B. J. West & M. Schlesinger (1990) The Noise in Natural Phenomena,
90   *Am. Sci.* **78**, 40.

WHEELER   J. A. Wheeler & W. H. Zurek (1983) *Quantum Theory and Measurement*,
83   Princeton Univ. Press, Princeton.

WIGNER   E. P. Wigner (1952) Die Messung quantenmechanischer Operatoren, *Z.*
52   *Physik* **133**, 101.

Notes on the exercises

Exercise 9.1A.2 Consider the ceiling in Fig. 9.1A.2. Alternatively redefine the angle.

Appendix 9.1.A: Cauchy and other long-tailed distributions

Many natural phenomena exhibit large fluctuations. One catalogue is the volume *Fractals in Physics: Essays in honour of B. B. Mandelbrot* ([AHARONY]). Typically, fractals are the hallmark of processes involving large fluctuations, for example, the dynamics of a system near a critical point ('second order phase transition'). It turns out that fractal systems are more prevalent than one would expect, considering the non-generic nature of critical points, and the concept of 'self organized criticality' has been proposed to explain why large fluctuations are so common.

Exercise 9.1.A.1. For the Curie-Weiss model of magnetism, the Boltzmann factor for states with k spins up (out of a total of n) is

$$\binom{n}{k} e^{-U(k)/T} \qquad \text{with} \quad U(k) = -\frac{1}{2n}(2k-n)^2 - h(2k-n)$$

For $n \to \infty$ the critical point is at $T = 1$, $h = 0$. Show that away from the critical point, fluctuations in k are Gaussian. Show that at the critical point they are not, and much larger fluctuations are common. Note that unlike many physically realistic systems the model does not exhibit power law fluctuations. Why not?

Whatever the justification for long-tailed distributions, they do occur in Nature. Although their physical importance has only recently been realized, mathematicians have been studying such distributions for quite some time, witness the association of the name Cauchy with the distribution function, Eq. (9.1.5). In this Appendix I will give the flavor of the mathematical theory; for details, extensions and applications the references in the notes should be consulted.

A classification of distribution functions naturally arises when one considers *sums of independent random variables*. Let X be a random variable, for example, the money transferred as the result of a roll of dice, the vector distance travelled by a molecule between collisions, or the thickness of a tree ring from a season's growth. Typically the quantity of interest results from many repetitions of the process. Let X_k, $k = 1, \ldots, N$ be N random variables, all having the same distribution function as the original X and all independent of one another (independence is often a touchy issue). Let $S_N = X_1 + X_2 + \ldots + X_N$. The quantity S_N is typically the thing you care about. Using the assumed independence one can calculate its distribution function for any finite n. In general if X and Y are independent random variables and $S = X + Y$, then

$$\Pr(S = s) = \int dx \, \Pr(X = x) \Pr(Y = s - x) \tag{9.1.A.1}$$

Let ϕ denote Fourier transform, so that, for example,

$$\phi_X(k) = \left\langle e^{ikX} \right\rangle = \int dx \, e^{ikx} \Pr(X = x) \tag{9.1.A.2}$$

By noting that Eq. (9.1.A.1) is a convolution, or by using the independence of X and Y in the form $\langle \exp[ik(X + Y)] \rangle = \langle \exp(ikX) \rangle \langle \exp(ikY) \rangle$, it follows that

$$\phi_S = \phi_X \phi_Y \tag{9.1.A.3}$$

which is the starting point for study of sums of random variables. For the sum S_N above, the Fourier transform of its distribution function is simply $\phi_{S_N} = [\phi_X]^N$. If the first few moments of X are finite, one can write

$$\phi_{S_N}(k) = \exp\left[N \log(\phi_X(k))\right]$$
$$= \exp(ikN\langle X \rangle) \exp\left(-(k^2/2)N\left[\langle X^2 \rangle - \langle X \rangle^2\right]\right) \tag{9.1.A.4}$$
$$\times \exp\left(-(k^3/6)N\left[\langle X^3 \rangle - 3\langle X^2 \rangle \langle X \rangle + 2\langle X \rangle^3\right]\right) \times \cdots$$

By rescaling S_N, namely subtracting $N\langle X \rangle$ and dividing the result by $\sqrt{N}$, one gets rid[17] of all terms higher than quadratic in k, so that the variable $[S_N - N\langle X \rangle]/\sqrt{N}$ has a Gaussian Fourier transform. Since the Fourier transform of a Gaussian is a Gaussian, the rescaled object in the $N \to \infty$ limit is normally distributed (a Gaussian). That's the central limit theorem, and it's the basis for much of our intuition about random processes in physics.

---

[17] This 'getting rid of' higher order terms is the job of proving the central limit theorem. Roughly speaking, the linear term in k goes away when one considers $S_N - N\langle X \rangle$; one then rescales the k variable by $\sqrt{N}$. This leaves the quadratic term as the dominant one and as $N \to \infty$ cubic and higher order terms drop out.

If the original X was itself a Gaussian, then there is no $O(k^3)$ (or higher) term in Eq. (9.1.A.4). Thus if X and Y are Gaussian, then $X+Y$ is Gaussian. From Eq. (9.1.A.4), it is clear that the mean of $X+Y$ is $\langle X \rangle + \langle Y \rangle$ and the standard deviation is the sum of the standard deviations. The family of Gaussian distributions is thus *stable*, i.e., the sum of two of them is a Gaussian with parameters related to the parameters of the summands. Moreover, the family of Gaussians is an *attractor*, in that sums of other distributions ultimately look like a Gaussian.

But if X does not have a sufficient number of finite moments this all falls apart.[18] Nevertheless, you still have the notion of a stable distribution and the stable distributions are attractors within the appropriate class. The relation, $\phi_{S_N} = [\phi_X]^N$, still holds, but the subsequent expansion (the second equality in Eq. (9.1.A.4)) need not. In particular, suppose the Fourier transforms have the form

$$\phi_X(k) = \exp(-a|k|^\alpha)\,, \qquad \phi_Y(k) = \exp(-b|k|^\alpha) \tag{9.1.A.5}$$

for $0 < \alpha \le 2$. Clearly the Fourier transform of $X+Y$'s distribution function will have the same α, but will have the parameter $a+b$ replacing the a and b of X and Y, respectively. Thus the class of distributions with Fourier transforms of the form Eq. (9.1.A.5), having the same α, are stable in this sense of addition.

Working with the expression for the distribution

$$\Pr(X = x) = \frac{1}{2\pi} \int dk \exp\left(-ikx - a|k|^\alpha\right) \tag{9.1.A.6}$$

can be messy. For example, even showing that $\Pr(X = x)$ defined by Eq. (9.1.A.6) is positive (necessary for a probability), is a task. The cases $\alpha < 1$ and $\alpha > 1$ require separate handling. The case $\alpha = 1$ is our case of interest, the Cauchy distribution, and for completeness we recall the equation from Section 9.1

$$C_a(x) = \frac{a/\pi}{x^2 + a^2} = \frac{1}{2\pi} \int_{-\infty}^{\infty} dk \exp(-a|k| + ikx) \tag{9.1.A.7}$$

Thus positivity is explicit. The (only) other easy case is $\alpha = 2$, for which we get a Gaussian, showing that they are a special case of the present classification. By non-trivial fussing you can show that for $|x| \to \infty$

$$\int dk \exp\left(-ikx - a|k|^\alpha\right) \sim \frac{\text{const}}{|x|^{1+\alpha}} \tag{9.1.A.8}$$

Thus for $\alpha < 2$ these distributions do not have a second moment, and for $\alpha \le 1$ no first moment as well.

In the sources given in the notes, further examples are explored (for example, complex parameter 'a'). They also provide an idea of the extent to which these distributions are basins of attraction (for sums). Further sources are also cited for this well developed and at times subtle mathematical topic.

In the text I mentioned one of the significant properties of the long-tailed distributions connected with the way in which large deviations are *least unlikely* to occur. In particular, if S_N is the sum of N Cauchy distributed random variables all with parameter a, and if it is known that for a particular occurrence S_N takes the value A, with $A \gg Na$, then with overwhelmingly probability (as $N \to \infty$) only one of the summands has a large deviation

---

[18] The central limit theorem needs a bit more than the existence of a second moment.

and the others are small. This is in sharp contrast to the Gaussian, normal distribution, for which the deviation is most likely to be shared.

A graphic exhibit of this is provided by the *Lévy flights*. Exactly as for Brownian motion, you plot S_N, say in two dimensions. If you use random variables with second moments and if you scale your discretized random walk properly,[19] then you will get a jittery line in the plane. This is the content of the theorem that Brownian motion paths are continuous but nowhere differentiable. But when your steps are taken from one of the long-tailed distributions, the limiting object is *not* a continuous curve. It is a jumpy object, that clusters for a while, then takes a big jump, then clusters, then jumps, then after a lot of this takes a really big jump. In the limit, it does this on all distance scales. This gives the 'path' the structure of a random fractal, with hierarchical structure and scale invariance. Mandelbrot has proposed this as a model for the fractal structure of the distribution of galaxies. In Fig. 9.1.A.1 I show three such images for 50, 250 and 1200 points. For all of these the Cauchy parameter a was one. The largest steps in the figures were respectively 300, 20 000 and 600 000.

An example of the Cauchy distribution can be realized by considering the spot where the light from a suspended mirror hits a wall. See Fig. 9.1.A.2. The mirror is a distance a from the wall. It is free to rotate about a vertical axis and is equally likely to be at any angle (between $-\pi/2$ and $\pi/2$). The point on the wall where the light hits is $a \tan\theta$, where θ is the angle the mirror actually makes with a line parallel to the wall. Call the random variable giving the position where the light hits X; then the probability distribution for X is

$$
\begin{aligned}
\Pr(X = x) &= \Pr(x = a\tan\theta) \\
&= \frac{d\theta}{dx}\Pr(\Theta = \theta) \\
&= \frac{a}{a^2 + x^2}\frac{1}{\pi}
\end{aligned}
\qquad (9.1.A.9)
$$

by elementary probability manipulations (and with an obviously defined random variable Θ, which is uniformly distributed on $[-\pi/2, \pi/2]$, giving rise to the last factor in the last line, above).

Fig. 9.1.A.1. Three excerpts from a Lévy flight with Cauchy distributed steps. The left figure shows 50 steps, the middle 250 and the right figure (which uses a smaller symbol for landing points) 1200.

---

[19] time step size $\sim$ (spatial step size)^2

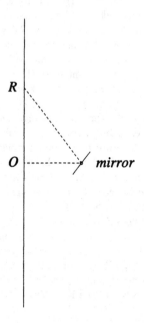

Fig. 9.1.A.2. Light originates from the source O, travels to the right and strikes the mirror. The mirror's orientation is uniformly distributed in angle so that the position of the reflection R on the wall is Cauchy distributed.

This example allows us to see another way in which the long-tailed distributions differ from the more familiar Gaussians. Suppose you checked the mirror many times, with the angle selection taking place independently each time. You collect the data on the reflection point, perhaps hoping that this could help you estimate a. What should you hope for the distribution of the *average* of your measurements? For better behaved random variables, the random variable $[X_1 + \ldots + X_N]/N$ will tend to zero, and things will settle down. Here, however, $[X_1 + \ldots + X_N]/N$ has exactly the same distribution as the original X, so it will neither shrink nor grow with N.

One of the major scientific puzzles of our time is $1/f$ noise (references in the notes). This is the observation that the power spectrum of many natural phenomena, from current fluctuations in resistors to luminosity in quasars to the height of the river Nile, exhibit increasing intensity with decreasing frequency, according to the approximate law, $1/$ frequency. There is little doubt that this phenomenon is intimately related to fractals and to large fluctuations. There is also little doubt that there's a long way to go in terms of dotting *i*s and crossing *t*s. I bring up the topic of $1/f$ noise both because the anyone who has read this far into the Appendix may be interested in related topics and also to make the point that there is still a great deal that is unknown about fluctuations in macroscopic systems. My claims about the ubiquitous existence of Cauchy distributed noise should be evaluated while bearing in mind that an apparently ubiquitous phenomenon whose existence is well known, remains unexplained.

Exercise 9.1.A.2. Let X be a Cauchy distributed random variable with parameter a. Show that the random variable $Y \equiv 1/X$ is Cauchy distributed, with the same parameter, a.

10

Experimental tests

In Chapter 6 I presented a proposal for how and why grotesque states do not occur in Nature. In subsequent chapters I explored consequences and found subsidiary requirements, such as Cauchy distributed kicks. To find out whether all or part of our scheme is the way Nature works, we turn to experiment. *How* to turn to experiment is not so obvious, since the basic dynamical law, $\psi \to \exp(-iHt/\hbar)\psi$, is the same as for most other theories.[1] Our basic assertion concerns not the dynamical law but the selection of states. Therefore it is that assertion that must be tested. For example, one way is to set up a situation where the states we demand, the 'special' states, *cannot* occur. Then what happens? Another part of our theory is the probability postulate, and this deals not only with the existence of special states but with their abundance. It enters in the recovery of standard probabilities but has far reaching consequences that may well lead to the best tests of the theory. Such tests arise in the context of EPR situations.

The experimental tests fall into the following categories.

Precluding a class of special states. This should prevent a class of outcomes. If the changes in the system (due to precluding the class of special states) do not change the predictions of the Copenhagen interpretation, then this provides a test. In particular, with a class of special states precluded, our theory forbids the associated outcome. Section 10.1.

Footprints of specialization—footprints of the future. If the least unlikely way to specialize involves action on the state prior to the experiment proper, there may be ways to determine that such action took place. This possibility arises in EPR type experiments. Section 10.2.

---

[1] Most versions of the Copenhagen interpretation invoke some black magic during a measurement, but when push comes to shove, i.e., when you can look more closely at the details of the apparatus, they allow detailed quantum evolution to take precedence over the black magic.

Evidence of auxiliary features. In Chapter 9 we found that specializing events necessarily involved long tailed (in particular, *Cauchy*) distributions. Is there other evidence of the large fluctuations behind those specializing events? Section 10.3.

Deviations from standard probabilities. When you measure an observable for a state that is in an eigenstate of that observable, our theory predicts an error rate due to cutoffs in the 'kick distribution.' Is this measurable? Section 10.3.

In the coming sections we take up these possibilities. In general, it's necessary to stick your neck out, to make more than minimal assumptions about the specializing influences, in order to make concrete predictions. However, I will try to be clear just how deeply embedded in the theory each assertion is.

After a bookful of moralizing on human limitations, the reader will not be surprised that I consider the providing of experimental checks essential. There is also a virtue beyond the fundamental one. Once an experimental test is offered, even a gedanken-experiment, one can be confident that the theory has physical content.

Finally, note that the experimental tests described here are tests of the quantum side of this book. Experiments associated with classical arrow of time issues were described in Section 4.6.

10.1 Precluding a class of special states

Consider the emission and detection sequence pictured in the upper portion of Fig. 10.1.1. A particle leaves the source and its wave packet spreads. As that spreading wave packet passes the detector there is a chance that the detector fires. Conventionally, the probability of firing is proportional to the square of the amplitude of the wave function that crosses the detector multiplied by the efficiency of the detector. Suppose that the packet emerges as a spherical wave, that the detector has linear dimension D, that the distance between the source and detector is R, and that the efficiency of the detector is $\mathcal{E}$. Then, up to πs and such, the probability of firing is $\mathcal{E}D^2/R^2$.

Suppose that this detector absorbs as it detects and suppose all forces between it and the particle detected are short ranged. Then according to the ideas presented here the entire wave function of the particle must find its way to the detector. Under these circumstances one expects things to look more like what is depicted in the lower portion of the figure; namely, the wave packet, for one reason or another, does not spread widely and manages to be focused onto the detector. In this section we will look

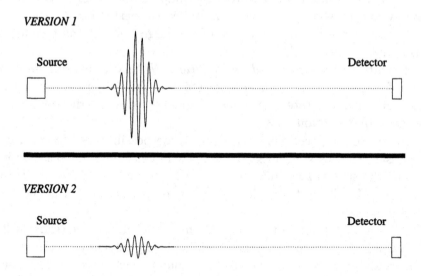

Fig. 10.1.1. Two versions of an emission/detection event. In the upper portion, a particle is emitted from the box labeled 'Source.' Its wave packet spreads as it moves toward the detector. As it passes the detector, there is overlap of the particle wave function with the detector. Local interactions between detector constituents and the particle give rise to absorption and detection of the particle. The probability of detection will be proportional to the square of the overlap times the efficiency of the detector. In the lower figure, the entire wave packet is heading for the detector. When a detection is recorded there will be no residual portion of the wave function continuing off to the right, past the detector.

closely at that 'one reason or another,' and try to prevent all focusing influences. This should result in the failure of the detector to trigger, despite the fact that the quantity $\mathcal{E}D^2/R^2$ is significantly different from zero.

Preventing a detector from firing

As in Fig. 10.1.1, the particle is emitted from the source and may be absorbed in the detector. Everything is contained in a chamber, which, as we shall see, must maintain a good vacuum. According to the theory of this book the particle's entire wave function must evolve to the region of the detector for the capture. According to other theories this is not necessary. We will deplete the physical environment so as to prevent the wave function from evolving—by pure quantum evolution—entirely to the region of the detector. This should change the predictions of our theory, but not those of others.

To prevent evolution of the wave packet to the detector we must have an idea of what environmental forces can guide it when the environment is *not* depleted. The forces available consist of the particles and fields in the emitter, the particles and fields in the chamber between source and detector, and the particles and fields in the detector. The fields in the chamber can come from its particles or its walls. There are also exotic possibilities, such as neutrinos, that uncontrollably pass through the chamber.

I will assume that all forces associated with the emitter and detector are short range. Any (long range) fields that their constituent atoms emit will be treated separately, basically in the same way as fields produced by the walls of the chamber.

If all other forces could be neglected and if the flight time from source to detector could be known, then by making both source and detector sufficiently small, detection could be prevented. To see this, recall the following inequality[2]

$$\Delta x_1 \Delta x_2 \geq \frac{\hbar T}{m} \qquad (10.1.1)$$

Δx_j, $j = 1, 2$, are the spreads in the particle's wave function as it leaves the source (at time t_1) and arrives at the detector (at time t_2), and $T = t_2 - t_1$. The particle is assumed to be free between those times. Suppose the source has characteristic linear dimension S and the detector characteristic linear dimension D. Then, for example, to say that the particle is ultimately detected requires that $\Delta x_2 \leq D$. Therefore by making

$$SD < \hbar T/m \qquad (10.1.2)$$

detection becomes impossible.

As to the mechanism by which these blobs of matter ordinarily focus the wave packet, I cannot be specific. It is clear that matter in the vicinity of an emission or detection event reacts to that event. For an appropriate *detection*, as a particle enters a detector it distorts the lattice and makes a significant disturbance—it must in order for the detector to work. Finally things settle down and the incoming energy is thermally dissipated. Consider what this disturbance looks like when the time evolution is run backwards. Moments 'before' the (reversed) detection, the lattice begins to vibrate, culminating in emitting a highly directional packet toward the source. That this can happen for both emission *and* detection is a

---

[2] For information on the derivation of Eq. (10.1.1), see Exercise 5.3.4.

statement about the existence of special states, but the fact that the environment participates in focusing is not in doubt. For emission as well, recoil tracks testify to the participation of the source.

Remark: A bit of elaboration on why the double ended scenario is a statement about the existence of special states. Consider the wave function evolution in the many worlds view, using the wave function in 'VERSION 1' of Fig. 10.1.1. Some is absorbed, some passes through. If you'd run the whole thing backwards, including detector particles, you'd send the wave function back to the source. Suppose you ran backwards only the part that was absorbed. (That is, take the total wave function and subtract from it the part that was not absorbed.) By the superposition principle, when you run this[3] backward, it will send out a packet in the direction of the source. However, this will not in general reconverge on the source. It would only do so if we used *all* the wave function. Without the entire particle wave function, the portion (backwardly) emerging from the detector would head for the source and be partly (backwardly) absorbed (anti-emitted), partly pass through. *With* the entire wave function the backward propagating absorbed and unabsorbed portions interfere destructively, except for the part going to the source. On the other hand, when the emission and subsequent evolution involve a 'special' state (VERSION 2), the absorption by the detector is absorption of the entire wave function, so that when run backwards the entire wave function returns and is anti-emitted by the source.

To get a feel for the conditions imposed by the above inequalities, let's put in numbers. Suppose we want to detect a slow neutron by having it disrupt a plutonium nucleus. Then $m = m_N$, the neutron mass. Let the detector be a clump of plutonium nuclei, essentially a grain of size D. Take the neutron to have the velocity of a thermal neutron at temperature 300 K. This is about $5 \times 10^{-6}c$. If the distance from source to detector is $R \sim 10^2$ cm and $S \sim 10^{-5}$ cm, our condition becomes

$$D \leq 4 \times 10^{-2} \text{ cm} \qquad (10.1.3)$$

This is not unreasonable. In practice, slow neutrons come out of big piles, but the relevant dimension should be the size of a grain of coherent emitter, and the meter-sized rod should not matter. One must also check that source and detector are sufficiently macroscopic that the uncertainty of *their* zero point motion does not contribute to the spread. Here is an order of magnitude check: if the source has density 1 gm/cm^3, and its wave function spread is its own size (10^{-5} cm) then the velocity imposed on it by the uncertainty principle is about 10^{-12} cm/s.

Thus source and detector must be small. Now we look into what happens in between. We consider four issues separately: 1, atomic/molecular scattering; 2, bunching of scatterers; 3, neutrino/exotic scattering; and 4, photon scattering. There is also scattering off the chamber walls, which we will take up later when dealing with a related question.

---

[3] ... including detector particle coordinates.

Atomic/molecular scattering

In thinking about the gas that lies between the source and the detector, you quickly realize that you can't get rid of it. Even at 10^{-12} torr, which is a very good (but attainable) vacuum, you still have 35 000 particles/cm^3 at room temperature (cf. Exercise 5.0A.1). Recall that our goal is to have a situation where, *no matter what all those particle do*, they cannot focus the wave packet onto the detector (and then our prediction will be: no detection, in contradistinction to other theories). Given the oft advertised capabilities of coherence, you might think it impossible ever to say that something is *impossible*, especially with upward of 10^4 particles per cubic centimeter. Nevertheless, a bound *can* be placed on the ability of a collection of particles to maintain narrow collimation for a wave function. ('Collimation' refers to the situation depicted in Fig. 10.1.1, VERSION 2.)

In the usual treatment of scattering theory, a plane wave $\exp(i\mathbf{k}\cdot\mathbf{x})$ becomes, after scattering, $\exp(i\mathbf{k}\cdot\mathbf{x}) + [\exp(ikr)/r]f_k(\theta)$. For wave packets, this means a piece with amplitude proportional to f is withdrawn from the onward moving packet. The proportionality factor, however, depends on the properties of the specific wave packet. Say the packet has transverse linear dimension L which is much larger than the effective size of the gas molecules off which it scatters. Up to factors of 2, the molecule removes from the forward moving beam a transverse area σ_T, the total cross section, as calculated in the usual way.[4] For a small gas molecule we take $\sigma_T \sim 10^{-16}$ cm^2. The amount (of $|\psi|^2$) removed or scattered from our wave packet will therefore be σ_T/L^2 where again, for simplicity, we assume that the wave packet is smooth and fairly uniform within the L^2 cross section. Let us first calculate the usual mean free path. Take the density of scatterers to be n gas molecules per cubic centimeter (about 35 000 for $T = 300$ kelvin, $P = 10^{-12}$ torr). Then the number of scatterings within a volume $L^2\ell$ will be $nL^2\ell$. When the product of that number with the fraction removed, (σ_T/L^2), is unity, ℓ will be the mean free path. Thus $(nL^2\ell) \cdot (\sigma_T/L^2) = 1$ or

$$\ell = \frac{1}{n\sigma_T} \qquad (10.1.4)$$

which is the usual result. Under the conditions mentioned, $\ell = 3 \times 10^9$ m.

For coherent scattering we look instead at the *amplitude* removed from the wave function and, as a worst case scenario, assume it adds coherently. Thus instead of having (the dimensionless) σ_T/L^2 sum to one, we take

---

[4] Recall, $\sigma_T = \int |f|^2 d\Omega$.

$\sqrt{\sigma_T}/L$ to sum to one. The criterion for 'complete' coherent deflection is thus $(nL^2\ell_c) \cdot (\sqrt{\sigma_T}/L) = 1$ or

$$\ell_c = \frac{1}{nL\sqrt{\sigma_T}} = \frac{\sqrt{\sigma_T}}{L}\ell \qquad (10.1.5)$$

with the 'c' standing for coherent. Evidently ℓ_c depends on the properties of the wave packet. We estimate the numbers involved using values on the order of the quantities S and D developed above. For simplicity we replace S and D by their geometric mean. Then $L = \sqrt{SD} \sim 6 \times 10^{-4}$ cm. This gives $\ell_c \sim 6 \times 10^4$ m, down by a factor 2×10^{-5} from ℓ.

This shows both the effectiveness of coherence and the fact that you *can* make a good enough vacuum to eliminate coherent deflection.

Bunching of scatterers

In the estimate just made we assumed that the particle density was uniform along the tube defined by the beamline and the cross section of the wave packet. If the scattering particles bunch significantly, all bets are off.

If the gas particles are well approximated by an ideal gas subject to no external forces, it is not difficult to estimate density fluctuations. In particular we should estimate fluctuations in the number of particles seen (by the wave packet) along the tube described above. Call this number N_0. In the example given above N_0 is on the order of 10^4 or 10^5 (i.e., $L/\sqrt{\sigma_T}$). For an ideal gas (in a rough container) the density function is Poisson and to an excellent approximation the fluctuations in the relevant number are normally distributed. A significant fluctuation would have to be on the order of N_0 itself;[5] this would have probability $\sim \exp(-N_0/2)$. More to the point is that with normally distributed variables I would not expect to be able to recover probabilities (in the sense of item 3 at the end of Section 6.3). For example, suppose that by emerging in different directions the emitted particle encountered different gas densities. Then my expectation, based on the results of Chapter 9, would be that Cauchy distributed variables entered. Furthermore, although it is not possible to have *nothing* in the chamber, it would be possible to have those last 35 000 atoms/cm^3 be nearly all helium or argon, hence 'ideal' (hence possessing Poisson distributed particle distributions).

There remains the possibility that bunching could be driven by something else, for example electromagnetic fields in the chamber. This will be discussed in *Photon scattering*, below.

---

[5] ... in which case the normal distribution is not such an excellent approximation, but still good enough for our purposes.

Neutrino/exotic scattering

A popular gee–whiz theme concerns the number of neutrinos passing through our bodies each second or the number of light years of lead it would take to stop a neutrino. Therefore the bad news is that it is impossible to build a chamber that excludes neutrinos. The good news is that it doesn't matter.

We estimate maximum possible neutrino influence as we did for atomic scattering in our rarefied chamber. That is, we examine the amount of wave function withdrawn in a single scattering event and place a bound on the effect of many such events by assuming that all withdrawn wave function adds coherently. In this case we have a large number of scatterers, but as will be seen in a moment, the minuscule scattering cross section consigns neutrino scattering to insignificance.

Let ϕ be the flux of neutrinos. For solar neutrinos[6] this is about 10^{11} cm^{-2} s^{-1}. Let σ be the cross section for scattering of the neutrino and the particle in the experiment. Typically this is on the order of 10^{-43} cm^2. We again call L the characteristic dimension of the wave packet, v its velocity and R the distance from source to detector. The amount of flux withdrawn is thus

$$\frac{\sqrt{\sigma}}{L} \frac{L^2 R \phi}{v} \tag{10.1.6}$$

For the values used in our previous estimates (and with $R \sim 100$ cm) this gives (the dimensionless number) 10^{-16}. Not to worry.

Remark: This small number leads to a different question. Deep in a mine in South Dakota are many tons of cleaning fluid, dedicated to the detection of solar neutrinos. One of the puzzles of our day is why so few are seen; nevertheless, some *do* register. According to the theory in the present volume, for detection, the entire neutrino wave function must arrive at a ^{37}Cl nucleus in the cleaning fluid. What interactions could possibly focus them on such a small spot? They are born, supposedly as spherical waves, in the interior of the sun. Given their lack of interaction, one expects it to be impossible to focus them on a point on earth. It's interesting that coherence is up to the job; at least the bound I have been using for coherence leaves the required focusing well within the realm of possibility. Consider the following question. How much must the wave function of a new-born neutrino spread before it encompasses enough matter for that matter to coherently scatter all of the wave function? Suppose it has spread to a wave packet of radius L. The amplitude withdrawn is on the order of $\sqrt{\sigma}/L$, as before. The number of scatterers in the volume is ρL^3, with ρ the number per unit volume. Setting the product of the number of scatterers and the withdrawn amplitude to unity, we get $L = \rho^{-1/2}\sigma^{-1/4}$. For $\rho \sim 10^{24}$ particles/cm^3, this gives about a tenth or a hundredth of a centimeter. Of course it's unlikely that scattering can be as efficient as this upper bound, but it shows

---

[6] The number and nature of solar neutrinos is a subject of great current interest. For our purposes an order of magnitude up or down won't matter.

there's plenty of room to spare. What makes this estimate work is the appearance of the fourth root of σ, showing that the neutrino's unresponsiveness is neutralized.

What else is there? Photons are discussed in the next subsection. I wouldn't expect cosmic rays to be important, but if they were suspected of being troublesome one has the option of going underground, thereby avoiding all but neutrinos. As to other objects, a variety of exotic particles have been proposed as solutions of the 'missing mass' problem. Some of these are supposed to pervade our surroundings. Could these direct our spreading wave packet to the detector? Since the nature of the particle varies with the suggestion a full answer cannot be given. What these proposals must have in common is that the particle is something we wouldn't have noticed. In other words it must be unreactive, even by the standards of neutrinos. In that case, I would say that it is not about to focus our expanding wave packet either.

Photon scattering

The walls of the chamber necessarily emit and absorb photons. If one takes these photons to have a black-body spectrum, the number of photons per cubic centimeter is about $24T^3$, with T in kelvins. At room temperature this will exceed 10^8. However, the photon-neutron cross section is down by many orders of magnitude relative to the neutron-atom cross sections discussed above. In any case, with a bit of cooling of the walls this effect should be negligible.

The effect that does concern me, however, arises from the interaction of the photons with the residual gas atoms. This could lead to the sort of bunching effect discussed in *Bunching of scatterers*, above. Although the photon-atom cross section is low, on the order of the Thomson cross section $(6.65 \times 10^{-25} \text{ cm}^2)$ or less for non-resonant scattering, still, there are many photons interacting with many atoms. Furthermore, the photons come with a tendency to bunch, due to their bosonic nature. Finally, to the extent that the photons are themselves emitted by transitions with characteristic Lorentz line shape, they already partake of the Cauchy distribution in their energy spectrum.

I have not managed to rule out the effect of photons on the atoms and in turn on the neutron (if that's the particle in the emission-detection experiment). Better vacuum and colder walls should help, but at this time I do not have quantitative bounds on this possibility.

Can you make nothing happen?

Suppose it's all true. Suppose you *can* build a detector that doesn't detect because special states are unavailable. Now build many such detectors, surround the source with them and put it all in an appropriate vacuum chamber. The particle is emitted. Where can it go? A number of people have asked me this question, and I believe that they are right in that the detection scenario will be changed, perhaps the emission scenario as well.

What happens will depend on the details, but most likely the wave packet will bounce off its surroundings until it manages to focus on one of the detectors. It is then absorbed/detected. There are other possibilities. If the edges of the detectors allow passage of the particle, several of them could absorb the particle coherently, following which its wave function would go to the detector that actually fires. (This obviously depends on the detectors and the builder of an experiment designed to see if 'nothing' happens should not have trouble eliminating this possibility.)

Yet another scenario, again depending on the details, is that the emission itself is suppressed. A similar phenomenon occurs in 'cavity quantum electrodynamics,' where an absence of quantum states for an emitted photon prevents the atomic decay in which that photon would have been emitted. In the experiments that demonstrate that phenomenon (see the notes below) it is the absence of particular microscopic states for the photon that control the emission. For us it is the absence of the complex states (the 'special' ones) that prepare the particle for its detection, but the result would be the same.[7]

If the detection does occur by multiple bounces off the walls it might be possible to know this from the time between emission and detection, although, again, whether knowledge of this time is feasible will depend on the details of the experiment. This point also brings out the necessity of recognizing bounces off the chamber walls in the basic non-detection experiment proposed in detail above. In that case it may be possible to have the chamber walls considerably farther away from the source than the detector, allowing easier discrimination based on arrival time.

Remark: Although this section has been oriented to laboratory experiments, it is conceivable that in the extreme vacuum of intergalactic space, particle scattering could be reduced by the inability to satisfy future constraints. (Alternatively stated: dynamical scenarios that *do* satisfy the boundary conditions are those for which inevitably delocalizing scatterings do not take place.)

10.2 EPR-type experiments: footprints of the future

Consider Bohm's version of the EPR experiment. A pair of distinguishable spin-$\frac{1}{2}$ particles is initially in a spin-0 metastable state within a source, S. The metastable 'molecule' formed by these two particles is

---

[7] Here too our prediction for the detection differs from that of Copenhagen.

called $\mathcal{M}$. The source, $\mathcal{S}$, is to be thought of as mesoscopic, and $\mathcal{M}$ is embedded in it.

Detection and spin measurement devices are located on the x-axis, so that only decays in which the particles exit collinearly along the x-axis are recorded. Furthermore, to keep things simple, the left detector is only sensitive to particle 1 and the right detector to particle 2. The initial wave function of $\mathcal{M}$ can be taken to be

$$\Psi_i(1,2) = \frac{1}{\sqrt{2}}\left[|+\tfrac{1}{2}-\tfrac{1}{2}\rangle - |-\tfrac{1}{2}+\tfrac{1}{2}\rangle\right]\Phi(1,2) \qquad (10.2.1)$$

with $\Phi(1,2)$ a function of the position coordinates of particles 1 and 2.[8] This is an entangled state.

Conventional analysis of the experiment

We first proceed with no explicit mention of the degrees of freedom of the apparatus. In this analysis, the material source, $\mathcal{S}$, in which we assumed $\mathcal{M}$ was embedded, plays no role, and the analysis would be unchanged if $\mathcal{M}$ were an isolated object. For convenience, assume that the measurement on the left (particle 1) takes place first. When particle 1 is measured, the detector is presented with a superposition of $|+\rangle$ and $|-\rangle$, each with a vector-valued coefficient of norm one half. Therefore with equal probability it gives spin-up or spin-down. With regard to the other particle, the wave function assigned to particle 2 depends on the outcome of the first measurement, namely

$$\psi(2) = \begin{cases} e^{i\phi}|-\rangle_2 & \text{if the particle 1 measurement gave 'up'} \quad (\phi \in \mathbb{R}) \\ e^{i\phi'}|+\rangle_2 & \text{if the particle 1 measurement gave 'down'} \ (\phi' \in \mathbb{R}) \end{cases}$$

$$(10.2.2)$$

This is because spin-up for particle 1 represents having the piece $|+\tfrac{1}{2}-\tfrac{1}{2}\rangle$ of Ψ_i be selected. For that piece, particle 2 has spin down. This automatically gives conservation of the z-component of angular momentum. However, *the total angular momentum of the particles constituting $\mathcal{M}$ is not conserved.* The states $|+\tfrac{1}{2}-\tfrac{1}{2}\rangle$ or $|-\tfrac{1}{2}+\tfrac{1}{2}\rangle$ that represent the possible outcomes of the experiment are themselves superpositions of states of total angular momentum one and zero. When the apparatus destroyed the phase correlation between the two states (ϕ and ϕ' are not be correlated) the non-conservation of angular momentum (of $\mathcal{M}$ alone) was forced. One then recovers conservation of total angular momentum from angular momentum transfers at the detectors. Since these are macroscopic, the

---

[8] Total-spin quantum numbers of the individual particles are suppressed. The labels in Eq. (10.2.1) are the z-components of spin: $|m_1 m_2\rangle$.

transfer is unnoticed. If one would go back to S, its angular momentum would be unchanged. Similarly, if one measured the x or y component of spin angular momentum for the decay particles, this would have no effect on S.

There is no principle preventing the source from picking up the decay angular momentum—or linear momentum for that matter. But if such were detectable, it would already constitute a measurement. Suppose that one of the decay particles leaves an observable track as it passes through S, and that the track is such that it can be used to deduce the spin orientation of the particle. This would mean that although in principle the total wave function could be a superposition of all possible outcomes, once the track is created, one of those outcomes has been selected. On the other hand, with the usual assumptions, the 'measurement' occurs only at the detectors further downstream. In that case there is no effect on S.

Analysis of the experiment with special states

According to the theory presented in this book, a particular measurement in which the detectors fired, recording a particular pair of spin results, represents a wave function that evolved entirely to that state. The microscopic state of the apparatus, etc., was unusual ('special') since the majority of microstates lead to macroscopic superpositions (grotesque states). Where does the 'specialization' take place?

Remark: Bear in mind that 'specialization' is not evolution under different dynamics. Usually the microstate of the environment-cum-apparatus is unimportant in that the actual evolution of the system is that associated with the (vast) majority of environment microstates. Sometimes the environment microstate leads to a *different* evolution of the system from that of the majority. This is 'specialization.'

The question of *where* specialization takes place is meaningful since special states are rare, and, as repeatedly emphasized, we look for the *least unlikely* among them. Presumably two rare occurrences at two locations will be less likely than a single one—assuming a single one can do the job.

Specializing with only one rare event can occur at the site of the decay. Moreover, this would be a way to be sure that the z-component of angular momentum of the two decay products be (anti-)correlated. Thus we have a 'two-state-discrimination' situation, as described in Chapter 9, and could get the desired result with a single rotation in the (real) two-dimensional Hilbert space defined by the initial state and the possible outcomes. We further assume that the operator implementing this rotation is local in coordinate space. This implies that when the decay products emerge from S they are already oriented so as to give particular definite results for the measurement that will take place. Because for this experiment the demand for spin orientation can take place a significant time before the measurement, one could call an observation implying the existence of this orientation, 'footprints of the future.' This characterization may not seem overly dramatic if one considers delayed-choice embellishments of the EPR experiment.

Remark: Having the spins oriented as they emerge from S does *not* contradict the quantum predictions of the Bell inequalities. Bell's theorem is predicated on a hidden variable that determines how the system would respond to a variety of measurements. In my description, the particles *do* leave S with a fixed value of the quantity to be measured, but only for the measurement that will actually be done, not for all or several possible measurements.[9] Implicit in this, and indeed in the entire theory, is the fact that past, present and future are intertwined, and the system's microscopic degrees of freedom take values consistent with events that lie in the future.

Since the passage from $\Psi_i(1,2)$ (of Eq. (10.2.1)) to $|+\frac{1}{2}-\frac{1}{2}\rangle$, or to $|-\frac{1}{2}+\frac{1}{2}\rangle$, takes place within S, it follows that the angular momentum book-keeping must balance within S. We do not look to the large magnets of a Stern-Gerlach apparatus or to some other equally unaccountable piece of equipment for the possible change in angular momentum. Rather it must reside in S at the end of the experiment.

Can this residual angular momentum be detected? Let $\mathbf{J}^S = (J_x^S, J_y^S, J_z^S)$ be the angular momentum of S. For each measurement of the sort we consider, when J_z (of particles 1 and 2) is measured, J_z^S should be unchanged. However, J_x^S and J_y^S can change, generally by $0, \pm\hbar$. It is this change we would like to measure. Repeating this experiment many times and finding such changes (perhaps looking for $\sqrt{N}$ effects—see Section 10.4) would demonstrate that angular momentum had been transferred in S. This is the experimental test.

However, one should not underestimate the difficulties of this approach. For the original EPR proposal (involving position and linear momentum, and embellished with a source analogous to our S) one can argue (C. G. Kuper & T. Mor, private communication, 1995) that if the source is big enough and interactive enough to leave a record of the specialization then it would at the same time destroy the EPR correlations. That argument is based on the position-momentum uncertainty relation. For our case (spin) there are additional observables, so that specializing J_z need not affect J_y. Thus, all you need is a non-zero value of J_y^S to tell you that angular momentum transfer took place. Such non-zero J_y^S need not tell you which way the particle J_z measurements will turn out, so that this measurement need not destroy the later EPR effects

---

[9] A philosopher might describe this as the denial of counterfactual reasoning.

(as the spin-dependent track measurements discussed above would). The experimenter would therefore have to walk a narrow line: a delicate measurement of y-component of angular momentum that at the same time, in principle, does not yield (or destroy) J_z information. One should also bear in mind that in this case the specializing must also disentangle. In Section 7.2 we looked into disentangling mechanisms for states of the form Eq. (10.2.1). We produced what I consider to be an artificial mechanism for achieving this end (cf. Eq. (7.2.6)) and argued that more likely mechanisms exist involving, for example, photons. Nevertheless, in the absence of a more detailed picture of the disentangling, it is not clear whether one can take advantage of the additional observables in the spin problem. One can also contemplate yet more exotic experiments involving delayed choice, but before doing so, one should work out the more straightforward version. There is also a significant experimental challenge in detecting angular momentum changes on the order of $\hbar$. Here too, I will not examine this issue until a clearer theoretical proposal exists.

EPR overview

From general considerations a prediction emerges for EPR-type experiments that is significantly different from the Copenhagen picture. The prediction is that at the time of breakup of $\mathcal{M}$ the spins are already oriented for the measurement that will be performed on them. What is it in the present theory that leads to this? After all, we have claimed that special state counting should lead to the usual results.

In fact we have made an assumption about the physical mechanisms of specialization that goes beyond the bare postulates given in Chapter 6. That assumption is that the interactions or forces that provide the 'kicks' that specialize are themselves local in space. This is *not* a local hidden variable (as I've emphasized before); the particles do not have any secret dial on which the outcomes of possible experiments are written. Rather, they are subject to forces caused by other objects in their environment, and I am assuming that those forces are due to locally unlikely states of those particles in the environment. There *could* be a long range force, it is just that my idea of what the special states look like tends more to near-encounters of particles or local fields in the neighborhood of $\mathcal{M}$.

Assuming the above to be true, there remains the question of determining that indeed the particles were so oriented. The suggestion above is that by embedding $\mathcal{M}$ in a larger object, $\mathcal{S}$, there would be measurements on $\mathcal{S}$ that would, on the one hand, not destroy the EPR correlations but, on the other, allow determination of the early orientation of the spins. The proposed experiment would be extremely delicate, because of the smallness of the effect and the need to maintain EPR correlations. I have shied away from considering the correlated photons in down-conversion EPR experiments, mainly because of the large crystal involved in this process. However, it is conceivable that a way to establish early spin orientation could be found there.

In summary, because we have physical input here that goes beyond the Copenhagen interpretation there does not seem any reason in principle why this effect could not be measured.

10.3 Other possibilities

Evidence of auxiliary features: Cauchy noise

In Chapter 9 we found that the recovery of standard probabilities required that the system sustain 'kicks' in Hilbert space and that those kicks were distributed according to the Cauchy distribution, a long-tailed distribution with no moments.[10] Can this agitation be detected in other contexts as well?

I bring this up not because I have the answer but because it would be a good thing to look for. The 'kicks' required in Chapter 9 did not necessarily involve energy transfer, so there is no evident connection to an energy scale. It is thus not clear whether there should be contributions to the specific heats of materials. On the other hand, the Lorentz line shape, possessing the same form as the Cauchy distribution, suggests a role for photons, in which case we would be talking about the photons' energy. More sensitive tests of 'kicks' than specific heat are noise or fluctuation spectra. The most famous anomalous fluctuation phenomenon is in the time domain, namely $1/f$ noise. Whether this has anything to do with our Cauchy distributed kicks, I don't know. In many circumstances long-tailed distributions do show up, from condensed matter to the physics of stars. Generally these too tend to be in the time domain, and they have been related to $1/f$ noise.

Perhaps the most familiar situation in which long-tailed distributions arise—in space as well as time—is in critical phenomena. At a critical point one does get large deviations and fractal structure. Furthermore, it is likely that manifestations of such behavior do not require the good fortune of finding your system exactly at a critical point. There is a concerted theoretical effort to understand why fractal structures are so common. One explanation of that propensity is that there is a tendency for open systems to evolve naturally to critical points. This same theoretical effort (coming under the rubric 'self

---

[10] Reminder: $C_a(x) = (a/\pi)/(x^2 + a^2)$.

organized criticality') has also claimed success in understanding $1/f$ noise.

Optimistically, these developments offer the possibility of relating our Cauchy distributed kicks to the associated physical phenomena. But even without a direct connection, they give credibility to the occurrence of non-Gaussian phenomena in Nature.

In Section 10.4 are references to physical applications alluded to above.

Reduction of fluctuations

Both major themes of this book tell you that there is less than meets the eye. The number of microscopic states associated with a macroscopic state is far smaller than the usual specification of that macrostate might lead you to suppose. In the first portion of this book we argued that by and large this reduction in available states would not be evident. Let us look again to see if more subtle tests might reveal cryptic constraints.

The essence of our argument was a kind of ergodicity. When the future constraint is many relaxation times away, the system 'forgets' its future and the actual points selected in phase space are pseudo-random. In practice this means that the points selected by the future constraint have no more structure than a random selection of points. But choosing random (phase space) points and watching them evolve is the standard way to make statistical predictions. This was why the forward evolution of the first few pictures in Fig. 4.3.2 (cat-map movie with two-time boundary conditions) looked completely normal.[11] It is also the solution to Exercise 4.3.2, in which one is asked to show that specific heats and other ordinary thermodynamic quantities would be unchanged by future conditioning.

But the naive version of the ergodic hypothesis implicit in this argument may be false. Classical dynamics of physical systems may include things like KAM tori within which systems can be effectively trapped. If *these* structures are homogeneously distributed in phase space, then, no problem. One can again appeal to pseudo-randomness. But they may not be uniformly distributed. Furthermore, quantum dynamics is notable for its failure to behave in an ergodic fashion. If for any of these reasons the particular constraint[12] does not do its selection homogeneously in phase space or Hilbert space, then there can be observable effects.

There is a related issue concerning the absolute normalization of entropy. The contemporary understanding of entropy is as the logarithm

---

[11] And in Appendix A to Section 5.3 there was analytic confirmation.

[12] ... whether that constraint arises because of a collapsing universe or because of special states (which may be the same thing).

of the number of microscopic states associated with a macroscopic state. Whether one does this with coarse grains (as in the present volume) or by another method, whether quantum or classical, there is a counting of states at the root of the modern definition. By claiming that the vast majority of microscopic states are in fact *not* available to the system, it would seem that we are implying an enormous change in all entropies, in all entropy calculations and comparisons with experiment.

This is not so drastic as one might suppose. It was only with the advent of quantum theory, with its basic dimension for phase space, $\hbar$, that one had any claim to providing an overall scale factor for entropy. As we argued above using ergodicity, if the reduction of microstates is homogeneous, one is again merely talking about a scale factor for entropy. However, since we now do have notions of an absolute scale factor for entropy, there is the possibility of direct comparison. I think the most fruitful place to look for this would be in fluctuations of small systems. For large systems one is back to the insensitivity to the quantum scale factor. For systems near the micro/macro boundary one might have reliable theoretical calculations of phase space dimensions along with the possibility of doing measurements. Again though, one must have good understanding of the extent to which the system is ergodic, since a tendency to hang about in a small region of phase space would also lead to measured entropy decrease.

Remark: If special states are distributed as in the decay model of Section 7.1, there would be *no* decrease in entropy. This is because the special states exhaust all dimensions of the Hilbert space.

Remark: If the special states arise semiclassically (i.e., the 'kicks' analyzed in Section 9.1 are collective motions for which the semiclassical approximation is good), then ergodicity at the classical level would provide homogeneous, pseudo-random distribution of phase space points.

Deviations from standard probabilities

In Section 9.1 we found that when the Cauchy distribution parameter γ is strictly positive, the ratio of up to down spins for a spin inclined at an angle θ to the z-axis was *not* $\cos^2 \frac{1}{2}\theta / \sin^2 \frac{1}{2}\theta$. Rather, for the initial wave function $\left(\begin{smallmatrix} \cos \theta/2 \\ i \sin \theta/2 \end{smallmatrix} \right)$ the probability of getting UP is (keeping the previous equation numbers)

(9.1.13) $$\Pr(\mathrm{UP}) = \frac{F(\theta)}{F(\theta) + F(\theta + \pi)} = \frac{\cos^2 \frac{\theta}{2} + \sin^2 \frac{\theta}{2} \tanh^2 \frac{\gamma}{2}}{1 + \tanh^2 \frac{\gamma}{2}}$$

which for small γ gives

(9.1.14) $$\Pr(\mathrm{UP}) \cong \cos^2 \frac{\theta}{2} - \frac{\gamma^2}{4} \cos \theta$$

In particular, an incoming up-polarized beam will be misread as down with frequency $\gamma^2/4$.

Before discussing experiments, the nature of γ needs to be clarified. Some modes[13] of a system may yield Cauchy distributed kicks, some may not. And even for those that are Cauchy distributed asymptotically, they need not be exactly Cauchy at small kick size. The role of γ in Section 9.1 is not so much as a strict Cauchy parameter; rather γ is the value beyond which one is in the asymptotic regime of the distribution. Therefore although I would not necessarily expect Eq. (9.1.14) to be true in detail (e.g., the '4' in $\gamma^2/4$ is not firm), I *would* expect the leading deviation to be proportional to γ^2. (γ^1 is ruled out by the $f(\phi) = f(-\phi)$ symmetry, for the 'f' of Eq. (9.1.2).)

Although it was convenient to phrase the discussion in Chapter 9 in terms of a spin-$\frac{1}{2}$ Stern-Gerlach experiment, this was merely a metaphor for any two-state discrimination situation. For a test of the above formulas, I believe the spin experiment would be a poor choice. First, assuming γ to be small (which it had better be, or else we're in deep trouble) one would need extreme precision in preparing the angular orientation of the incoming beam. Second, the 'WAY' theorem (Appendix A to this section) already guarantees that there be a non-zero error rate. A spin experiment measures one component of angular momentum. However, the apparatus has conserved angular momentum components that do not commute with the component being measured; in this case the WAY theorem tells you that there will be errors with a rate that is roughly $\hbar^2/8L^2$, where L^2 is the mean square angular momentum of the apparatus. This is a small quantity, with strong suggestions of the macro/micro difficulties that afflict all of quantum measurement theory. Nevertheless, in the absence of preconceptions about γ it would be best not to introduce additional uncertainties.

Thus the ideal experimental check for $O(\gamma^2)$ deviations from standard probabilities would be a measurement of a two-level system in which there were no mischievous commuting observables and in which extreme precision is possible in the preparation of the initial state. For the Stern-Gerlach apparatus there is also the problem that in any case the measurements are only accurate *modulo* the validity of the adiabatic approximation. The theoretical sorting efficiency[14] of any proposed apparatus would therefore have to be very good for the experiment to be reliable.

---

[13] 'Mode' here refers to a general collective coordinate, *not* just a linear oscillator.
[14] ... meaning the probability of errors in discriminating between the two states, as calculated in the usual way.

Microscopic/mesoscopic frontiers

In the past few years experimentalists have watched individual atoms decay, manipulated them, taken pictures of surfaces showing individual atoms, and generally have control over dimensions previously considered microscopic. There is also good theoretical understanding of many of these operations. Although for the moment I cannot suggest explicit ways to use them for experimental tests of my theory (for reasons to be given below) this may well turn out to be the practical route to a test— people are already devoting effort to exactly the kind of control I would require, but for reasons of their own.

The approach to experimental tests suggested in Section 10.1 is to de- plete the environment until some of the usual special states cannot exist. If you deplete without changing the Copenhagen predictions, then there will be a difference between the two theories. In contrast to Copenhagen, I predict that outcomes for which special states are unavailable do not occur. Although the dramatic experiments to which I referred in the previous paragraph do focus on individual atoms, nevertheless, they are performed in an environment that is rich in other degrees of freedom. Thus when one watches a single atom decay (a 'quantum jump'), the manipulation and measurement involve intense irradiation of the atom. Not only is this radiation rich in degrees of freedom but, as discussed in Section 7.4, it is an excellent place to look for real world examples of special states.

Furthermore, should one succeed in identifying special states for the quantum jump there would be additional benefits. Knowing what those states are may mean knowing how to eliminate them. For example, I expect that the many laser beams that push, pull, stimulate and monitor the atom provide the degrees of freedom for the special state. My guess is that all can be turned off temporarily without affecting the experi- ment, except the beams that monitor.[15] By some version of Murphy's law those monitoring beams are probably also the raw material for the special states. However, as for the scattering experiment of Section 10.1, I would hope to be able to walk the fine line of maintaining the Copen- hagen prediction while eliminating the special state. If such elimination is possible, it would clearly provide another experimental test. Since ex- perimental artistry is at such a high level in this context there may be a real chance for a significant test.

At the mesoscopic level and in connection with certain optical sys- tems, a number of tests of superposition have been proposed, involving what people sometimes call 'Schrödinger cats.' The SQUID has figured

---

[15] Note that in the actual experiment some beams do double duty.

prominently in other such proposals. Generally there is also good quantum theoretical understanding of these systems, and it is this competence that makes the tests so interesting. Unfortunately, I do not know if this can provide a test of the ideas in this book. In these proposals there is no spatial separation of the 'cats.' The state is grotesque, but not spatially grotesque. Perhaps this is easy to remedy, but since the rationale I have given for the avoidance of grotesque states involves spatial localization, at the moment the issue remains open.

The most ambitious attempts to control relatively large microscopic systems arise from contemporary efforts to build 'quantum computers.' The idea is that coherent evolution of a quantum system can be thought of as a parallel computation in the high dimensional tensor product of the many degrees of freedom of the coherently evolving system. Algorithms have been proposed to take advantage of this. For the physicist the challenge is to build a system that is large enough to make the computation interesting, and under sufficient control to retain coherence. 'Retaining coherence' represents the level of control sought in Section 10.1. The builder of such an apparatus requires that no external force cause loss of phase coherence among portions of the wave functions in the many regions of the Hilbert space. It is likely that an apparatus that is sufficiently isolated so as to implement a quantum computation could not be 'specialized' during its isolation. Reading out the results of the quantum computation involves 'measurements,' i.e., interactions with the larger environment, so at that end of the process there is the possibility of specialization. Initial preparation is another potential source of specializing.

Abstractly stated, the scheme for using a quantum computer to test my theory is as follows. Call the initial state of the computer ψ_0. It evolves to $\psi_T = \exp(-iHT/\hbar)\psi_0$, with little incoherent contribution. The cleverness of the computational theorist is in designing H and ψ_0 and in finding a measurement (say a projection, P) so that $P\psi_T$ will answer an interesting computational question. Generally one wants $\|P\psi_T\|^2$ close to 0 or 1. For my purposes I prefer a projection, Q, such that $\|Q\psi_T\|^2 \approx \|(1-Q)\psi_T\|^2 \approx 1/2$. If the outcomes represent macroscopically (including spatially) different states I require that ψ_T be an eigenstate of Q. For the proposed quantum computers, 'specializing' in my sense should be difficult because of the experimental control over the microstate. Thus according to me you would not get half and half for the outcome. What you actually *do* get will depend on the details of Q (cf. the scattering experiment of Section 10.1, where the default is nondetection). On the other hand, the Copenhagen interpretation would continue to give 1/2-1/2.

> *Appendix for this section (at the end of the chapter):*
> 10.3.A   The 'WAY' theorem: a limitation on idealized measurement

10.4 Notes and sources

Section 10.1. The scattering experiment was considered in [SCHULMAN 88, 91]. Discussion of EPR type experiments appears in [SCHULMAN 94, 95].

A review of the continuing search for an explanation of the solar neutrino deficit is [BAHCALL]. Suppression of radiative emission due to an absence of quantum states for an emitted photon is discussed in [HAROCHE].

Section 10.2. My conclusions regarding angular momentum transfer in the conventional analysis of the Bohmian-EPR experiment (that it occurs in the spin-measuring apparatus) agree with those of [POPESCU].

When an experiment has a small outcome of varying sign, performing N such experiments leads to an $O(\sqrt{N})$ effect that may be measurable. Such $\sqrt{N}$ enhancement has been used by [AHARONOV], and indeed he suggested this possibility for the EPR experiment described in the text.

Section 10.3. Observations of long-tailed distributions in physical phenomena have been reported by [SHLESINGER], [BELL], [SCHER] and [MONTROLL 83, 87]. A theoretical study using the kind of Lévy flight found in those phenomena is [BERGERSEN]. A short review of self-organized criticality is [BAK].

A tendency to localize is one of the major findings of the field of 'quantum chaos.' For example, there is the 'standard map,' describing the dynamics of a rotator (a single angular degree of freedom, θ) that is kicked at integer times. If p is the conjugate momentum and n the time at which the rotator is kicked for the nth time, then successive coordinate and momentum values are $p_{n+1} = K \sin \theta_n$ and $\theta_{n+1} = \theta_n + p_{n+1}$. This defines the kick and uses a parameter K, such that for sufficiently large K the momentum grows indefinitely. (There is a rich structure to this map; see [LICHTENBERG].) In the quantum version, the system gets stuck. The momentum does *not* grow indefinitely; this open system does not explore all the phase space you might have thought available to it. At the root of this phenomenon is a kind of Anderson localization. See [CASATI 83, 85] and [PRANGE]. Another kind of localization—i.e., failure to explore phase space—is studied by [BOHIGAS] and was alluded to in Chapter 5.

The original paper on the 'WAY' theorem is [WIGNER]. The error estimate for spin measurements that I give in the text is due to [YANASE]. This paper, as well as [ARAKI], are reprinted in [WHEELER]. Additional discussion can be found in [EARMAN] and [STEIN].

Several kinds of microscopic/mesoscopic manipulations were mentioned in the text. I am sure that there is a larger list than I am now about to give, but the following can serve as an entry to the literature. The decay of individual atoms and the techniques involved in achieving that are reported in [WINELAND], [BERGQUIST] and [DIEDRICH]. Theoretical papers on the SQUID are [CALDIERA], [LEGGET] and [CHAKRAVARTY]. Optical 'Schrödinger cats' are suggested in [SONG] and [YURKE]. Background on quantum computing can be found in [BROWN]. Additional references are given in Chapter 1. For a critique see [LANDAUER].

References

AHARONOV
88
Y. Aharonov, D. Z. Albert & L. Vaidman (1988) How the Result of a Measurement of a Component of the Spin of a Spin-1/2 Particle Can Turn Out to be 100, *Phys. Rev. Lett.* **60**, 1351.

ARAKI
60
H. Araki & M. M. Yanase (1960) Measurement of Quantum Mechanical Operators, *Phys. Rev.* **120**, 622.

BAHCALL
93
J. N. Bahcall & H. A. Bethe (1993) Do solar-neutrino experiments imply new physics? *Phys. Rev.* D **47**, 1298.

BAK
90
P. Bak (1990) Self-Organized Criticality, *Physica* A **163**, 403.

BELL
78
T. L. Bell, U. Frisch & H. Frisch (1978) Renormalization-group approach to noncoherent radiative transfer, *Phys. Rev.* A **17**, 1049.

BERGERSON
91
B. Bergersen & Z. Racz (1991) Dynamical Generation of Long-Range Interactions: Random Lévy Flights in the Kinetic Ising and Spherical Models, *Phys. Rev. Lett.* **67**, 3047.

BERGQUIST
86
J. C. Bergquist, R. G. Hulet, W. M. Itano & D. J. Wineland (1986) Observation of Quantum Jumps in a Single Atom, *Phys. Rev. Lett.* **57**, 1699.

BOHIGAS
93
O. Bohigas, S. Tomsovic & D. Ullmo (1993) Manifestations of classical phase space structures in quantum mechanics, *Phys. Rep.* **223**, 43.

BROWN
94
J. Brown (1994) A quantum revolution for computing, *New Scientist*, Sep. 24, p. 21.

CALDEIRA
81
A. O. Caldeira & A. J. Leggett (1981) Influence of Dissipation on Quantum Tunneling in Macroscopic Systems, *Phys. Rev. Lett.* **46**, 211.

CASATI
83
G. Casati (1983) Quantum Dynamics of Classical Stochastic Systems, in *Trends and developments in the eighties*, ed. S. Albeverio & P. H. Blanchard, World Sci., Singapore.

CASATI
85
G. Casati (1985) *Chaotic Behavior in Quantum Systems, Theory and Applications*, NATO ASI, Ser. B, Phys., vol. 120, Plenum, New York.

CHAKRAVARTY
84
S. Chakravarty & A. J. Leggett (1984) Dynamics of the Two-State System with Ohmic Dissipation, *Phys. Rev. Lett.* **52**, 5.

DIEDRICH
89
F. Diedrich, J. C. Bergquist, W. M. Itano & D. J. Wineland (1989) Laser Cooling to the Zero-Point Energy of Motion, *Phys. Rev. Lett.* **62**, 403.

EARMAN
68
J. Earman & A. Shimony (1968) A Note on Measurement, *Nuov. Cim.* B **54**, 332 .

HAROCHE
89
S. Haroche & D. Kleppner (1989) Cavity Quantum Electrodynamics, *Phys. Today* **42**, Jan., p. 24.

LANDAUER
95
R. Landauer (1995) Is quantum mechanically coherent computation useful?, in *Proc. of the Drexel-4 Symposium on Quantum Nonintegrability— Quantum-Classical Correspondence*, ed. D. H. Feng & B. L. Hu, International Press.

LEGGETT
84
A. J. Leggett (1984) Schrödinger's Cat and her Laboratory Cousins, *Contemp. Phys.* **25**, 583.

LICHTENBERG 83   A. J. Lichtenberg & M. A. Lieberman (1983) *Regular and Stochastic Motion*, Springer, New York.

MONTROLL 83   E. W. Montroll & M. F. Shlesinger (1983) Maximum Entropy Formalism, Fractals, Scaling Phenomena, and $1/f$-Noise: A Tale of Tails, *J. Stat. Phys.* **32**, 209.

MONTROLL 87   E. W. Montroll & B. J. West (1987) On an enriched collection of stochastic processes, in *Fluctuation phenomena*, 2nd ed., E. W. Montroll & J. L. Lebowitz, eds., North-Holland, Amsterdam.

POPESCU 95   S. Popescu (1995) Private communication.

PRANGE 85   R. E. Prange, D. R. Grempel & S. Fishman (1985) Quantum Chaos and Anderson Localization, in *Chaotic Behavior in Quantum Systems: Theory and Applications*, ed. G. Casati, NATO ASI, Ser. B, Phys., vol. 120, Plenum, New York.

SCHER 91   H. Scher, M. F. Shlesinger & J. T. Bendler (1991) Time-Scale Invariance in Transport and Relaxation, *Phys. Today* **44**, Jan., 26.

SCHULMAN 88   L. S. Schulman (1988) Detection with Compulsory Nonabsorption, *Phys. Lett.* A **130**, 194.

SCHULMAN 91   L. S. Schulman (1991) Definite Quantum Measurements, *Ann. Phys.* **212**, 315.

SCHULMAN 94   L. S. Schulman (1994) Particle detection via special states, in *The interpretation of quantum theory: where do we stand?*, ed. L. Accardi, Encic. Ital., Rome (Proc. conf. Columbia Univ., New York, 1992).

SCHULMAN 95   L. S. Schulman (1995) Consequences and Inconsequences of Cryptic Constraints, in *The Dilemma of Einstein, Podolsky and Rosen—Sixty Years Later*, A. Mann & M. Revzen, eds., *Ann. Israel Phys. Soc.* **12**.

SHLESINGER 89   M. F. Shlesinger (1989) Lévy Flights: Variations on a Theme, *Physica* D **38**, 304.

SONG 90   S. Song, C. M. Caves & B. Yurke (1990) Generation of superpositions of classically distinguishable quantum states from optical back-action evasion, *Phys. Rev.* A **41**, 5261.

STEIN 71   H. Stein & A. Shimony (1971) Limitations on Measurement, in *Foundations of Quantum Mechanics*, ed. B. D'Espagnat, Academic, New York.

WHEELER 83   J. A. Wheeler & W. H. Zurek (1983) *Quantum Theory and Measurement*, Princeton Univ. Press, Princeton.

WIGNER 52   E. P. Wigner (1952) Die Messung quantenmechanischer Operatoren, *Z. Physik* **133**, 101.

WINELAND 87   D. J. Wineland & W. M. Itano (1987) Laser cooling, *Phys. Today*, June, p. 34.

YANASE 61   M. M. Yanase (1961) Optimal Measuring Apparatus, *Phys. Rev.* **123**, 666.

YURKE 86   B. Yurke & D. Stoler (1986) Generating Quantum Mechanical Superpositions of Macroscopically Distinguishable States via Amplitude Dispersion, *Phys. Rev. Lett.* **57**, 13.

Appendix 10.3.A: The 'WAY' theorem: a limitation on idealized measurement

If you can measure an observable M and if there is an overall constant of the motion for the system plus measurement apparatus, then M commutes with the constant of the motion. This was discovered by Wigner (**W**) in 1952 and elaborated by

Araki, Yanase (**A**, **Y**) and others. I seldom see reference to this remarkable result. Perhaps its impact has been limited by its being based on a mathematical idealization. Its validity depends on something being strictly zero; with the least deviation, the implication is lost or severely softened. In any case, it shows that the picture of a measurement that was promulgated by von Neumann must be imperfect. For example, it implies that you cannot make an idealized measurement of the z-component of angular momentum in a Stern-Gerlach experiment if your total system Hamiltonian commutes with the x-component of total angular momentum—which in fact one expects.

The proof is only a few lines,[16] although those lines involve a lot of notation. Let the system be in a Hilbert space $\mathcal{H}_1$ and have an initial wave function $\sum_{\mu,\rho} a_{\mu\rho} u_{\mu\rho}$, with $\{a_{\mu\rho}\}$ coefficients and $\{u_{\mu\rho}\}$ normalized eigenfunctions. The label μ refers to the eigenvalues (which we assume to be discrete) of an observable (self-adjoint operator), M. The other labels, ρ, are carried along and help keep track of changes in the wave function before and after the measurement. The value of μ is assumed not to change during the course of the measurement. Let the apparatus (environment, etc.) state vector be in a Hilbert space $\mathcal{H}_2$ and initially be in a state Ω_0. These assumptions are formally stated as

$$M u_{\mu\rho} = \mu u_{\mu\rho}$$
$$\langle u_{\mu\rho} | u_{\mu'\rho'} \rangle = \delta_{\mu\mu'} \delta_{\rho\rho'}$$
$$\Psi_{\text{initial}} = \Omega_0 \sum_{\mu,\rho} a_{\mu\rho} u_{\mu\rho} \tag{10.3.A.1}$$

Now let a time t go by. The total Hamiltonian H generates a unitary evolution, $U = \exp(-iHt/\hbar)$. The wave function becomes

$$\Psi_{\text{final}} = U \Psi_{\text{initial}} = \sum_{\mu,\rho'} \Omega_{\mu\rho\rho'} a_{\mu\rho'} u_{\mu\rho'} \tag{10.3.A.2}$$

This is the standard von Neumann analysis. At the end of the measurement the apparatus states $\Omega_{\mu\rho\rho'}$ correspond to particular values μ, of the observable M. In effect the μ value of the original system is 'measured' by looking at $\mathcal{H}_2$, where presumably the states $\Omega_{\mu\rho\rho'}$ are clearly distinguished from one another and may even describe the observer. In particular, this distinguishability is expressed as

$$\langle \Omega_{\mu\rho\rho'} | \Omega_{\mu'\rho''\rho'''} \rangle = 0 \quad \text{if} \quad \mu \neq \mu' \tag{10.3.A.3}$$

Now suppose there is an observable L with the following properties: (1) $[L, H] = 0$ and thus $[L, U] = 0$; (2) for times before and after the apparatus and system were in contact, L can be broken into a sum of operators acting on the system and on the apparatus. Formally: $L = L_1 \otimes 1_2 + 1_1 \otimes L_2$, with L_1 and L_2 self adjoint operators and the subscripted 1s are the appropriate identity operators.

The essential step in the argument is to consider the inner product

$$\left\langle u_{\mu'\rho'} \Omega_0 \,\middle|\, L \,\middle|\, u_{\mu\rho} \Omega_0 \right\rangle = \left\langle U \left(u_{\mu'\rho'} \Omega_0 \right) \,\middle|\, UL \,\middle|\, u_{\mu\rho} \Omega_0 \right\rangle$$

$$= \left\langle U \left(u_{\mu'\rho'} \Omega_0 \right) \,\middle|\, L \,\middle|\, U [u_{\mu\rho} \Omega_0] \right\rangle = \left\langle \sum_{\rho'''} u_{\mu'\rho'''} \Omega_{\mu'\rho'\rho'''} \,\middle|\, L \,\middle|\, \sum_{\rho''} u_{\mu\rho''} \Omega_{\mu\rho\rho''} \right\rangle \tag{10.3.A.4}$$

---

[16] If one considers measurements of continuous eigenvalues things can get touchier.

The first equality is true, because for any two vectors applying a unitary operator to both of them does not change the inner product. The second equality reflects the commutation of L and H. The third—where the indices proliferate—is what happens during the measurement. The additive structure of L allows the extreme left and right sides of Eq. (10.3.A.4) to be broken into sums of terms, as follows.

Left side:

$$\left\langle u_{\mu'\rho'}\Omega_0 \,\middle|\, L_1 \otimes 1_2 + 1_1 \otimes L_2 \,\middle|\, u_{\mu\rho}\Omega_0 \right\rangle$$

$$= \left\langle u_{\mu'\rho'} \,\middle|\, L_1 \,\middle|\, u_{\mu\rho} \right\rangle \langle \Omega_0|\Omega_0\rangle + \left\langle u_{\mu'\rho'} \,\middle|\, u_{\mu\rho} \right\rangle \langle \Omega_0|L_2|\Omega_0\rangle \tag{10.3.A.5}$$

Right side:

$$\left\langle \sum_{\rho'''} u_{\mu'\rho'''}\Omega_{\mu'\rho'\rho'''} \,\middle|\, L_1 \otimes 1_2 + 1_1 \otimes L_2 \,\middle|\, \sum_{\rho''} u_{\mu\rho''}\Omega_{\mu\rho\rho''} \right\rangle$$

$$= \left\langle \sum_{\rho'''} u_{\mu'\rho'''} \,\middle|\, L_1 \,\middle|\, u_{\mu\rho''} \right\rangle \left\langle \sum_{\rho'''} \Omega_{\mu'\rho'\rho'''} \,\middle|\, \sum_{\rho''} \Omega_{\mu\rho\rho''} \right\rangle \tag{10.3.A.6}$$

$$+ \left\langle \sum_{\rho'''} u_{\mu'\rho'''} \,\middle|\, \sum_{\rho''} u_{\mu\rho''} \right\rangle \left\langle \sum_{\rho'''} \Omega_{\mu'\rho'\rho'''} \,\middle|\, L_2 \,\middle|\, \sum_{\rho''} \Omega_{\mu\rho\rho''} \right\rangle$$

For $\mu \neq \mu'$ there is a great simplification. All terms containing 'L_2' drop out because the functions u are orthogonal for unequal μ. But the important point is that on the right side the 'L_1' drops out also. This is because it multiplies an inner product of Ωs for differing values of μ, which are orthogonal. This is where the assumption on the effect of the measurement is made. There remains only a single term, and it must be zero:

$$\left\langle u_{\mu'\rho'} \,\middle|\, L_1 \,\middle|\, u_{\mu\rho} \right\rangle = 0 \quad \text{for} \quad \mu \neq \mu' \tag{10.3.A.7}$$

It is now a standard exercise to show that if the operator L_1 does not connect any states with differing values of μ, it commutes with M (whose eigenvalues the set $\{\mu\}$ are). Thus $[M, L_1] = 0$.

How then does one measure the z-component of spin? The answer is that if you allow a small—very small—piece of wave function to come out wrong, if you allow a little bit of down-spin to result from the measurement of an initial up-spin, then everything goes through without the implied commutativity of M and L_1. This little bit of 'wrong' wave function provides the error rate.

The reason you can get away with this small correction is that in Eq. (10.3.A.5) the 'left side' is microscopic in size, basically $O(\hbar)$ for angular momentum. On the other hand, in Eq. (10.3.A.6) there is a matrix element of L_2 between apparatus states, Ω. Any error that creeps in will be enhanced by a macroscopic factor, the value of L_2 in a macroscopic state. In fact this is the scale of the minimal error level found by Yanase.

This exposition has been included for several reasons. Most directly it impacts the possibility of the experimental test discussed in Section 10.3. Second, the result deserves more publicity. Third, although this effect and the 'ϵ' that shows up in finding special states seem to me logically distinct, they may after all turn out to be related. Furthermore, there is a debater's point connected to this ϵ, in that when I've been challenged over the epsilons that I seem to need for my special states, I can point to this one in conventional theory. Actually, my expectation is that the conventional epsilon is much worse than that which will be needed for special states. The WAY epsilon is $O(1/N^\alpha)$ with N the system size and α order unity. For the special state epsilons I expect the same kind of numbers that you get in decoherence calculations, $O\big(\exp(-\text{const} \cdot N)\big)$. Finally, the

WAY theorem shows that there are logical problems in the underlying structure of the conventional theory. The von Neumann analysis plus ensembles (or whatever) does not always work smoothly, and an 'ensemble' of spin-up particles does *not* always give spin-up measurements.

11

Conclusions and outlook

Less than meets the eye

If even some of the ideas presented here are correct, the world is different from what it seems. The major theses of the book, on time's arrows and on quantum measurement theory, are unified by the notion of *cryptic constraints*. We see, sense, specify, *macroscopic* states, but what we predict about these states depends on an important assumption concerning their *microscopic* situation. The assumption is that the actual microstate is equally likely to be any of those consistent with the macrostate. And I say, not so. For various reasons, both classical and quantum, many otherwise-possible microstates are eliminated. As presented in detail in previous chapters, such elimination impacts many areas of physics, from the cosmos to the atom. But we also make the point that this elimination can be difficult to notice and in particular there is no experimental evidence that confirms the usual assumption.[1] By an explicit example on a model (Fig. 4.3.2), we show that a future constraint eliminating 98% of the microstates can go completely unnoticed.

My expectation is that this fundamental change in the foundations of statistical mechanics is needed. Whether or not it takes the forms I've proposed will be determined by future investigations.

In the next section I will review open problems for the program implicit in this book. The tone will be that used in speaking with colleagues: an attempt to be frank about difficulties and a willingness to be wildly speculative. I see nothing wrong with either mode provided the reader

---

[1] Actually, there are beautiful experiments, such as the spin-echo effect, that rejuvenate degrees of freedom that appear to have 'thermalized.' Such an apparent—but false—thermalizing is an example of a cryptic constraint, but is not the sort with which we are mainly concerned.

understands the context and claims—after all, it is with 'wild speculation' that we first grope for truth and it is with sharp criticism that we can confirm its presence.

For the reader who has turned to this chapter to get an overview of the book, I can only say, return to Chapter 1. That's where the summary is. Here I will be assuming familiarity with what's in between.

Finally, a bit of advice: don't get hung up over the arrow of time revolution—cryptic constraints, inability to 'control,' determinism. That's a matter of getting used to, and it is conceptually less demanding than learning special relativity. What you should worry about, or criticize, is the technical stuff. Is it really possible for a special state to disentangle? What could give rise to the Cauchy distributed kicks I need? Is there an EPR experiment in which the 'footprints of the future' could be noticed? Can such footprints be seen in the dynamics of galaxies?

11.1 Open questions

1. Galactic dynamics and large scale structure. If ours is a roughly time-symmetric cosmology, then dynamical processes whose relaxation time is comparable to the big bang–big crunch time interval, will equilibrate differently from the way they would if there were no big crunch awaiting them. In Section 4.6 I talked about ways to observe this effect, but my aim was to show conceptual feasibility; my *hope*, however, is that a concrete, practical method will be found.

Other slow dynamical processes may be sought for observation of similar effects. Some nuclei have lifetimes on scales that could be candidates for the big bang–big crunch time interval; but it does not seem that observing small samples of them could give the desired information.

2. Gravity and thermodynamics. For the purposes of deriving our local thermodynamic arrow of time, I believe that treating gravity as an external force (cf. Chapter 4) is justified. But more ambitious quests may be formulated in which the entropy of black holes and, perhaps more important, the tendency of gravity to create chaotic, fractal structure, is included.

3. Special states: existence. Are there enough degrees of freedom around to do the specializing? Although in Chapter 7 I gave examples from models, I think you really need to learn from the answer found (in Chapter 9) to the 'abundance' question. Namely, the special states must be able to give large kicks (in Hilbert space) to the system, with the probability of a kick of a given size dropping off only as the square of that size. Perhaps this is due to ubiquitous low energy photons, or to low energy something else, so long as it has a Lorentz line shape. This is one of those good news/bad news situations. It makes it easier to find special states, but harder to rule them out, as in the sort of experimental test contemplated in Section 10.1. In any case, the choice is not mine, and ways to investigate this, theoretically and experimentally, should be found.

Another feature demanded in Chapter 9 is the smallness of the Cauchy parameter; that is, the distribution function for kick size achieves its asymptotic value for small values of the microscopic coordinate it affects. This is needed for agreement with experiment. If the kicks are due to the products of unstable or metastable transitions (as suggested by the photon comment in the last paragraph), then we are talking about a narrow linewidth, which I would say is consistent with the special states arising from collective motions of relatively large numbers of particles (at the mesoscopic level perhaps). This conforms to the picture that I now have of these states, but a more precise assertion awaits a better understanding of the sources of the Cauchy noise and the relation of the general arguments of Chapter 9 to the specific model calculations of Chapter 7.

4. Rationale for special states. In Chapter 8 the reader was told to ignore flaws in the argument: all I wanted to do was show you *something*, some condition that produced the particular cryptic constraint needed for the quantum measurement theory. Clearly, there is more to be done here. First, that argument itself needs to be improved, most urgently, in my opinion, in the treatment of identical particles. It would also be good to pin down the source of the wave function localization—gravitational bunching? particle production?

Second, can another rationale be found? Based on general ignorance of Nature, I argued that we should not be too insistent on having a rationale. After all, Boltzmann did not know why his *final* states were special (a different kind of 'special,' but highly selective nonetheless). But it would be nice to have other possibilities in mind, if for no other reason than to be insulated from the vicissitudes of contemporary observational cosmology.

5. Two-time boundary value problems. Two-time boundary value problems played an important part in the material presented in this book, both for understanding the thermodynamic arrow of time and for the quantum measurement problem. For quantum systems, such boundary value problems can be difficult. Partly the difficulty arises from the interplay of the microscopic and the macroscopic. The reason is that the kind of boundary conditions one finds natural are usually spatially motivated, usually macroscopically definable, usually involving coarse grains with spatial definition. This is in contrast to what is often the more convenient arena for quantum mechanics, namely a Hilbert space with a basis defined by microscopic and not necessarily local operators.

In Chapters 5 and 7 we made inroads on this problem, in particular our special state examples were seen to be the solution of a relevant kind of quantum two-time boundary value problem. However, many questions remain open, and progress in this area is important for our program, as well as possessing independent mathematical interest.

What *is* under good control is the stochastic two-time boundary value problem. Here the challenge would be to formalize many of the intuitively based assertions in Chapters 4, 5, 7, 8 and 9.

6. Cauchy noise and cryptic constraints. There are apparently lots of unusual microstates, at least if my theory is correct. More precisely: big rotations are possible in Hilbert space, although there is no energy scale that is automatically associated with these rotations (as far as I can tell). In Chapter 10 we looked for experimental consequences. Here I will make my mumbled speculations more explicit. In the time domain, there are more long term correlations than anyone knows how to account for, specifically, $1/f$ noise. The reader will by now realize that I draw an ironic pleasure from examples of global human ignorance, and $1/f$ noise is one of the points in my data set. There have been attempts at general answers, one even related to QED. I am proposing a long-tailed distribution in another arena, in some abstract Hilbert space coordinate. Nevertheless, it seems plausible to me that all these phenomena are part of a general tendency of Nature to yield large deviation results where we have mostly expected regularity. (N.B. This meandering talk is what I apologized for in the previous section. I do not claim to explain $1/f$; I only find the relation to the Cauchy noise suggestive.) One open question is: can anything solid be said, can these ruminations be turned into science? Also, whether or not there is any relation to $1/f$ noise, the presence of large fluctuations of the sort I require should have *some* observable consequences.

7. More noise. Of the various theories of quantum measurement that have been proposed, my ideas are most closely related to the many worlds picture (relative state, post-Everett, decoherence, whatever). Nevertheless, some of the nonlinear, wave-function collapse theories invoke stochastic forces to accomplish their ends. Recovery of standard experimental predictions imposes particular forms on those stochastic forces. Such recovery is what gave me Cauchy noise. Superficially, I don't see a relation between the two forms of noise. Moreover, the conceptual origin is completely different. Nevertheless, in listening to seminars by those advocating this approach, I've noted resemblances. Such a relation is not part of my measurement program and it's not part of theirs, but it would be interesting to know if it existed, if only at the mathematical level.

8. Abundance arguments: legal and illegal Hilbert space directions. Part of the argument in Chapter 9 required conditions on the permissible rotations ('kicks') in Hilbert space. When you made the required assumptions, things worked out beautifully, mathematically, that is. I do not have a solid physical justification for all those assumptions, although I could wave hands a bit. Nevertheless, the consistency and smoothness of the mathematics convinces me that I'm on to something. It would be good to improve this situation by arguing from grounds other than personal contentment, however aesthetically motivated.

9. Entropic manifestations of cryptic constraints. If a system is small, cold and isolated enough that you think you understand all its 'fluctuations,' then you should have an absolute notion of its entropy, *modulo* coarse graining ambiguities. If ours is a world with cryptic constraints, the number of available microscopic states should be reduced. This was discussed in Section 10.3, but without a concrete experimental proposal.

10. Experimental tests. When all is said and done in Chapter 10, I cannot walk up to Joe or Josephine Experimentalist and say do this and that and you'll know the answer. I can't even do this if Dr Experimentalist has a $3000000 budget and three years to spend on my project. There are both conceptual and experimental issues.

Nevertheless, I am optimistic about the possibility that a feasible experiment will emerge in the next few years—not necessarily by virtue of my own efforts, but because theoretical and experimental work are converging on problems of this sort. For example, there is contemporary research, both mathematical and physical, on 'quantum computing.' (See Section 10.3.) The physical challenge is to build a large enough system to compute without losing quantum coherence. These are some of the same issues confronted in Section 10.1 in trying to prevent 'specialization.' The kind of isolation from the environment required for building such a system will be similar to what I sought in designing the test of Section 10.1. Similarly, greater control of both optical and material systems at the nanometer level suggests that some of the less direct predictions of this theory, those phenomena associated with the special noise and with cryptic constraints, will also be subject to experimental investigation.

11.2 Notes and sources

Section 11.1. Articles relating $1/f$ noise to quantum electrodynamics are [HANDEL, VAN DER ZIEL] and [VAN VLIET]. These authors and others might dissent from my characterization of $1/f$ noise as 'unexplained.' In any case, I would only claim it is the *universality* that is unexplained; for many specific examples the mechanism is understood. For nonlinear wave-function collapse theories, see [GHIRARDI 86, 89] and [PEARLE]. For a review on quantum computing, see [BROWN].

References

BROWN 94   J. Brown (1994) A quantum revolution for computing, *New Scientist*, Sep. 24, p. 21.

GHIRARDI 86   G. C. Ghirardi, A. Rimini & T. Weber (1986) Unified dynamics for microscopic and macroscopic systems, *Phys. Rev.* D **34**, 470.

GHIRARDI 89   G. C. Ghirardi, P. Pearle & A. Rimini (1989) Markov Processes in Hilbert Space and Continuous Spontaneous Localization of Systems of Identical Particles, *Phys. Rev.* A **42**, 78.

HANDEL 80   P. H. Handel (1980) Quantum Approach to $1/f$ Noise, *Phys. Rev.* A, **22**, 745.

PEARLE 94   P. Pearle & E. Squires (1994) Bound State Excitation, Nucleon Decay Experiments, and Models of Wave Function Collapse, *Phys. Rev. Lett.* **73**, 1.

VAN DER ZIEL 89   A. van der A. Ziel, A. D. van Rheenen & A. N. Birbas (1989) Extensions of Handel's $1/f$-noise equations and their semiclassical theory, *Phys. Rev.* B **40**, 1806.

VAN VLIET 89   C. M. Van Vliet (1989) An Alternative Theory for Quantum $1/f$ Noise based on Quantum Electrodynamics, in *Tenth International Conference on Fluctuations in Physical Systems, Budapest*.

Author index

For each author cited, the chapter(s) in which that author's work appears are listed. The entries refer only to works cited in the 'Notes and sources' section of each chapter. Note that in those sections citations are arranged according to first author, whereas the following list includes all authors.

Abraham, R; 2
Accardi, L; 1, 6
Adler, R; 3, 5
Agazzi, E; 6
Aharonov, Y; 10
Aharony, A; 3, 9
Ajdarl, A; 3
Albert, D; 10
Alers, G; 9
Alpher, R; 3
Altenmuller, T; 6
Anandan, J; 1
Anderson, J; 3, 4
Andre, J; 3
Araki, H; 10
Arnold, V; 2
Arya, A; 6
Ash, R; 2
Avez, A; 2
Avron, J; 2
Ax, J; 3
Azbel, M; 7
Bahcall, J; 10
Bak, P; 3, 9, 10
Balian, R; 1
Ballentine, L; 1
Bazin, M; 3, 5
Bekenstein, J; 4
Bell, J; 1
Bell, T; 10
Bellman, R; 3
Benatti, F; 1

Bendler, J; 10
Bergé, P; 3
Bergerson, B; 10
Bergquist, J; 1, 5, 7, 10
Berman, A; 2
Bethe, H; 3, 10
Birbas, A; 11
Black, T; 1, 6
Blatt, J; 7
Blatt, R; 7
Bocchieri, P; 2
Bohigas, O; 5, 10
Bohm, D; 1
Bollinger, J; 2
Born, M; 9
Brehme, R; 3
Brewer, R; 2
Brodbeck, M; 1, 6
Brown, J; 10, 11
Broyles, A; 6, 7
Brush, S; 4
Buldyrev, S; 3
Caianiello, E; 3
Caldeira, A; 10
Casati, G; 10
Caves, C; 10
Cerdeira, H; 5
Chaitin, G; 3
Chakravarty, S; 7, 10
Chalmers, D; 3
Chester, G; 2
Christensen, K; 9

Christenson, J; 3
Chuang, I; 1
Clarke, J; 9
Clauser, J; 1
Cloudsley-Thompson, J; 1
Cocke, W; 4, 5
Coleman, P; 3
Coleman, S; 2, 3, 6
Colin, L; 3
Cooke, K; 3
Coopersmith, M; 2
Courant, R; 3
Cover, T; 2
Cronin, J; 3
Davies, P; 4
De Alfaro, V; 2
De Groot, S; 3
Dehmelt, H; 7
Della Riccia, G; 6
Dennett, D; 1, 6
DeWitt, B; 1, 3, 6
Dicke, R; 2, 3
Diedrich, F; 10
Dirac, P; 3
Doering, C; 3, 5–7
Dominguez-Tenreiro, R; 3
Dorsey, A; 7
Driver, R; 3
Dudewicz, E; 9
Dutta, P; 9
Earman, J; 10
Ezawa, H; 1

333

Index